**Stargazing Basics**
**Getting Started in Recreational Astronomy**

How do I get started in astronomy?
Should I buy binoculars or a telescope?
What can I expect to see?

This wo
absolut
techniq
observi
takes y
and mo

In this b
affordal
the sky,
photos
through
and ref

Whethe
a newc
steps in

**PAUL** 
and edu
schools
his kno

# Stargazing

# Basics

## Getting Started in Recreational Astronomy

Paul E. Kinzer

CAMBRIDGE UNIVERSITY PRESS
Cambridge, New York, Melbourne, Madrid, Cape Town, Singapore, São Paulo, Delhi

Cambridge University Press
The Edinburgh Building, Cambridge CB2 8RU, UK

Published in the United States of America by Cambridge University Press, New York

www.cambridge.org
Information on this title: www.cambridge.org/9780521728591



First published 2008

Printed in the United Kingdom at the University Press, Cambridge

*A catalog record for this publication is available from the British Library*

*Library of Congress Cataloging in Publication data*
Kinzer, Paul E., 1960–
Stargazing basics: getting started in recreational astronomy / By Paul E. Kinzer.
p. cm.
ISBN 978-0-521-72859-1
1. Amateur astronomy–Observers' manuals. 2. Astronomy–Amateurs' manuals.
3. Astronomy–Observers' manuals. I. Title.
QB63.K466 2008
520–dc22

2008024378

ISBN 978-0-521-72859-1 paperback

To Wina and Bjorn

# Contents

# Acknowledgements

Many people helped in making this book a reality, but there are several I would like particularly to thank: Dr. John Dickey, Professor of Astronomy at the University of Tasmania, for reading an earlier draft and offering kind words and encouragement; Larry Gautsch, Meredith Houge, and Daniel Kelliher, for reading later drafts and offering many helpful comments; my nieces Gretchen and Kendra Dischinger, and my son, Bjorn Mortenson, for agreeing to appear in photographs; my wife, Wina Mortenson, for her patience and pushing; her brother Jon, for help with photo choice and processing; and, finally, Vince Higgs, my editor at Cambridge University Press, for taking an interest in the first place, and for quick and clear guidance along the way.

# Introduction: why another stargazing guide?

I had been an armchair astronomer for years: I loved reading about science, and I had some general knowledge about the Universe and how it works. However, I knew little about the actual night sky as it appears above the Earth, and I had never owned a decent telescope.

Then, in 1995, I started a nature program for urban nine- to twelve-year-olds. For the program, I bought a good Nikon "spotting scope": a small telescope made primarily for viewing birds and other Earth-bound objects (see Figure 1). On an overnight camping trip, I aimed the telescope at the Moon, so the kids could see the detail: craters, mountains, and "seas." The children were amazed, since most of them had never really seen the Moon before. I was impressed, because the telescope gave excellent views: the images were very sharp. But I had seen the Moon through binoculars, and it was familiar enough that it was not as thrilling for me as it had been for the children.

After the kids were asleep in their tents, I used the telescope again. The sky was very dark once the Moon had set, and I looked at fields of stars in the Milky Way for some time. Then I aimed the scope at a bright yellow star. I was not expecting much: stars are so far away that they appear

**Figure 1 Spotting scopes, though designed for terrestrial (Earth-bound) use, can be useful for wide views of the nighttime sky. This model, with its angled eyepiece, provides comfortable upward viewing.**

**Figure 2 Even a small telescope can show the detail seen here when viewing Saturn; and the "live" view seems much more three-dimensional.**

only as sharp points in even the largest telescopes. But this "star" had *rings*: I was looking at the planet Saturn!

I had known, of course, that Saturn was up there in the sky somewhere. I had also seen more detailed views of it in satellite photographs. But here it was, real and bright and sharp; a ringed ball floating in space right before my eyes! I had no idea that so small a telescope could show it so clearly; and I had just stumbled upon it, like a jewel on the beach.

Well, that was it: I was hooked. As soon as I could afford it, and after much research – often confusing and contradictory – I bought a larger telescope, on a sturdy mount that was better for astronomy. Though I have been finding more jewels ever since, I am still continually amazed at just how much I can see through a fairly small scope.

Until I actually started doing it, there was something about stargazing that I would not have guessed. It is *much* more satisfying to find objects myself than it is to look at a photograph in a book or magazine. To see the real thing. To know that the light hitting my eye actually left the object up there, and that that light may have traveled countless trillions of miles and, possibly, millions of years, seemingly just for me to find it. Of course, the light would have come even if I was not there to catch it, but each new catch is, literally, wonderful to me.

Others agree. It is always fun to see the happy, dumbstruck look on people's faces when *they* see Saturn (or Jupiter, or a close-up view of craters on the Moon, or a galaxy) for the first time through my telescope. They are filled with excited questions afterwards; and, at least for a little while, they are hooked, too.

Many of these people ask a particular question: can I recommend a single, basic book to help someone who is an absolute beginner get *started* in recreational astronomy?

Well, I know of many great books. Some were especially written to explain the make-up of the Universe. Others were written to help people learn the constellations and the night sky, and how they change through the night and year. Still others were written to help people choose and buy astronomy equipment. And a few do all of these things, thoroughly and well.

However, it has seemed to me that many people who have looked through my telescope want something that is more basic than what these books offer. They want a *simple* outline that will provide them with all the necessary information to make a good beginning, both on what is "Up There" and on the techniques and equipment used to view it; but not so much information that they will be overwhelmed or intimidated. (There is an amazing amount of jargon in astronomy publications, which often goes unexplained.)

In other words, I believe there is a need for a sort of road map that will allow people to get to a place where they can dip their toes into recreational astronomy before deciding whether to take the plunge. For someone who is enthusiastic, yet knows little, getting started can be a daunting job. It is for these people, and I think there are a lot of them, that I have written this guide.

## What this guide is

*Stargazing Basics* is meant to be a starting point. I hope it will provide enough information to allow just about anyone to get started in recreational astronomy, or stargazing. The guide is divided into three sections:

First, Part I, "Stargazing techniques and equipment," starts with a brief description of how the night sky "works," and goes on to explain what the novice will need to view it; whether with eyes alone, binoculars, or a telescope. Since this is a guide for beginners, the equipment discussed is of small to moderate size, and relatively inexpensive.

Part II, "What's up there?", is a brief tour of the Universe that is observable to the beginner. We start nearby, in astronomical terms, and head outward.

*More info* boxes will appear along the way in both of the first two sections. These contain information which may add meaning to the main text, but which might be distracting if contained within it. The first two parts also have *Resources* sections at various points. These are references to equipment, books, magazines, websites, and organizations that can provide more information and assistance.

Finally, as you read these first two parts, you will notice that many words and phrases appear in **bold** print the first time they appear. These terms are described in alphabetical order in Part III, "A stargazing glossary." They are terms that may be commonly used in other publications, but not always clearly defined for a beginner. They may also be explained in the main body of the guide, but I thought it would be helpful to point out that clear, simple definitions are readily available for easy reference at a later time. If you see an unknown term, try looking it up in the Glossary.

## What this guide is not

This guide is not "a complete guide." It may lead you toward completion, but it is not meant to be the only resource you need.

This guide is not really meant to be taken into the field. Simple star charts are included, but anything more would have made the book too expensive. Besides, there are already many good books specifically designed to show the detailed layout of the night sky, and it is possible to get up-to-the-minute sky maps from the Internet, which show the night sky as it will look from your exact location on Earth.

This is also not a complete guide to equipment buying. Astronomy can be both the least and most expensive of hobbies, and this guide emphasizes the least. If you are wondering whether you should buy that wide-field apochromatic refractor, or a 20″ Dobsonian reflector instead, then this book is not for you!

This guide is, finally, not very long. Too much information can be both confusing and intimidating to someone who is unfamiliar with a topic. I have tried to include only what the beginner may need or want to know at the very beginning.

# Part I
# Stargazing techniques and equipment

If you asked most people to name the one thing that is most needed by someone who is just starting out in **astronomy**, they would almost certainly say "a **telescope**." And if you then asked them to name the most important thing to look for when choosing that telescope, the answer – if there was one – might be "power" (as in **magnification**). The first answer is definitely questionable, and the second answer is simply wrong.

Many people believe both answers, and who can blame them? How would they know any differently? Few people have had any education in the understanding of amateur astronomy and its equipment, and cheap-telescope distributors know it.

Here is a typical experience from not so long ago:

A certain young person looked through a telescope on a camping trip one night, and became interested in the idea of pursuing this hobby (as many people do after looking through a *good* telescope). It seemed, to this person, only natural to think that the first thing to do was to buy his own telescope. He remembered seeing them in department stores, camera dealers, and at a gift shop in a museum. He went to one of these places and asked someone to help him, only to find that there was no one there who could. So all he had to go on was the information provided on the telescope packaging. All the boxes had large, bold print, boasting of things like "magnifications up to 600x!"

He bought the one with the "highest power" and took it home. That night, he set it up in his suburban backyard. He looked up at the sky, but wasn't sure where to aim the new scope. "Where's the **Moon**?" he wondered. Four nights ago, at the campground, he had looked through a telescope at the Moon. His friend had said it was just past full.

Oh, well. He had also looked at the **planet Jupiter**, and it had been amazing. He was almost certain it was that bright "**star**" up there. But where were all the other stars? He knew the city lights washed some of them out, but only now did he see how few were visible. Anyway, he tried to aim the telescope at what he thought was Jupiter, but the **finder scope** was next to impossible to adjust, or even to see through, and the telescope would not

**Figure 3 An example of a telescope that is not very "powerful," but still very useful for astronomy.**

stay where he aimed it because the **tripod** was too shaky. When he finally got the object in view, it would not come into clear focus, no matter what he did. It was a dim, fuzzy ball. It was not worth looking at, which was lucky, because just being near the telescope seemed enough to make the image bounce all over. And it kept moving out of the **field of view**. He felt like he had done something wrong, and was so frustrated that he packed the telescope up and never used it again.

This has been the first – and last – experience that many people have had with astronomy. It is, more or less, a description of *my* experience as a teen. Many of the problems I had were caused by my lack of knowledge, all of which will be explained by the end of this book. But if I had managed to get a steady, sharp view of Jupiter for even a few seconds, I might have tried harder to overcome my ignorance. Unfortunately, the "starter" telescopes sold at many stores are often just cheap junk.

However, there is good news: new technologies, and many new manufacturers and distributors, have led to keen competition, which has resulted in an improvement in quality, and the lowering of prices. Even department stores now often carry inexpensive, yet decent, beginner's equipment. (But beware: there is still plenty of junk!)

**Figure 4 A very suspect claim. The telescope that came in the box with this printed on it, purchased at a well-known national chain store in the USA in 2007, could not produce *sharp* images even at its lowest magnification, 40×.**

This first section of the guide discusses the selection of a first telescope. Having a little knowledge ahead of time will be a big help in avoiding disappointment.

However, there is much to getting started in stargazing that can come before the selection of a telescope. In fact, a telescope is not even a necessity for exploring and enjoying the night sky. Binoculars, and even your unaided eyes, can show you many wonderful sights! If I had known this as a teen, I could have saved a lot of money and disappointment. I might also have stuck with the hobby, instead of waiting decades to rediscover it.

# With the naked eye alone 1

*Note: It may be that the information presented in this first section would more properly fit in "Part II: What's up there?" But I wrote the guide with the idea in mind that people would probably read it from beginning to end. It seemed to me, then, that a very basic description of the night sky – as a whole – would be very helpful right at the start.*

People have been looking up and pondering the night sky since – well, since there have been people. Every culture in the world, both past and present, has studied and tried to explain what is up there. However, until **Galileo Galilei**, in 1609, took a small telescope and aimed it upward, all others who had studied the stars used just their eyes.

Exploring the night sky with a telescope is made much easier if a person has at least some knowledge of the layout of the heavens. But it is also a great pleasure in itself to just lie back and look up at the wide expanse of the "starry bowl." It looks essentially the same as when ancient cultures named the stars and **constellations**, many of which have had the same names since before recorded history began (see Appendix 2 for a complete list).

**More info**

It is not strictly true that astronomers before Galileo used only their eyes to study the heavens. They had various devices for accurately measuring and mapping the positions of the stars. They made charts with the information gathered, which were used in surveying, and, especially, navigation. But none of these devices magnified the view.

## The Celestial Sphere

Because this is a *simple* guide, and because there are excellent books devoted entirely to teaching the positions, names, and lore of the stars and constellations, I will not go into such detail here. However, a basic understanding of how the sky "works," how it changes over time (and how it does not), is very helpful to the beginning stargazer.

Amateur astronomers divide the Universe into two broad regions: the **Solar System** (containing the planets, their moons, and other objects under the gravitational influence of the **Sun**); and the **deep sky**, which contains everything beyond the Solar System. (It should be said that the term **deep-sky object** is often used by astronomers to describe all the objects

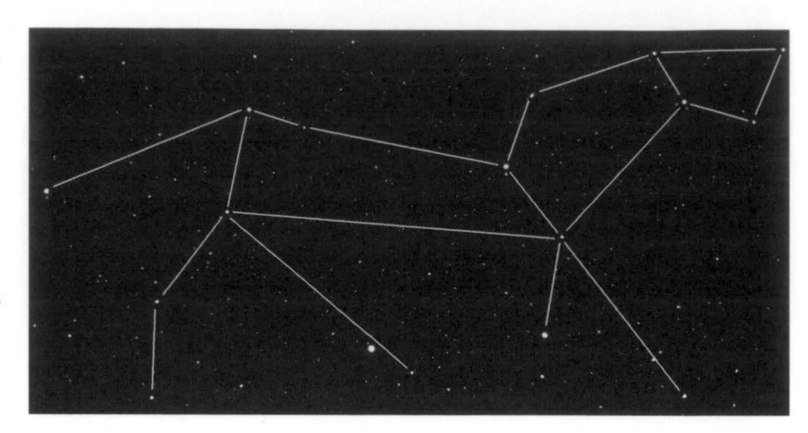

**Figure 1.1 The constellation Leo, the Lion, actually looks like the creature it is meant to represent. This is definitely not the case with some other constellations. In this photograph, taken in 2008, one of Leo's rear legs seems to be stepping past the bright planet Saturn.**

outside the Solar System *other than* stars. Still, stars *are* part of the deep sky. This possibly confusing point is made clear in Part II.)

The reason for separating these groups is this: all deep-sky objects are so far away that they seem, in a way, to be frozen in place in the sky. They may rise and set as the **Earth** spins, and they may also change position as the Earth **orbits** (travels in a near-circular path around) the Sun, but *they do not noticeably change position relative to each other.* In other words, the **Big Dipper** is the Big Dipper, whether it is this year or next, winter or summer, midnight or 3:00 am (or noon, for that matter). There is, it seems, a large unchanging bowl of stars hanging above our heads at night. The ancients named it the **Celestial** (or Heavenly) **Sphere**.

In contrast, all objects in the Solar System are close enough to us that they do not keep the same positions upon the bowl of heaven. Before the use of telescopes, only seven *planets* (the word originally meant *wanderer*) were known: the Moon, the Sun, **Mercury**, **Venus**, **Mars**, Jupiter, and **Saturn**. They were called wanderers because they, unlike the stars, crawled *across* the celestial dome. The motion cannot usually be seen from moment to moment, but it is there.

The ancients did not know it, but *we* now know that the planets, including the Earth, and the thousands of other objects within the Solar System, all orbit the Sun. We also know that they are far closer to us than any of the stars or other deep-sky objects.

Why is all of this important? Because it means that we can make maps of the deep sky, and that these maps can be used for decades to find interesting and beautiful

**More info**

I find it curious that it was not until the advent of the telescope that the planet **Uranus** was discovered. It is very dim, but it is visible to the naked eye, if you look in the right spot under dark skies. And since it moves across the sky, I am surprised that no one noticed this planet before **William Herschel** saw it through his telescope in 1781.

## More info

Here is a basic description of why the sky changes through the night, as well as through the year. The Earth turns on its **axis** once every day. That means it spins around like a top. So objects in the sky seem to rise in the east and then set in the west, though it is actually the Earth that is turning.

The Earth also orbits the Sun. It completes a circle around our own star once in a year, or 365.24 days. A circle is divided into 360 **degrees**, so we travel about one degree a day around our orbit. (This is probably not a coincidence. The practice of dividing circles into 360 degrees dates back thousands of years, to the Babylonians, or even further. Their calendar also divided the year into 12 months, each 30 days in length, for a total of 360 days.) This means that, if you were to look at the sky *at the same time on the clock* over several nights, the bowl of the sky would seem to revolve about one degree to the west each night, or about 30 degrees in a month. Furthermore, the spinning of the Earth on its axis causes the sky to turn 15 degrees in an hour.

This is all confusing at first, but once you understand it, you will have a much easier time searching for things in the sky. And, of course, this is just an explanation for the movement of the deep sky; Solar System objects each move in their own paths (but more on that later).

The drawing below (Figure 1.2) shows that some objects, like the constellation **Orion**, are visible in the evening sky in December, but are in the sky during the day, and overwhelmed by the light of the Sun, in June. Others, like the "Teapot" shape seen in the constellation **Sagittarius**, appear at night in June, but are invisible in December.

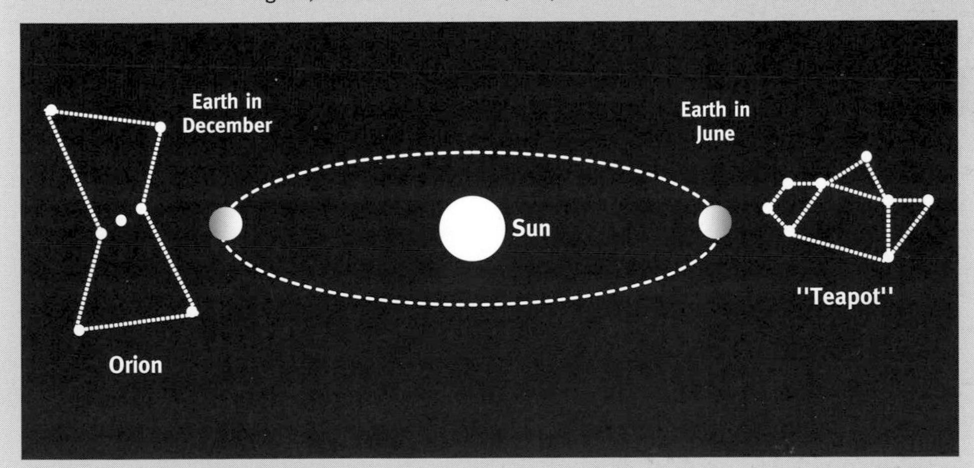


**Figure 1.2 Orion and the Teapot.**

things to study. But finding objects in the Solar System must be done differently, since they change position so quickly. Luckily, the most interesting planets to look at are bright enough that they are usually fairly easy to locate.

## Making the most of the "wide-eye" view

If you want to look with unaided eyes at the whole sky (or, for that matter, any part of it), here are some tips:

- *The first thing to do is to get as far away from artificial light as you can.* **Light pollution** is a serious and growing problem, and it gets more difficult every year to find truly **dark sites**. Still, it can be done, and must be, if you want to pursue astronomy at its best, whether with the eye or the largest telescope in the world. There are certainly some things that can be studied from the brightest of places (the Moon and the brighter planets), but to enjoy the sweep of the whole night sky you should go where the most can be seen. (To learn more about light pollution, and the fight against it, visit the **International Dark-Sky Association** at www.darksky.org.) If it is possible, go to a higher elevation as well, since this will put you above some of the **atmosphere** that blocks and distorts the incoming light.
- *Plan your viewing for a night when the Moon will set early or rise late*. The Moon is fascinating, but only for a short while as a naked-eye object, and its light washes out that of less bright things (meaning everything else in the night sky!)
- *Dress appropriately*, both for the *weather*, and for the possibility of unwelcome guests: *insects*. Layers work well. Since the temperature will probably drop as the evening progresses, it is good to have more clothing to put on. And being virtually motionless for long periods of time requires you to dress more warmly than you might think. Being still also makes you a prime target for bugs, so go prepared!
- *Bring something comfortable to sit on, or better yet, for reclining, or lying down*. A lawn chair, an air mattress, a blanket or sleeping bag – all of these are great choices.
- *Be prepared to let your eyes adjust to the darkness*. It takes about half an hour for the human eye to become totally dark-adapted. If you look into headlights, or a bright flashlight, even for a second, be ready to start that half hour over again.
- *Try, for at least part of the time, to not look at a particular object in front of you, but to take in as much as is in your field of view as possible*. I used to tell the children I worked with to use their "wide eyes" when looking at the whole sky. Naked-eye stargazing is best for wide swaths of sky.

To make the experience more interesting, you may want to take a couple of items along. First, in order to know just what you are looking at, you may want a **planisphere** (see Figure 1.4 on page 15). This is an inexpensive device that shows you what will be visible in the deep sky at any hour of any date of the year. Of course, it can only contain a few hundred objects, at best, on its small surface. Practicing with it ahead of time can really make the changing (as in rotating) sky more easily understandable. Astronomy dealers sell them, and they can also be found, often for under $10, at book stores in the science section, or at museum gift shops. It is also possible to design and print out your own (see the *Resources* section on page 15).

A red **LED** (light emitting diode) flashlight is another handy item to have, if only to read your planisphere. Red light does not affect your night vision nearly as much as ordinary white light. Such LEDs are available from astronomy dealers, but I have also seen them at department stores. It is also possible to tape red plastic over a regular flashlight, but if it is too bright, even red light can ruin your night vision.

By the way, these tips will also be helpful when using **binoculars** or a telescope.

### What might you see?

On any given night, you can study the *constellations*. Astronomers have divided the entire sky into 88 distinct sections that are recognized around the world. Most of these constellations were given to us by ancient cultures, but some came from explorers of later centuries. (The "official" names and boundaries, used and recognized internationally by professional astronomers, were not set until 1930.) The stories behind them are fascinating, and learning some of the lore adds something to the actual seeing. Of course, other cultures have divided the sky into their own constellations, and looking for these is also an interesting possibility. However, if you get to know the "map of the sky" that is used by the professionals, it will make finding objects up there *much* easier, whether with the naked eye, or a powerful telescope.

Under even fairly dark skies, you are likely to see a few **meteors** on any given night (and many more during a **meteor shower**). You will be able to see the brighter parts of the **Milky Way** (our own home **galaxy**, see Figure 1.3). If the Sun is "active," you may see **aurora** (the Northern or Southern Lights), if you live far enough north or south. If you know where to look, you can see **nebulae**, up to five (or six?) planets, and even another galaxy or two.

Of course, what you will see mostly is stars: thousands of stars. Though some brightly lit cities now only have a few visible, it is still possible to view virtually countless stars, if you can get to the right place.

However, if you are lucky, the best thing about getting out under a dark sky and looking up will not be what you *see*, but what you *feel*. Even

**Figure 1.3** The clouds of "steam," seemingly rising here from the teapot shape in the constellation Sagittarius, are actually countless stars contained in the disc of our own galaxy, the Milky Way (the center of which is just off the "spout" of the teapot). These clouds of stars are clearly visible under dark skies.

though I have been exploring it for years, there are still times when I am nearly overwhelmed by the experience of the night sky. The Universe is vast beyond real comprehension; but sometimes, for no reason I can explain, it – or something in me – seems to open up, and I am chilled with joy. I can never predict when these moments of epiphany will strike, and they never last long, but I look forward to them: they bring me back to childhood, when everything seemed new and amazing.

## Resources

### Books

*The Stars: A New Way to See Them*, by H.A. Rey. Houghton Mifflin, 1976.

There are many good books devoted to teaching an understanding of the constellations and the workings of the celestial sphere, but I know of no better one for the beginner than this. Rey was the creator of the "Curious George" picture books for children. He was also an avid stargazer himself, and an excellent illustrator. This book was originally published more than 50 years ago, and some say it was written for children. But I did not read it until I was an adult, and it helped me get my mind around some concepts which other books had not. It is full of information that is explained simply, and very clearly and cleverly illustrated.

### Websites

The Internet changes rapidly, and I hesitate to include specific sites, but some seem destined to last because they are so popular and useful. One of these is a site called cleardarkskies.com. It was created to give amateur astronomers in northern North America weather forecasts for the skies above their observatories or preferred viewing sites. It can also help anyone in this region find dark sites from which to observe, and to know what the weather will be like at that location up to 48 hours in the future.

**Figure 1.4 Everything you need to begin at the beginning: A good planisphere; *The Stars: A New Way to See Them*, by H.A. Rey; a red LED flashlight; and a soft lawn for reclining under wide, dark skies.**

Another great site is heavens-above.com. This website is full of fun stuff. Once you enter a town near your astronomical viewing location, you can call up a **star chart** for that site, for any time or date you like. And unlike a normal planisphere, this chart will show the location of the planets and the Moon, as well as a map of the entire visible sky. The site also has great information on Earth-orbiting **satellites**, **comets**, and more.

It seems a little unfair to single out only these websites, since there are literally thousands more that would be useful to stargazers. Just do a web search for an object or area of interest, and you will get more information than you can possibly use. This is one more reason for the increasing popularity of the hobby. Here are a few more sites to get you started, each of which should be around for some time. I include them because they are written, to an extent, with beginners in mind, and because each contains regularly updated information and links to more information:

antwrp.gsfc.nasa.gov/apod/ (Astronomy Picture of the Day)

www.SkyandTelescope.com

www.SpaceWeather.com

## Software

Planispheres are designed to be used in particular regions of the world. Going east or west does not matter, but shifting north or south causes changes in what can be seen in the sky, and in what should be shown on a planisphere. If you live in Sweden, you cannot use a planisphere meant for people in New Zealand. *Planisphere* is a free, downloadable program that you can get by going to the web address nio.astronomy.cz/om/. It allows you to create a planisphere for your own place on the planet, though it does not cover the whole world.

If, by the time you read this book, this particular website has disappeared (as so many of them do), try doing a web search for "free planisphere software." That is how I found this excellent program.

# Binoculars: the next step 2

The human eye, under truly dark skies, can see a few thousand stars. We can see so many because, where there is no artificial light to interfere, the eye's **iris** contracts to its smallest size, allowing the **pupil** to open to its widest **diameter**. This allows the maximum amount of light to enter and shine upon the **retina**, the light-gathering area in the back of the eye.

Yet why do we only see a few thousand stars, when there are hundreds of billions in our own galaxy, and billions of galaxies? Because the human eye's pupil can only open to a diameter of about seven millimeters (a little more than a quarter inch). As we get older, this usually decreases to about five millimeters (about a fifth of an inch).

Why is this important? Because, contrary to what most people think, it is this small opening that primarily determines how much we can see in the night sky, *not* magnification. Most stars and other objects are simply too *dim* to be seen by the human eye. A larger **aperture** (usually expressed as the diameter of a light-gathering surface, but actually its **area**) is the key thing, for two reasons: it allows dimmer objects to be seen by gathering more light over a wider area, and it also causes objects to be seen more clearly (see *More info – Comparing apertures*, on page 20). The first seems obvious, the second less so; but both are true. **Nocturnal** animals have large eyes for good reasons!

**Figure 2.1 Binoculars come in a wide range of shapes and sizes. Any of these would enhance the view when compared to the eye alone, but some choices are better – much better – than others.**

However, so far anyway, *we* are stuck with the eyes we were born with. So how do we increase our apertures? With telescopes! The large **lenses** or mirrors of telescopes gather light and concentrate it for delivery to the eye. And the best and easiest telescopes to begin with are binoculars.

## Why binoculars?

Binoculars are a good beginning choice for stargazing for many reasons:

1. *Many people already have them* for other purposes; so, for these people, there is no cost involved.
2. If you do not have them, a good pair is *not terribly expensive*.
3. They are *very portable;* just put them around your neck, and you are ready to go.
4. They are *easy to use*: just lift them to your eyes and focus.
5. Many people find that *using two eyes is more comfortable*, and some report that they can actually see more stars when using both eyes.
6. They show a ***correct image***. Astronomical telescopes may show images that are upside-down, inverted left to right, or both, depending on the equipment used. This can take getting used to. Binoculars, because of their special **prisms**, show images of things as they actually appear in the sky, or on star charts.
7. They give *low magnification for steadier views*. If you magnify an object by a given amount, you magnify any vibrations by an equal amount. Low magnification is why binoculars can be steadily hand-held.
8. Though they have low magnification, the large apertures of binoculars can take you from seeing a few thousand to *more than 100,000 stars*.
9. Binoculars are also a good starting place because *you will never move entirely beyond them*. No matter how large a telescope you may someday own, you will always want a good pair of binoculars. Why? First, because they are so easy to use, you will use them more often than a telescope, which takes time to set up. But, more importantly, because there are some things that are better viewed through binoculars than with either the naked eye or a telescope.

Some objects in the sky are too small or dim to be seen without some aid, but fill too much sky to be viewed with the higher magnification of a telescope. Binoculars are the perfect step between the eye and the astronomical telescope.

**More info...**

Here are just a few binocular targets (items seen well or best through binoculars): Some comets, the whole Moon, the **Andromeda Galaxy**, many **open star clusters**, some large, bright nebulae, and Milky Way star fields.

## Selecting binoculars for astronomy

If you already own a pair of binoculars, they may be very useful for stargazing, or they may not. Binoculars are almost always described and labeled with two numbers: 10 × 25, 7 × 35, 8 × 42, 10 × 42, 10 × 50; these are some common combinations. The first number is the *magnification*, or how many times larger an object will appear when viewed. The second number is the *aperture*: the diameter, in millimeters, of the **objective lens**. There are at least two important features you should consider when choosing binoculars for astronomy: *aperture* and ***exit pupil***.

In astronomy, a larger aperture means brighter, sharper images; so it would make sense to get the largest objective lenses possible. However, since using binoculars requires you to hold them steadily for long periods of time, weight is also important. Large apertures mean more weight. So a personal balance must be found.

The second thing to consider is the exit pupil. This is the size of the circle of light that leaves the **eyepiece**. It can be measured, but if you know the numbers, it is easier to just figure it out: divide the aperture by the magnification. So, for 7 × 35 binoculars, you get 35 (the aperture in millimeters) divided by 7 (the magnification) equals an exit pupil of 5 mm.

The exit pupil matters because it is best that the light entering the objective lens is used most efficiently by the eye. If the exit pupil is too large, you will have to move your eye around to see the entire view; if it is too small, the light-collecting surface of your dark-adapted eyes will not be entirely used. Since the human pupil can range in size from about five to seven millimeters, it makes sense to have an exit pupil on your binoculars of about the same size. An exit pupil of about five millimeters is a good choice for just about everyone.

The numbers 7 × 35 fit the ratio necessary for a 5 mm exit pupil, and it is a very common binocular ratio. But 10 × 50 also fits, and is a better choice for astronomy, since the larger objective lenses gather more than twice as much light. However, 10 × 50s are bound to be heavier, and holding them steady may be a problem for some.

**More info**

The exit pupil is only an important consideration at fairly low magnifications. When using a telescope at higher powers, the exit pupil has to become smaller: dividing a given aperture by a larger magnification gives a smaller result. So there is a trade-off that must take place.

### Some other considerations

***Eye relief***. Depending on the **optics** used, binoculars (and telescope eyepieces) have different

### More info – Comparing apertures

Apertures are circles, and the area (A) of the circle is what is important. It is measured by taking the **radius** ($r$) of the circle (half the diameter) and multiplying it by itself (or **squaring** it), and then multiplying by **pi** ($\pi$, about 3.14). So the mathematical formula (the only one in this book!) is $A = \pi r^2$. The squaring leads to results that may be surprising: If you double the diameter of an aperture, you multiply its light-gathering area by four times. So, increasing the aperture by what seems a small amount can increase its light gathering ability by quite a bit. This is useful to know and use when choosing binoculars or a telescope. For example, using the two binoculars described in the text, an aperture with a diameter of 35 mm gives the following: the radius, 17.5 mm (half of 35) times itself equals 306.25; this times pi gives an area of about 962 square millimeters. The 50 mm binoculars have a radius of 25 mm; squaring this gives 625; multiplying by pi shows an area of 1962.5 square millimeters. So, increasing the aperture by 15 mm more than doubles the light-gathering area of the 35 mm aperture.

*eye relief.* This is the distance, usually measured in millimeters, at which you must hold your eye from the eyepiece to get a proper view. Many binocular adverts or brochures will list this distance, and some will say that 14 mm is "ample" or "good." But not if you wear glasses! I do, and I think 20 mm is an acceptable distance. Personally, I am still more comfortable taking my glasses off. However, if you plan to do the same, beware: some binoculars will not come to focus for those of us with particularly weak vision. Binoculars with long eye relief are sometimes referred to as **high eyepoint**.

*Type of prisms used.* Binoculars virtually always use one of two types of prism: **roof** and **porro**. Both types are designed to do the same things: provide a correct image, and shorten the distance from objective lens to eyepiece by reflecting the light along folded paths. *Roof* prisms are used in some of the least and some of the most expensive binoculars made. They allow for smaller, lighter designs; but it is expensive to make excellent roof prisms. As a result, binoculars using them that are large enough for astronomy are both rare and extremely expensive (see Figure 2.2).

Binoculars using *porro* prisms are the most familiar; and though somewhat bulkier, they are made in many sizes, can provide excellent images, and have prices much more within reach. A good pair of 10 × 50s can be bought for under $200, and can sometimes be found for quite a bit less.

*Type of glass in prisms.* Brochures or ads will often state that a pair of binoculars has **BK-7** or **BAK-4** prisms. This is a description of the chemical make-up of the glass from which the prisms are made; BAK-4 is typically the better choice, and should be looked for, because in most binoculars it produces brighter images across the entire field of view.

*Tripod adaptability.* Holding binoculars in your hands makes them very easy to use, but there are times when it is also good to be able to rest

**Figure 2.2 These two pairs of binoculars are comparable in size (10 × 50) and quality, but the pair on the right, with roof prisms, is priced three times higher than the pair on the left, which uses porro prisms.**

your arms, as well as get steadier views. This is why you should look for a pair that is adaptable for tripod use (most good ones are). A special adapter is required, but is not very expensive (see Figure 2.3).

*Lens coatings*. Any good pair of binoculars sold today will come with **coated lenses**. Coating a lens with special chemicals reduces internal reflections, which allows more light to reach the eye, and also reduces "ghost" images from around brighter objects. **Multi-coated** lenses have more than one type of coating, for further reflection reduction. **Fully coated** means that all lens surfaces that are exposed to air (and there can be several) are coated. **Fully multi-coated** (**FMC**) covers it all.

*High-quality optics*. Maybe this should go without saying, but I am including it because it is very difficult to simply look *at* a pair of binoculars to tell their quality. You must look *through* them. Is the focused image sharp, and sharp across most of the field of view? (Even great binoculars may blur some near the edges.) If not, try testing a different pair of the same model. The first pair may be a "lemon." If things still don't look pristine, try a different model or brand.

*Find a balance*. Of course, getting a pair of binoculars with all these features may put them out of your price range. You will need to do some research, and make some choices.

## Things to avoid

***Zoom** binoculars*. Zoom lenses change magnification by moving **lens elements** internally. Their designs are usually not good for stargazing.

*Water- or weather-proof binoculars*. Since astronomy, almost by definition, takes place in dry conditions, it is not necessary, or desirable, to get binoculars with rubber- or plastic-armor coatings. It only adds weight to

**Figure 2.3 Steady views are helpful for seeing detail, so a tripod adapter, and binoculars equipped to use them, are good investments.**

them. Of course, if you are buying "multi-purpose" binoculars, to be used during the day as well, then that's a different story.

*"Ruby" coatings*. These red-tinted lenses look flashy, but the coatings are not ideal for astronomy. On well-made binoculars, they are meant to block the glare seen on bright, daylit objects. When used for astronomy, these coatings can block incoming light, and even change the color of objects seen. On cheap binoculars, ruby coatings are often just a cosmetic gimmick.

## Big binoculars

There *are* binoculars that are especially designed for astronomy. They feature larger objective lenses, and, usually, higher magnification. Most are also, of course, heavier, and not really meant to be hand-held. They range in size from 9 × 63 to 25 × 150 or more. The wide, bright, seemingly

Figure 2.4 *Comparing binoculars*: From top to bottom, here are the pairs of numbers that describe the magnifications and apertures of each of these pairs of binoculars: 8 x 20, 10 x 25, 7 x 35, 10 x 42, and 10 x 50. The top two are not appropriate for astronomy: they are poorly made, with low-quality roof prisms, and each pair has an exit pupil of only 2.5 mm. The bottom three are much better choices: they all have good- to excellent-quality porro prisms; they have larger objective lenses; the exit pupils are at least 4.2 mm; and those circles between the lenses are covering threaded tripod-mounting holes. Note also the reflections off the objectives. The middle pair of the bottom three has the fewest, and this is a sign of excellent multi-coatings, which is also a likely pointer to overall quality. They are, in fact, the most expensive pair of the five.

three-dimensional views they give to both eyes can be fantastic. Some of the smaller models are quite reasonably priced, and seem to get cheaper all the time.

## Are binoculars the right choice for you?

The simple answer to this question is "yes".

Though there are some who may get complete satisfaction out of viewing the whole sky, with no more aid than their own eyes and possibly a planisphere, I believe most people will find that investing in a decent pair of binoculars is well worth it. They are so valuable and easy to use that, with virtually no practice, a whole new world will open up within a few seconds of putting them between your eyes and the night sky. That is a big claim, but it really is true. Especially with larger apertures (50 mm or more), and with the

**Figure 2.5 What's up in the sky? Binoculars are the easiest way, and often the best way, to find out.**

steadiness provided by a tripod, exploring the sky with binoculars can be truly awe-inspiring.

It may also be true for many who read this book that a good pair of binoculars is all the optical equipment that will ever be wanted or needed. Sweeping through the Milky Way; identifying features on the Moon; studying **sunspots** (with proper **filters**!); searching for "faint fuzzies" like star clusters, galaxies and nebulae; getting a closer, brighter look at a visiting comet; all of these can provide great enjoyment and challenge through a pair of binoculars.

Setting up and using a telescope takes at least a little effort. If you are not yet sure that astronomy is something you will vigorously pursue, then I strongly advise you get a good pair of binoculars first. You may decide they are all you want, or all you can afford. Even if you do go on to buy a telescope, your binoculars will always be useful for wide, beautiful, easily obtained viewing.

## Resources

### Binocular dealers

Any large camera shop or sporting-goods store probably has an adequate selection of binoculars for stargazing. Some department stores also carry decent brands. There are also many mail-order companies that feature them. These companies run advertisements in the magazines described below.

## Magazines

There are two major magazines in the USA that cater specifically to the amateur astronomer: *Sky & Telescope* and *Astronomy*. Both feature articles on the latest astronomical discoveries, as well as information on amateur equipment and events. They have monthly features, including a sky chart that shows what is currently visible in the evening sky. They are excellent resources for people just starting out in the hobby (if only for the advertisements), though some of the articles may be intimidating to anyone without much knowledge of the field.

## Books

Once you own a pair of binoculars, or even before, it is a good time to get an astronomy field guide. There are many great choices, but below are a few that I think would be useful for a beginner.

1. *Cambridge Guide to Stars and Planets*, by Patrick Moore and Wil Tirion. Cambridge University Press, 2000.

   This is a good, basic guide to what is up in the sky. The book arranges its star charts by constellation, which I believe helps one better learn the layout of the sky.

2. *A Field Guide to the Stars and Planets* (Peterson Field Guide Series), by Jay M. Pasachoff. Houghton Mifflin, 2000.

   This is a more complete guide, with much more detailed star charts. As with the guide above, the charts were compiled by Wil Tirion. It divides the sky into regions, and arranges its maps in a way that might be confusing for a novice, by spiraling around the sky from north to south.

3. *Stars and Planets*, by Ian Ridpath. Dorling Kindersley Handbooks, 2000.

   This is an excellent book, published by the company famous for creating the Eyewitness series of illustrated non-fiction. It is very well organized, and wonderfully illustrated.

4. *Sky & Telescope's Pocket Sky Atlas*, by Roger W. Sinnott. Sky Publishing, 2006.

   This book is different from most. Rather than arranging charts alphabetically by constellation, it divides the sky into 80 charts (and four close-up charts of smaller regions) arranged in a simple, user-friendly way. It is also not actually pocket-sized, which is probably a good thing since it allows for fairly large charts that cover wide areas of the sky. It is a sort of bridge between a traditional field guide and larger star charts, which are discussed later.

All of the above guides are reasonably priced, and designed to be taken into the field.

Several books which specialize in stargazing with binoculars are also available. Here are just a few of them:

1. *Exploring the Night Sky With Binoculars*, by Patrick Moore. Cambridge University Press, 2000.
2. *Touring the Universe Through Binoculars*, by Phillip S. Harrington. John Wiley & Sons, 1990.
3. *Discover the Stars: Starwatching Using the Naked Eye, Binoculars, or a Telescope*, by Richard Berry. Crown Publishing, 1987.

# “But I want a telescope!” 3

Of course you do, but there are good reasons for waiting until this point to get to telescopes. Once you have learned something about the sky, by getting out under the stars with a planisphere, or a field guide, or one of the monthly magazines – whether with binoculars or just your eyes – then you have probably done two things. First, you have discovered whether or not you want to go further in pursuit of this hobby, and you have done so at a relatively small cost. You may have decided that your eyes, or binoculars, are enough. Many people do.

Second, you have probably gone some way toward making the use of a telescope easier, and so, more enjoyable. If you have a basic understanding of how the night sky “works,” and how to find specific things in it, then you will avoid much of the frustration that beginners often feel when first trying to find something in a telescope. It is much easier to view something through the telescope if you have some idea of where it is going to be, and when. And no matter how much you may read about it, there is no substitute for watching the sky change, over a stretch of hours, and a succession of weeks or months.

For these reasons, I strongly suggest that, if you have not already done so, you get out to a dark site and study the sky before you buy your first telescope.

## The beginner’s telescope

My belief is that there are three primary concerns a beginner should have when buying a first scope: *cost, “size,”* and *ease of use*. We should look at each in detail.

### Telescope cost

You will, of course, be the one to choose how much you spend on a complete, ready-to-use telescope package. I have somewhat arbitrarily chosen a *maximum*

**Figure 3.1 The author's scopes, with apertures from 80mm (3.1 inches) to 317mm (12.5 inches).**

of $500 here in the United States, and based my choices on current (2008) prices. These prices may have changed by the time you read this, but the trend in telescopes for several years has been *falling* prices, so you may actually be able to go beyond what I am going to describe. Prices will, of course, vary from this in other parts of the world, but the *choices of available telescopes for a given price* should, I think, be the same. The italics explain the reason I think it makes sense to choose an arbitrary, maximum price: to make it possible to compare different telescope types within a moderate boundary.

Five hundred dollars is much more than I spent on my first, perfectly adequate, telescope. It should be enough for just about anyone. Personally, I do not think it makes sense to spend more, initially, on a pastime you may not yet be sure of. If you do decide to spend more, you should plan to do research beyond the scope of this book. The $500 maximum leaves out many choices of size and type of telescope and accessories, but that is a *good* thing: there are so many choices out there that it can get very confusing.

**More info**

$500 may not seem like a "moderate boundary" to some, but astronomy equipment can get just about as expensive as the imagination can reach. Even some quite small (though wonderful) telescopes can cost many thousands of dollars.

There is also a minimum that you should be willing to spend, if you want an instrument that is to be of any use. You will

**Figure 3.2** When this 80mm telescope was new, in the mid 1990s, it retailed for nearly $500. These days, in 2008, you can get a slightly larger (90mm) scope package from the same manufacturer for less than half that price.

probably not be able to find a decent, complete telescope set-up for less than $200, unless you find a real bargain (which is possible, though not likely). A price range of $300–$500 is more realistic, for a scope that may be all you ever need. Something less expensive may *look* impressive, but probably will not be an impressive performer (see Figure 3.3).

### Telescope "size"

This is much more complicated, and will take some explaining. There are two features to look for when determining any telescope's "size:" *aperture* and

**Figure 3.3 Although these two scopes look similar, the one on the left, which cost $100 and came in the box shown in Figure 4, is virtually useless, from almost any perspective (image clarity, sturdiness, focusing... I could go on and on). The scope on the right – or a current model of similar quality – though still not as sturdy as it could be, is definitely useable, and you can find it for about $250.**

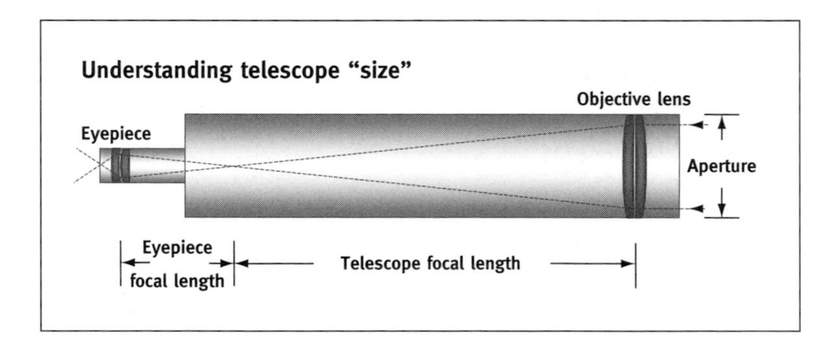


**Figure 3.4 A simple drawing that shows what factors determine a telescope's "size."**

***focal length***, and dividing the second by the first gives another important factor: ***focal ratio***.

Remember, a telescope's *aperture* is the diameter of its main light-gathering surface (either a lens or a mirror). It is the aperture that determines how bright and clear an image can appear through a telescope.

The *focal length* is the distance between the light-gathering surface and the **focal point**, the place at which the light rays, which have been bent by the lens or mirror, come to a point (or, more accurately, a small plane or spot). This measurement determines how "powerful" a telescope can be, when combined with a given eyepiece. Eyepieces have focal lengths of their own, and when you divide a telescope's focal length by an eyepiece's focal length, you get the magnification.

Here is an example. Let's say you have a telescope with a focal length of 1000 mm. If you insert an eyepiece with a focal length of 25 mm into the focuser, images will be magnified by 40 times (1000 divided by 25=40). If you changed to a 4-mm eyepiece, the magnification goes up to 250 times (1000 divided by 4=250). It is focal length alone which determines magnification.

*Which is more important: aperture or focal length?*

Aperture, without a doubt. Why? Recall that it is the aperture that determines both brightness *and* clarity. (The measure of the clarity a telescope can be expected to have is called **resolution**, or resolving power.) So, in a telescope with a small aperture, images appear dimmer and blurrier than in a telescope with a larger aperture. At low powers, this may be acceptable (as in binoculars); but the higher the magnification, the dimmer and fuzzier the image seems.

Here is a good rule of thumb that is often used: useful magnification is equal to two times the aperture, when measured in millimeters; or fifty times the aperture, when measured in inches (since 1 inch is about

**Figure 3.5 If all other things were equal, the telescope on the right, with an aperture of 100 mm (about 4 inches) should not only produce 56% brighter images, but also *sharper* images, than the 80 mm scope on the left.**

 **More info**

Sixty millimeters is the aperture of many "department-store junk" telescopes. They often have a focal length of 900mm, and might come with an eyepiece with a very short focal length. This means they truly can reach "powers" as high as they claim, but their aperture makes the magnification a useless, very dim, frustrating blur. There ought to be a law!

25mm). Of course, this assumes you have decent optics in your telescope. So, if you have a 60-mm telescope, the highest magnification you should expect from it is about 120×.

Notice that focal length does not matter here: a beginner's telescope with good optics and an aperture of 100mm (4 inches) should not be expected to reach powers higher than 200×, whether it has a focal length of 500mm and is used with a 2.5-mm eyepiece, or a focal length of 1500mm with a 7.5-mm eyepiece.

*Magnification limits*

At this point, since I am discussing magnification, I should also mention this: the Earth's atmosphere can really cloud things up. Even the best telescopes with the largest apertures have their images blurred and distorted by **turbulence** in the atmosphere. When things are truly bad, stars appear to twinkle, even to the naked eye. In a telescope, images flare and spike and quiver. Astronomers call this bad "**seeing**." But even under very good conditions, atmospheric interference usually prevents useful magnifications above about 300×. With great optics under perfect skies, higher powers are sometimes possible for fleeting moments, but the beginner should not expect this. The good thing is, you will almost never want more than 300×, and you will almost always want much less.

*Focal ratio*

If you divide a telescope's focal length by its aperture, you get its *focal ratio*. For example, one of the telescopes above, with the 100-mm aperture and the 500-mm focal length, would have a focal ratio of five, usually written f/5; this might be called a "fast" scope. The 100-mm scope with the 1500-mm focal length would be f/15, and might be called "slow."

These speed references are terms from the world of photography, and have nothing to do with the brightness of objects viewed with the eye through a *given eyepiece*. There is a misconception out there, often found on the Internet, repeated by sales clerks, and even in books, which insists that, like camera lenses, telescopes with higher focal ratios do not let in as much light as those with lower focal ratios. This is simply not true. Using the two 100-mm aperture telescopes mentioned above as examples, the scope with a focal length of 500mm (giving a focal ratio of f/5) would give a magnification of 20× if you inserted a 25-mm eyepiece (500 divided by 25 equals 20). The same eyepiece, when used in the 100-mm scope with a focal length of 1500mm (f/15) would give a magnification of 60×. The magnification is different, *but the brightness is the same*. Others may tell you otherwise, but brightness is determined by aperture alone.

**Figure 3.6 Two telescopes with different apertures, focal lengths, and focal ratios. The scope on the left, with an 80-mm aperture and 910-mm focal length (giving a focal ratio of about 11.4, when you divide the focal length by the aperture), is a good choice for a small, all-in-one scope for a beginner. It should be able to span a useful range of magnifications somewhere between 30× and 160×. The scope on the right, with an aperture of 100 mm, and a focal length of 500 mm (for a focal ratio of f/5) might seem like it could be used at higher magnification because of its larger aperture. But at f/5, and with such a short focal length, it is much better suited for wide views, and can give stunning images at powers as low as 16×.**

*Choosing the right "size"*

For *aperture*, this is easy. Once you have chosen the type of telescope you want (we will get to that soon), get the largest aperture you can afford. Telescopes can get huge: there are two in Hawaii with mirrors more than thirty feet (10 meters) across. However, for under $500, physical size will probably not become a burden.

An aperture of 80 mm is, I believe, the *minimum* you should get (because of the **central obstruction** in some types of scope, 90–100 mm would be better). With anything much less than that, you would probably be better off sticking with binoculars. However, an aperture of 80 mm will give images more than two and a half times brighter than 50-mm binoculars, and will allow magnifications up to 160×, enough to see good detail on the Moon and some on the planets, and more than enough for most deep-sky objects you will be able to view.

For *focal length*, things get more complicated. A telescope with a longer focal length allows the use of eyepieces that are easier on the eye at higher magnifications (more on that later). Also, it is usually easier and cheaper to make good optics when using slower focal ratios (which means longer focal lengths). Even in great telescopes, **aberrations** (optical problems) creep in at short focal lengths. For these reasons, it might seem a long focal length is best.

On the other hand, telescopes with shorter focal lengths are often, well, shorter. This tends to make them lighter and more portable, which means they require a smaller, cheaper **mount**. Also, a shorter focal length means lower magnifications are possible, allowing wider views.

So, how to choose? Strike a balance. At any focal length less than 500mm, I think you would be better off with one of the less expensive pairs of "big" binoculars, which might be just fine. At anything more than 1,500mm, you begin to seriously limit yourself to using higher magnifications. You would have what amateur astronomers call a **planetary scope**. This also might be just fine. But somewhere in between these two will allow you to view a wide range of objects at a variety of magnifications, in a package portable enough to be easily used.

### Ease of use

This is the last broad consideration when choosing a telescope, and it is another area where personal choice comes into play. Some telescopes are easier to set up and use than others. This is even more true of whatever it is they are mounted on. Some are easier to aim than others. Some are prone to one aberration, others to another.

The question comes down to this: what features are most important to you, and what are you willing to put up with to get them? One package may offer the brightest views, but be heavier and more awkward. Another may be light and portable, but have limited choices of magnification. After you learn what options are out there, it will be up to you to choose what you think will best meet your needs. However, be aware that if you find your telescope a pain to use, you probably will not use it.

I have my own personal biases, which are based on what I think an average beginner (me, some years back) would be happiest with. I will make recommendations, and I will give reasons for them; but in the end, you must decide what is best for you.

## Telescope types available to the novice

There are three telescope designs, easily available to the beginning stargazer, which fit the constraints of cost, size, and ease of use that I have described. They are the **achromatic refractor**, the **Newtonian reflector**, and the **Maksutov–Cassegrain** (or **"Mak"**). Each has pluses and minuses, and each has its own devoted fans. Competition has led to widespread availability of all three types, in a wide variety of sizes.

### The achromatic refractor

When most people think of a telescope, a refractor is the type that comes to mind, and with good reason. It was a refracting telescope (a telescope that refracts, or bends, light through lenses) that was the first to be built

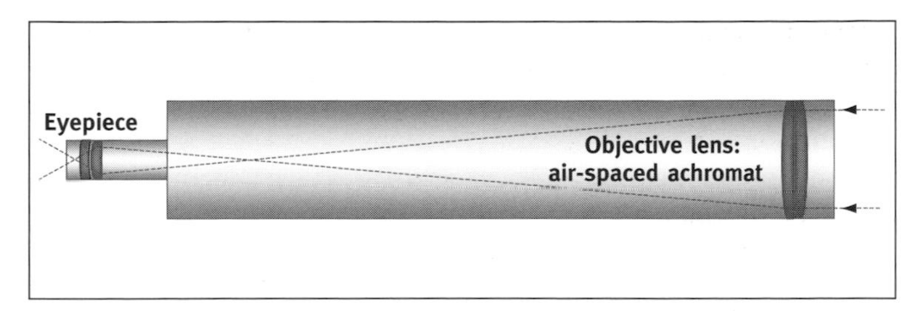


**Figure 3.7 Inside an achromatic refractor.**

and used by Galileo to look at the night sky. It was the only type to exist for some time. Also, the refractor has long been a popular choice among amateurs.

There are several different types of lenses used in refractors, but for the beginner, the choice will almost certainly be an objective lens made with an **air-spaced achromat**. Galileo's telescope used a simple objective lens (just one piece of glass). It worked, but simple lenses have a problem: while bending the light in a way that magnifies, they also act as a sort of prism, breaking the light that enters the scope into its component colors. This is called **chromatic aberration**. It was a huge problem in early refractors, and took centuries to solve.

Someone finally figured out that putting together two lenses of different shape (and usually of different types of glass) greatly reduces the **false color** created by chromatic aberration. The lenses usually have a very short space between them. This is the air-spaced achromat (achromat means "without color").

**More info**

Early telescope makers found that one way to minimize chromatic aberration was to make telescopes with very long focal lengths. Telescopes reached lengths of more than a hundred feet (30 meters). They became so long that tubes could not be used because they would sag and block the light. Lenses were suspended from wires inside large halls.

*Advantages*

1. The refractor is virtually *maintenance free*. The objective lens, when treated well and protected from dust, should last a lifetime or more.

2. Refractors have *no central obstruction*. In other types of telescopes, a mirror is placed in the center of the light's path in order to reflect it into the focuser. This obstruction decreases **contrast** in images seen through the telescope. These mirrors, on some scopes, also need periodic adjustment.

3. It is *easy to aim* because its focuser sticks out of the bottom, or back, and is in line with the direction of the targeted object.

4. Refractors come in a *wide variety of focal ratios*.

**Figure 3.8 A classic "planetary" refractor. This scope, the author's home-built 120 mm (4.7-inch) refractor, with a focal length of 1500 mm (f/12.5) is beyond the price range of this guide; but not by much. A 120-mm refractor with a focal length of 1000 mm (f/8.3) is a possibility within our price range.**

5. ***Cooling time*** *is less of a problem*. In refractors, light travels down the tube just once, and heads out through the focuser. In other telescope designs, light is reflected back up the tube, and sometimes back down again. If the air inside the telescope tube is

at a different temperature from the air outside, **tube currents** can form which will distort images. The more times light travels through these currents, the more distorted the images become.

*Disadvantages*

1. *"Faster" achromatic refractors (those with shorter focal lengths compared to their apertures) still have problems with false color.* The lenses act like prisms, and blue or purple edges might appear around objects being viewed. But above f/8 (a focal length eight times the aperture), a good lens will show truly objectionable color only on quite bright objects. At f/12 or higher, color should not be noticeably bothersome.
2. Because the focuser sticks out of the bottom of the telescope, *viewing might get uncomfortably close to the ground.*

**More info**

One of the more important advances in amateur telescopes of the past several decades was the introduction of the **apochromatic** refractor. Various designs are available, but all use special lens arrangements and/or glass types virtually to eliminate false color. Many people consider them the best telescopes made. For the extremely high prices charged for them, they should be!

**More info – Changing times**

Until a few years ago, *cost* would have been considered a major disadvantage of the refractor, but scopes with decent optics are now being made and sold by many companies. Competition among these companies has been stiff, and prices have gone down considerably. In larger diameters (and apochromatic designs), refractors are still the most expensive type of telescope, per unit of aperture. But in the sizes appropriate for a beginner, price is not as big an obstacle as it once was.

### The Newtonian reflector

First developed by another pioneer in science, **Sir Isaac Newton**, the design of this telescope is fairly simple. Other than the lenses in the eyepiece, there is only one surface that has a curve in it (the **primary mirror**). Light enters the top of the telescope and travels to the bottom of the tube, where it is reflected off of the primary mirror, and focused back up the tube. Before reaching the focal point, the light is reflected off the **secondary mirror**, which is set at a 45 degree angle in the center of the tube, near the top. The light then leaves the scope through the side of the tube, where the focuser is placed.

The design is so simple that, prior to the arrival of large-telescope manufacturers, many amateur astronomers built their own reflectors, grinding the necessary curve into the glass of the primary mirror. Some still do.

*Advantages*

1. The single largest advantage of the Newtonian reflector is *cost*. Because of the simplicity of its design, it is by far the least

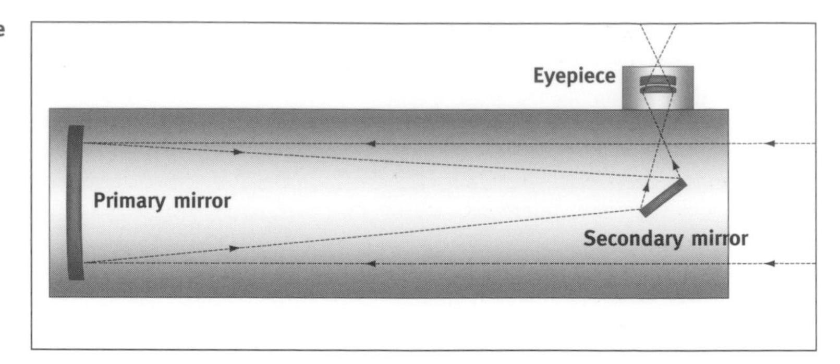


**Figure 3.9 The simple design of the Newtonian reflector.**

**Figure 3.10 To get the 114-mm (4.5-inch) aperture of this reflector in another telescope design would cost about twice the price of this scope.**

expensive type of telescope, per inch of aperture. For the price of some 4 inch apochromatic refractors, you could buy a 20 inch reflector. Of course, a larger aperture means brighter, clearer images; and the brighter the images, the more dim objects you can see.

2. Reflectors *do not suffer from chromatic aberration*, since the light is reflected off a single surface, instead of being refracted through a lens.
3. In the sizes available to the beginner, the *focuser* is usually placed at a *comfortable viewing position* near the top of the scope. (A ladder is needed to reach the focusers of larger, more expensive reflectors!)
4. Because one end of the tube is open, Newtonian reflectors *cool down more quickly* than some other types of telescope.

*Disadvantages*

1. Reflectors have their own type of aberration: ***coma***. Stars near the edge of the field of view seem to have tails, like little comets, or commas. The problem is worse in fast scopes (f/5 or faster).
2. Reflectors have a *central obstruction*. The secondary mirror blocks some of the incoming light, which lowers the contrast.
3. ***Diffraction spikes:*** the secondary mirrors in Newtonian reflectors are held in place by a **spider**, which is simply a structure with three or four thin vanes that stretch from the edge of the tube toward the middle, where the mirror is attached. This structure interferes with the view and causes *diffraction spikes*, small rays that appear on brighter stars in the pattern of the spider's vanes (see Figure 3.11).
4. Reflectors need some *cooling time*. Because the light entering a reflector travels down and then back up the tube, it is more affected by tube currents. Time is needed for the temperature inside and outside of the tube to equalize. The larger the scope, the longer it takes; partially because the main mirror itself becomes so thick in bigger reflectors that it needs considerable time to cool.
5. Reflectors need to be ***collimated***. The mirrors need to be periodically lined up with each other and with the focuser; otherwise, images will be distorted. To do so is a fairly simple process, but Newtonian reflectors need it more often than other designs.
6. The *coatings on the mirrors are not permanent*. Though, with care, they should last for many years, mirror coatings will probably need eventual replacement. There are companies that do this for a reasonably small price.
7. Because the focuser sticks out at a 90 degree angle to the direction of targets, Newtonian reflectors are *somewhat difficult to aim*.

Figure 3.11 Looking down the front, or top, end of a Newtonian reflector, with a view of the spider, secondary mirror, focuser tube, and, at the bottom, the primary mirror.

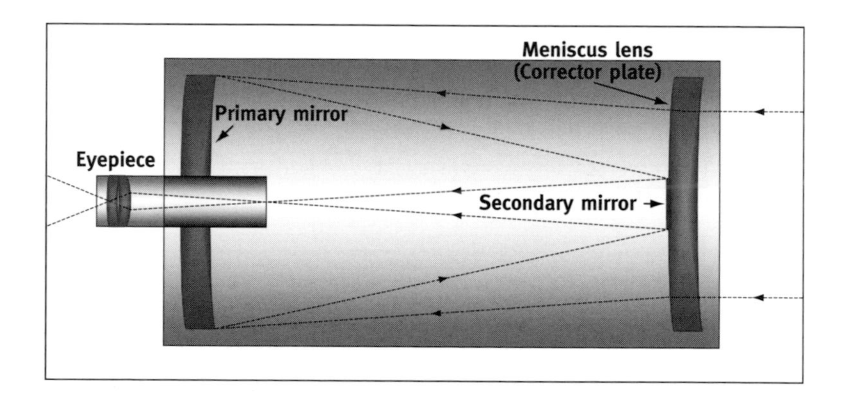


Figure 3.12 The complex light path through a Mak.

### The Maksutov–Cassegrain (or Mak)

Of the three available to the beginner, the *Maksutov Cassegrain* (or *Mak*) is the newest telescope design. **Dmitri Maksutov** developed it in the Soviet Union in the 1940s. It uses both a lens and a mirror (telescopes that use both are called **catadioptric**), and has four curved surfaces that influence the incoming light: the two sides of the lens, the primary mirror, and the secondary mirror (which is actually just a coating applied to the back of the lens).

What makes the design affordable is that all four of these curves are **spherical**. It is much easier to grind precise spherical curves into glass than other shapes.

Spherical mirrors produce severe coma, and this is why the Mak has a lens: the curves in the lens are designed to correct the image created by the mirror in order to reduce coma. In fact, the lenses on catadioptric telescopes are more often called **corrector plates**.

**More info – Some terminology**

A Cassegrain telescope is any reflecting telescope that has a hole in its main mirror at the back of the scope, where the focuser is located. The lens on a Mak is a **meniscus lens**. Meniscus means "crescent shaped;" and a meniscus lens is one that is curved outward (**convex**) on one side, and inward (**concave**) on the other. This gives it a sort of crescent shape.

**More info**

There is another type of Mak: the **Maksutov Newtonian**; which, as the name implies, combines some features of the Maksutov Cassegrain and the Newtonian reflector (and was also designed by Dmitri Maksutov). As I write, this type of scope is not readily available within our price range of $500, so it will not be discussed in detail. But by the time you read this, that may have changed, and you may want to do some research on this excellent design.

*Advantages*

1. *Physical size* is the main advantage of the Mak. Because it bends light up and then back down the tube, off of two curved mirrors, it is possible to contain a long focal length in a short tube. This allows the use of smaller, cheaper mounts. It also makes for a very portable package (see Figure 3.13).
2. The closed tube and thick corrector plate, which are both part of the design of the Mak, mean *sturdy, solid construction*. The closed tube also protects the mirrors' coatings.
3. The corrector plate does just that: a well-made Mak will have *less coma* than a Newtonian reflector.

*Disadvantages*

1. The closed tube means that a Mak will need *longer cool-down time*. And because the light travels through the tube three times before reaching the focuser, tube currents in an uncooled tube will affect images more than in a Newtonian reflector or, especially, a refractor.
2. Like the Newtonian reflector, the Mak has a *central obstruction that will affect contrast*. (It does not, however, have diffraction spikes, since it has no spider vanes: its secondary mirror is applied directly to the corrector plate.)

**Figure 3.13 (a) Because of the complex path and four curved surfaces of a Mak, this short (less than 12 inch) tube is able to house a 102-mm (4-inch) aperture scope with a focal length of 1300 mm. (b) The mechanics of a Mak: visible here are the meniscus lens/ corrector plate, the secondary mirror attached to the plate's center, the primary mirror, and the focuser housing coming through the primary's center.**

3. The design of the Mak makes it "slow." Most have focal ratios around f/13, or even higher. This makes them excellent for high magnification, but less well-suited to **wide-field** viewing.
4. At the beginner's level, *cost* is a factor with Maks. They are usually as expensive, per inch of aperture, as refractors; and much more expensive than Newtonian reflectors.

## As important as your scope: the mount

This is not an exaggeration. If using your telescope is not an enjoyable experience, you will not use it; and shaky images or difficult aiming and **tracking** do not add up to fun!

Sturdiness is vital. A mount that allows you to move smoothly and easily from one target to another, or to easily bring objects back to the center of the field (as the Earth's rotation causes objects to move out of view), is also necessary. Finally, a telescope is much more enjoyable to use (especially by groups) if it can automatically keep objects in the field of view for long periods of time. There is only one type of mount available to the beginner that meets *all* of these requirements; but before getting to it, we will look first at the alternatives.

### The alt-az mount

The term **alt-az** is short for altitude- (or vertical) azimuth (or horizontal). That is how these mounts move. A camera tripod is a basic alt-az mount, and

**Figure 3.14 Alt-az mounts. (a) When you adjust the tension of the lever on top of the tripod, and the handle coming out from it, then you can easily turn the scope on this mount by hand, and have it stay where you put it. (b) A 6-inch (150-mm) reflector with a Dobsonian mount. You can easily adjust the aim of this scope by pushing and pulling on the round knob on the bottom of the front end.**

the alt-az mounts designed for beginner scopes are not very different from camera tripods (see Figure 3.14 a.) They should, however, be sturdier, and should include controls that allow for slow-motion movements in both directions. Otherwise, the telescope is not worth using for astronomy. A shaky mount makes finding and focusing on objects almost impossible; and without slow-motion controls, keeping an object in view, as the Earth's rotation causes it to drift, just cannot be done. The junk telescope described at the beginning of the guide tells the effect of a *bad* alt-az mount: it gathers dust in a closet.

*The* ***Dobsonian*** *(or Dob) mount*. This is a simple, lightweight, inexpensive type of alt-az mount that is usually used with Newtonian reflectors (see Figure 3.14 b.) Popularized by amateur astronomer **John Dobson** in the 1970s, it is one of the reasons for the growth of astronomy as a hobby. It allowed the building, by amateurs, of huge yet inexpensive scopes. A good Dobsonian is sturdy and well balanced for easy movement.

***Fork mounts***. These are, basically, another type of alt-az mount. They have two tines that attach to the side of a telescope, and this is what gives them their name. They work best on telescope designs that use

**More info – Fork equatorial mounts**

A German equatorial mount is, basically, an alt-az mount that can be tilted to an angle at which the azimuth (horizontal) circle of motion can be matched to the equator. Fork mounts can be equipped with accessories that make them equatorial, too.

short tubes, and are usually associated with the **Schmidt–Cassegrain** telescopes (**SCTs**) sold by **Meade** and **Celestron** (the two largest American telescope manufacturers). Schmidt–Cassegrain telescopes are outside our price range.

### The German equatorial mount (GEM)

Because the Earth spins, stars in the night sky appear to rise in the east and set in the west;

**More info – Right ascension and declination**

Locations on a globe of the Earth are marked with perpendicular lines. The lines running from north to south between the poles represent **longitude**; those circling the Earth from east to west represent **latitude**. These lines are divided into degrees (360 degrees make a circle); which are divided into **minutes of arc** (60 minutes make a degree); which are divided into **seconds of arc** (60 seconds make a minute). The "zero line" for longitude runs from the north pole, through Greenwich, England, and on to the south pole. For latitude, the equator is zero; the poles are 90 degrees. If you know a place's latitude and longitude, you can find it, easily and exactly, on a map.

The globe of the heavens is measured in a similar, though slightly different, way. Celestial latitude is called declination (Dec), and is also measured in degrees; zero at the equator, 90 degrees at the poles. Heavenly longitude is called right ascension (RA). It is different from longitude in that astronomers divide it into hours, instead of degrees. These hours are then divided into minutes and seconds.

or, if they are near the **celestial poles**, stars seem to go in a circle around these points. These two spots in the sky are exactly aligned with the Earth's *axis*, the imaginary line that runs through the north and south poles of the planet, and about which it spins like a top, once a day. The north celestial pole is very near the star **Polaris**; the south celestial pole has no bright stars very close to it.

When you look through a telescope, the image you see will drift out of the field of view as the Earth spins. The higher the magnification, the quicker the drift. At high powers, an object in the center of the field can disappear from view in less than a minute. This causes a problem for alt-az mounts. They move vertically and horizontally, but drifting objects almost always move diagonally (the angle depends on your location, and the object's place in the sky). So, to keep an object in view, you need to adjust an alt-az mount in both directions. If you plan to use your telescope for bird-watching or other terrestrial viewing, then you will want a good alt-az mount, but they are not the best choice for stargazing.

An **equatorial mount** is designed to remedy this problem. It has an axis of rotation that is adjustable, and meant to be centered on the celestial

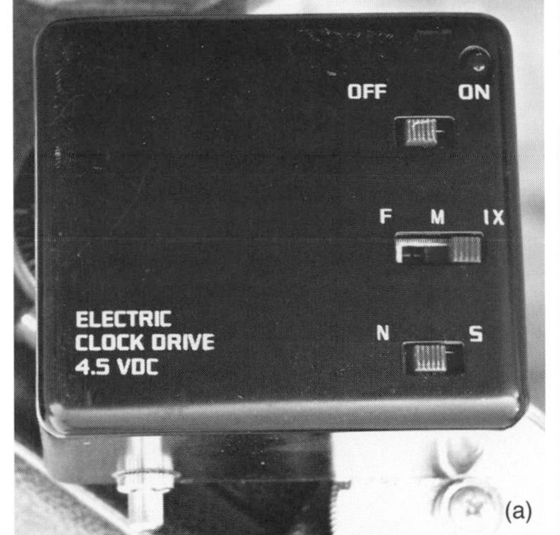


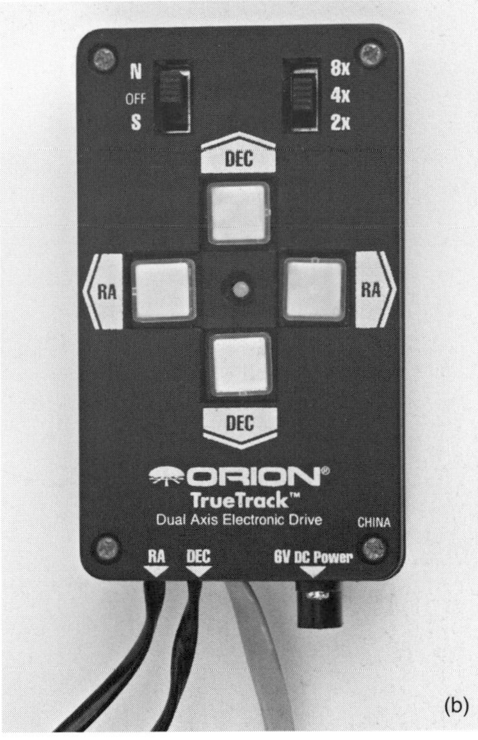


**Figure 3.15 Electronic drives can be as simple as (a) a small, battery-operated motor that is attached directly to the RA control shaft; or (b) much more deluxe, with hand-held, push-button, variable-speed control of motion in both directions in both declination and right ascension.**

pole. The most common design is called the German equatorial mount (or GEM). When you set one up, you do a **polar alignment**, so that the adjustable axis of your mount matches the axis of the Earth. Afterwards, once an object is centered in the field of view, the telescope will only need to be adjusted in one circular direction (called **right ascension** or **RA**) as the image drifts. *Exact* polar alignment is somewhat difficult, and not necessary most of the time. Without it, images will still drift a little. Minor adjustment is then made in **declination** (or **Dec**), the circle of motion **perpendicular** to right ascension. This may not seem like a big improvement over alt-az, but it makes things much easier if you want to study an object for a while, or share the view with others.

Things can be made even easier. Since adjustment (or tracking) only needs to be done in one constant, circular direction, a **clock drive** can be attached to the mount. This is a small motor, usually battery operated, which is attached to the RA axis. It makes the mount turn so that it closely matches the turning of the Earth. Once an image is centered in the field of view, and the motor is turned on, the mount will keep the image centered. This is called **motorized tracking.**

More advanced drives have two motors, so that periodic electronic adjustments can be made in declination and right ascension, if the image shifts a little. A **hand controller** is used to move the telescope precisely in any direction. A mount with this set-up has **dual-axis drives**. For mounts

**Figure 3.16 The German equatorial mount. (1) Counterweight and counterweight shaft. The counterweight is used to balance the telescope, so that it is easy to move (either by hand or with the drive), and so that it stays in place once the movement is done. To use it, you loosen the bolt visible on the top of it, and slide the counterweight on the shaft to find the proper, balanced point, and then tighten again. (2) Declination lock lever (the RA lever is on the other side of the mount, so not visible in the photo). When the levers are unlocked, it is easy to move a well-balanced scope close to the observing target by simply pushing and pulling by hand. Once close, you can lock the levers, and make final adjustments with the slow-motion hand controls, or with the electronic hand controller. (3) Right ascension motor housing. When this motor is running, the mount's motion matches that of the turning of the Earth, keeping objects centered in the eyepiece. (4) Azimuth adjustment knobs. (5) Altitude adjustments bolts. The azimuth and altitude adjusters are used only to align the mount with the celestial pole; they are not meant to help you in aiming the telescope. (6) Polar alignment scope. This is a small telescope that is mounted right inside the mount, and is used to help you get accurate alignment with the pole.**

with only one motor (a **single-axis drive**), periodic declination adjustments are done, easily enough, by turning a knob by hand.

A perfectly aligned mount with perfectly accurate gears and motors would keep an object perfectly centered for as long as you kept

**Figure 3.17 The CG-3 (or EQ-2) mount; this one is equipped with a single-axis drive. Note the slow-motion control cable angling downward to the right.**

the motor running. Since this is not realistically possible, adjustments will still have to be made every once in a while. But a casual polar alignment is enough for even a fairly cheap motor to keep adjustments down to only once every several minutes. With a little more careful alignment, an object can stay in the field of view, with no adjustments, for a half hour or more.

More advanced equatorial mounts have the option of mounting a **polar alignment scope** on them. This is a small, wide-field telescope, mounted along the RA axis, which helps in making a careful and accurate polar alignment. This is a great convenience if your telescope will be used by a group, or if you want to try some **astrophotography** (more on this later).

### Some commonly available GEMs

Though there are many different companies that sell German equatorial mounts, most models for the beginner seem to fit one of four basic designs. They come with many different names, but the three most useful designs are basically "clones" of those sold for years by Celestron: the CG-3, CG-4, and CG-5. The CG-4 and -5 are themselves clones of the Polaris and Super Polaris mounts once – but no longer – made by the Vixen Company of Japan. In fact, the CG-4 and -5 were, at first, made by Vixen for Celestron, but were later

**Figure 3.18 The mid-sized CG-4 (or EQ-3) mount.**

replaced by less expensive (and lower quality) copies. The fit and finish of these mounts has improved over the years, but quality control can still be a problem, so let the buyer beware.

I have included this history because the names may come up when searching for a specific mount, and it will help to be able to compare apples to apples, rather than oranges. And the names I have given are only those I have come across in the USA; in other parts of the world, other names are

**Figure 3.19 The CG-5 (or EQ-4 or EQ-5) mount.**

almost certainly used. Use the accompanying photographs to help with identification.

*The EQ-1 mount*. This mount, or one very like it, is often included with very small, light telescopes (especially Maks), and with junk telescopes. I think it is just too flimsy to be of much use.

*The CG-3*. Sometimes called the EQ-2, this is the lightest-weight GEM that I have found to be useful. It can support small-to medium-sized telescopes, up to about 90 mm (3.5 inch) in aperture, as long as they are not very long (the longer a scope, the more prone it is to noticeable shaking). This mount can also be equipped with a single-axis drive. Celestron has recently begun selling a much different-looking, though probably quite similar, mount with the same name, CG-3.

*The CG-4*. This is the clone of Vixen's Polaris mount, and its copies are sometimes called EQ-3 or the EQ3–2. It is the least expensive mount that is available with the capacity to be equipped with a polar-alignment scope and dual-axis drives, necessary for longer-exposure astrophotography. It is quite adequate for visual use with a medium-sized telescope; up to a 100-mm (4-inch) refractor, or an even larger scope of any type, if it comes in a short tube.

*The CG-5*. This clone of the Vixen Super Polaris is sometimes sold under the name EQ-4 or EQ-5, and others. It is the top-end mount that might be within my price range for the beginner. The latest models are quite sturdy, and will support loads of more than 20 pounds (9 kilograms). Nearly all of the **astrophotos** in this book were taken with cameras and telescopes mounted on this GEM. Both Celestron and Meade sell heavier-duty **computer-guided** versions of this mount (see below). Celestron still uses the name CG-5; Meade calls theirs the LXD-75. Though both are outside the price limits of this book, the fact that these companies use the design means the mount should be around for some time to come.

### Computer-guided mounts

One of the biggest changes in amateur astronomy in the past decade or so has been the rise of the computer-guided (or **"go-to"**) mount. To use them, the time and the viewing location (longitude and latitude, or postal code, or nearby city) are entered into the telescope's hand controller. The telescope is then pointed at two or three known stars, while these stars' names or identification numbers are entered into the hand controller. The built-in computer in the mount can then point the telescope at other objects (usually many thousands of them) contained within its memory.

There are telescopes for under $500 that come on computer-guided mounts. But, for the price of the mount, you trade aperture size, mount stability, and more. In our price range, the apertures available in a go-to scope may not even be able to show you some of what is in their computer's memory. And the tripods, finders, eyepieces, and other accessories that come with beginner go-to scopes just do not seem as robust as comparable non-go-to packages. However, this may soon change. Prices continue to fall, and decent-sized telescopes on good go-to mounts may soon be available for the absolute beginner. (On the other hand, the growing prevalence of go-to mounts has tended to drive down the prices of traditional non-go-to mounts, especially on the used market, making these simpler mounts even more attractive.)

Of course, there is also the question of whether a computer-guided telescope is a good idea for a beginner. Many beginners – who would otherwise have stayed away – have gotten started in the hobby, and kept on with it, through the use of a go-to scope. There are very dedicated fans of the various models. There are many advanced amateurs who would say that go-to mounts are *the* way for beginners to get started, since they bypass the need to understand the complexity of the night sky and how it changes from hour to hour and night to night.

There is sense in that argument, but it is my own belief that these mounts make things both too easy and too difficult: too easy, because they take away the necessity (and challenging fun!) of learning the night sky; too difficult, because they add yet another layer of learning to the process of just getting started, as well as an added possibility of getting something defective.

**Figure 3.20** Perhaps the largest-aperture go-to telescope within this guide's price range: a 130 mm (5.1 inch) Newtonian reflector with a focal length of 650 mm (for a focal ratio of f/5.)

There are more advanced go-to mounts that are very rugged, very stable, and very simple to use, almost as simple as turning them on and stepping to the eyepiece. But these are far beyond the price range set as a limit for this book. Beginners' go-to scopes (at least for now) tend to be finicky, frail, and not very accurate in their pointing. It takes time, practice, and some slight knowledge of the sky to get them started each time they are used (unless they are not moved). Also, no beginner's go-to scope that I know of can be operated manually. If the user becomes frustrated with the difficulty of getting an object centered in the eyepiece because of some glitch in the software, or because of a bad wire or battery, the scope will become a dust-gatherer.

Besides, there are many fascinating objects in the sky that are very easy to find without any special aid, or any great knowledge of the night sky. Getting started without a computer's help is not that hard.

There is an ongoing debate among amateur astronomers about this issue. In the end, it is another matter of choice. For myself, finding objects on my own provides much of the challenge and interest of stargazing. I have been at it for years, and have felt no need for help. In fact, much of the fun has been in stumbling upon things on my way to somewhere else, and then figuring out what they are by studying charts. I do not know how easy this would be with a go-to mount, or if it would even be possible. As I continue in the hobby, and search for ever dimmer and more elusive "faint fuzzies," I may change my mind; but for now, I see no need for a computer inside my mount.

## The busy end of the telescope: eyepieces, Barlows, diagonals, and finders

There are many accessories available for the eyepiece end of the telescope. Some are necessary; others are very helpful. They range in quality from exquisite to useless.

### Eyepieces

Astronomical telescopes come with focusers that use interchangeable eyepieces. These are lenses, or groups of lenses (anywhere from one to eight, or

**Figure 3.21 In the small area near the focuser of this scope you can see many of the items described in this section: an eyepiece, diagonal, Barlow lens, and finder.**

even more) bound together in a small tube. There are three standard eyepiece sizes, based on the diameter of the tube that slides into the focuser: 0.965 inches, 1.25 inches, and 2 inches (These are physical diameters of the eyepiece tubes, and should not be confused with their focal lengths, which can vary widely, and are almost always expressed in millimeters.). *You should avoid telescopes with 0.965-inch focusers.* Few eyepieces are available for them, and most of those are of poor quality. The 1.25-inch eyepieces are the most common, and all you may ever need. But if you can, get a telescope with a 2-inch focuser. They almost always include a simple adapter that allows the use of 1.25-inch eyepieces, and if they do not, the adapters are inexpensive. The 2-inch eyepieces allow lower magnification for wider views. You can use 1.25-inch eyepieces in 2-inch focusers, but it does not work the other way. You may not get any 2-inch eyepieces to start with (they are, all other things being equal, more expensive), but it is a good idea to have options for the future.

**More info**

This is a strange thing, but people in the telescope world – and not just in the USA – always seem to describe focusers using inches, and sometimes telescope apertures as well (especially larger ones); but other values, such as focal lengths, are almost always expressed in millimeters. If this is confusing to anyone unfamiliar with one or the other, just remember that one inch equals exactly 25.4 millimeters.

There are now so many choices available in eyepiece design and size that an entire book could be devoted just to them. However, many are far too expensive for the beginner; they will be left out of this guide. Others are so inadequate they are probably not worth having. Unfortunately, we cannot ignore *them*, because they are the eyepieces that sometimes come with a beginner's package (see Figure 3.23).

Some less expensive telescopes, especially the department store variety, include eyepieces of the **Huygens** or **Ramsden** design. If your telescope has eyepieces that have an "H" or an "R" written on them next to their

**Figure 3.22 Eyepieces come in a wide range of size, focal length, and quality. Here are two 2-inch (on the left) and four 1.25-inch good-quality eyepieces that would provide magnifications, in a telescope with a focal length of 1000 mm, of, from left to right, 15.4×, 31×, 40×, 59×, 67×, and 167×.**

**More info**

Eyepiece designs are usually named after the person who invented them. Huygens and Ramsden have been dead a very long time. Improvements have been made since.

**More info**

It is now possible to watch the Moon, Sun (with a proper **solar filter**), planets, and even some bright deep-sky objects live through your television. For not much more than the price of a decent Plossl eyepiece, you can buy electronic eyepieces that fit in a standard 1.25-inch focuser. These eyepieces are available in color and black-and-white versions (the color models typically have higher resolution and, of course, cost more). They are not much larger than a normal eyepiece, and have a cable that plugs directly into the video- or digital-recorder input on any television.

focal length, you will want to upgrade to better ones. In fact, you will probably want to avoid buying a telescope that includes these: if a manufacturer is willing to supply such eyepieces, the rest of the package is suspect as well.

Beginners' scopes might also include either **Kellner** or **modified achromat** eyepieces. While better than Huygens or Ramsdens, and even useable, the cheap ones usually included are still not very desirable.

The most common good eyepiece available for a reasonable price today is the **Plossl**. It is a decent all-around choice that provides quite sharp images. Plossls are also fairly inexpensive, especially if you get the "no-name" models that are often available from various telescope dealers or on the internet. Name-brand eyepieces may be of higher quality; but then again, they may not, or the difference may not be noticeable. The brand name *will* raise the price.

*Choosing eyepieces*

It may be that the telescope you select already comes with a couple of decent eyepieces, preferably Plossls. If it does not, or if it only comes with one (as many do), you will want to get one or more of your own. Two eyepieces are definitely enough to start with, and may be all you ever need, especially if you also get a **Barlow** lens (see below).

When choosing eyepieces, once you have decided on the type, you should keep two things in mind: *focal length* and *eye relief*.

*Focal length*. Remember, it is the focal length of both the telescope and eyepiece which determine magnification. You should choose the focal length of your eyepieces based on the magnification it will give you with your particular scope.

Here is an example: If you have a telescope (the type does not matter) with a focal length of 1000 mm, a 32-mm eyepiece will give a magnification of about 31× (1000 divided by 32 = 31.25). A 10-mm eyepiece will produce 100×. These are two useful "powers" to have.

If you have a telescope with a longer focal length, you would need eyepieces with proportionally longer focal lengths to get the same magnification. Shorter-focal-length telescopes require shorter-focal-length eyepieces. Beyond a certain point in either direction, this becomes difficult. To get wide-field views (say, 40×) with a telescope of 2000 mm

**Figure 3.23 It is not just focal length that is responsible for short eye relief. The cheap, 0.965-inch eyepiece of unknown design on the left has a focal length of 20mm, but it has much less eye relief, and a much smaller eye lens, than the 1.25-inch, 15-mm focal length Plossl eyepiece on the right.**

focal length, it would take an eyepiece with a focal length of 50mm. They are available, but usually in expensive, 2-inch sizes. At the other extreme, to get 200× using a telescope with a focal length of 500mm would require an eyepiece with a focal length of 2.5mm. Such an eyepiece would be virtually unusable, except for some of special design, and high price. This is because, typically, the shorter the focal length of an eyepiece, the smaller the eye relief.

*Eye relief.* The focal length of an eyepiece is also often an indicator of its relative eye relief. In other words, an eyepiece with a longer focal length will almost certainly have more eye relief than an eyepiece of the same type with a shorter focal length.

Some beginner scopes come with eyepieces that have focal lengths of 4mm, or even less. To use them, it is necessary to get your eye so close that it is, at best, very uncomfortable; and at worst, virtually impossible.

Eye relief is often stated in company advertisements or brochures for eyepieces. Anything less than 10mm is, for me anyway, getting too close. One problem I have noticed is that my lashes touch the **eye lens**, the lens that comes closest to the eye, and oil deposits are left on the glass.

What all of this means is that, if you have a telescope with a relatively short focal length, higher powers are going to be difficult to get, or uncomfortable to use.

There are special eyepiece designs that provide short focal lengths *and* long eye relief. They cost more, but may be worth considering, especially if you wear glasses.

**More info – Eyepiece filters**

Though not necessary to the absolute beginner, glass filters are available which are meant to be screwed onto the ends of eyepieces to enhance the view through the telescope. By blocking certain colors, some filters help the viewing of planets; others are meant to block light pollution; still others are meant to help with the false color which occurs in modestly priced refractors (these are often called **"minus-violet" filters**). You may want to choose eyepieces that have threads to accept these filters.

It should be remembered, though, that no filter actually makes anything brighter. They remove certain wavelengths of light so that others can be more noticed. Also, *NEVER use any eyepiece filter meant to block the light of the Sun!* The concentrated light reaching such a filter is extremely hazardous; it can crack such filters and cause immediate eye damage. There are filters meant for solar viewing, but a proper one is designed to be placed over the large end of the scope.

## Barlow lenses

The specialized eyepieces mentioned above usually attain longer eye relief by incorporating a special lens, called a Barlow, within them. A Barlow lens has the effect of lengthening a telescope's focal length (or, put another way, of halving the focal length of the eyepiece). Barlows are typically purchased as separate units and usually come with 2× or 3× designations.

Barlows are a good investment. If you have two eyepieces and a Barlow, it is like having four eyepieces. You insert the barlow into the focuser, and then the eyepiece into the Barlow. Using an example from above, a 1000-mm telescope and a 32-mm eyepiece using a 2× Barlow would give 62.5× (1000 × 2 divided by 32=62.5); a 10-mm eyepiece would yield 200×. This would result in an eyepiece collection that gave 31×, 62×, 100×, and 200×. This is a good, even spread over a wide range. You may never want more.

Barlows are a good choice for another important reason: *they increase magnification without decreasing eye relief.* That 10-mm eyepiece will become, in essence, a 5-mm eyepiece, but it will keep the same eye relief. For scopes with shorter focal lengths, this may be the best (only?) way comfortably to reach higher magnifications.

When buying a Barlow, avoid the cheapest ones. The view through a telescope is only as good as the weakest optics used. If you have a decent telescope and good eyepieces, spend a little extra for this accessory; it will be worth it.

**More info**

Notice that, if purchasing a 2× Barlow, you need to think carefully about the focal lengths you choose for your eyepieces. It would not make sense, for example, to buy a 10-mm and a 20-mm eyepiece, since the Barlow will, in effect, convert the 20-mm eyepiece into a 10-mm eyepiece.

## Diagonals

Because focusers on telescopes can end up in awkward positions when viewing (especially in refractors and Maks), it is often helpful to use

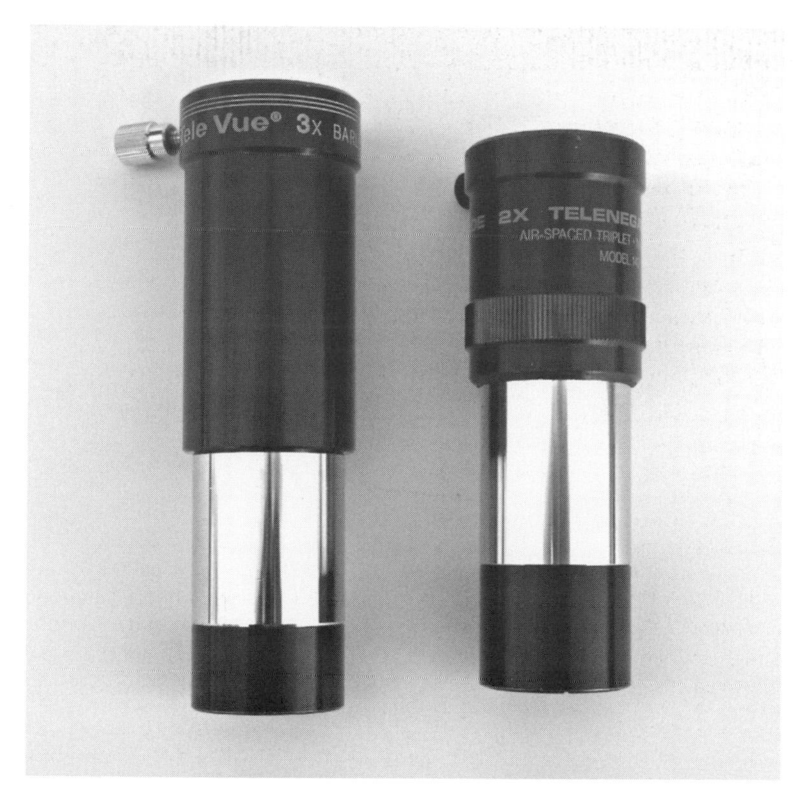


Figure 3.24 Two high-quality Barlow lenses.

a **diagonal**. This is a handy attachment that is placed between the focuser and eyepiece, and uses a prism or mirror to change the angle of view by 90 or 45 degrees. Some even use special prisms that provide a correct, upright, image (remember, astronomical telescopes usually show things upside down).

A diagonal is very helpful if two or more people of quite different height are going to use a telescope. It is often possible to find comfortable viewing angles for a variety of heights just by turning the diagonal in the focuser.

### Finders and finder scopes

Even the smallest beginner's telescope will probably come with an even smaller one attached to it. This is the **finder scope**. It is a low-power, wide-angle scope meant to help in the initial pointing of your main scope toward an intended target. Finder scopes usually have **cross-hairs** that help in centering; and some give a correct image, which makes using field guides or star charts easier. Since the field is so wide in the finder, it is easier to aim with this first. Finders need to be adjusted when you set up your telescope, so that the center of the eyepiece in your main scope matches up with the

**Figure 3.25 Top: a 1.25-inch, 45-degree "correct-image" prism diagonal that would be very useful for both astronomical and terrestrial use; middle: a 1.25-inch, 90-degree mirror diagonal; bottom: a 2-inch, 90-degree mirror diagonal.**

center of the field in the finder scope. A finder is only useful if it can be used to find things. Like binoculars, finders are usually described using numbers; anything smaller than $6 \times 30$ is not very useful.

Some telescopes now come with **red-dot finders**. These do not magnify an image through a lens (which is why they are also sometimes called unit power finders), but rather project a tiny red dot or circle against

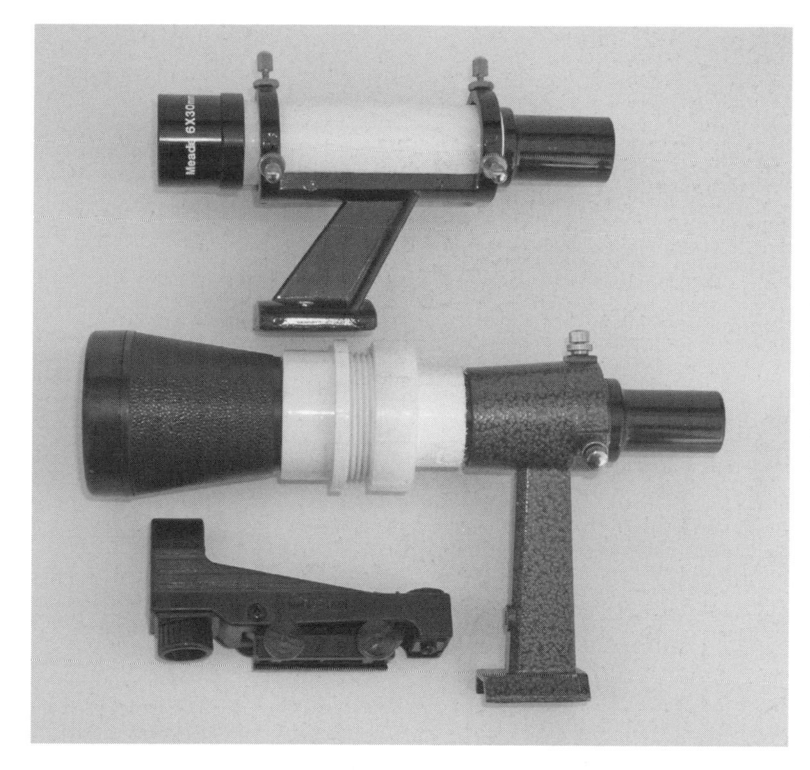


**Figure 3.26 Three quite different finders. Top: a 6 x 30 finder scope that comes as a standard on many smaller scopes; middle: a finder that started out as a 6 x 30, but which has been modified with the objective end of a broken pair of binoculars into an approximately 8 x 50 finder scope; bottom: a "red-dot" or unit power finder.**

a special piece of clear glass. Somehow, even if you move your head around, the dot continues to appear on the same spot in the background sky. They are very comfortable and easy to use. If a low-power eyepiece is used in the telescope when first aiming, these finders work well.

## Astrophotography

If you are someone who enjoys taking pictures, you may want to keep this in mind when choosing your astronomy equipment. Cameras work by focusing light through a lens onto a surface (either film or a computer chip) that collects and stores it. When the camera's release button is pressed, a shutter curtain opens and closes to allow the light for each photo to strike the film or chip. The larger the aperture of the lens and the longer the shutter is left open, the brighter the final photo will be. Photos taken in daylight, or with a flash, freeze any motion because the shutter is only open for a short fraction of a second. **Astrophotos** typically require the shutter to be open for at least several seconds, and often for many minutes. With a clock drive on an equatorial mount to stop the apparent motion of the stars, even a beginner's equipment is enough to capture great shots of celestial objects, providing

**Figure 3.27 An approximately two-hour exposure of star trails; interesting here, but beautiful when seen in the colors that the stars recorded on the film.**

the exposures are relatively short, and the magnification not too high. Getting a package that can be used for astrophotos will probably be cheaper than upgrading later.

Of course, the simplest astrophotos do not require a telescope at all. A camera with a normal or wide-angle lens, mounted on a sturdy tripod, is all you need, as long as the camera has a "b" setting, allowing the shutter to remain open indefinitely. With **fast film** (ISO 400 or higher), brighter constellations can show up with exposures of 20 seconds or less, and the turning of the Earth will not blur the image much. Longer exposures will show **star trails**, curved streaks on the film caused by the Earth's rotation, and these can be very pleasing.

### Photos with a telescope

There are three main techniques used for taking astrophotos using a telescope that might be possible for a beginner: **afocal**, **piggy-back**, and **prime focus**.

An *afocal* astrophoto is the easiest to take, and only works for very short exposures of bright objects like the Sun, Moon, and planets. You simply focus the lens of a tripod-mounted camera on the image in the eyepiece of your separately-mounted telescope, and shoot. It is also possible to buy adapters that allow cameras with their lenses to be directly attached to telescopes with eyepieces. **Digital cameras** work especially well with this technique, since even inexpensive point-and-shoot models

**Figure 3.28 Afocal astrophotography: A point-and-shoot digital camera attached, with a screw-on adapter, to an eyepiece inserted into a telescope's focuser.**

can take very quick shots of dim objects, and the results can be immediately viewed. Eyepieces with long eye relief work the best. No other special equipment is needed.

For a *piggy-backed* astrophoto, you mount a camera on top of your polar-aligned, motorized telescope. Many telescopes come with piggy-back mounting adapters already part of the package, but they can also be bought later for a small cost. You then use your telescope and its mount to **guide** the photo (if you expose for more than a minute or two, or are using a **telephoto lens**). This is done by centering a star in the field, and keeping it centered by using the hand controller. This works best if you use an eyepiece with cross-hairs that are lit up with an LED. This is called an **illuminated reticle eye-piece**. Piggy-backing works well for wider angle shots; of constellations, the Milky Way, or comets. A camera with a "b" setting and a **cable release** are necessary for piggy-backing. It is best if the camera's shutter does not need a battery to hold it open for long exposures, since these pictures sometimes need several minutes or more. This is a technique that does not work well with inexpensive digital cameras, since they are not capable of very long exposures.

*Prime-focus* astrophotos are those taken with a camera attached directly to the telescope, with no lens on the camera, and no eyepiece in the telescope. This is, strictly speaking, not really a beginner's technique. It requires the use of a camera with interchangeable lenses; almost always a 35-mm **single-lens reflex (SLR)**. Old, used, (cheap!) manual cameras work best, since features like light meters or auto-focus are not necessary. What are necessary are adapters to attach the camera to the telescope; these are available from telescope dealers, and are not very expensive. The telescope essentially becomes a very powerful lens for the camera. The Moon is an excellent beginner's target for prime-focus shots using film; the Sun (with a proper solar filter) is another. Dimmer targets need longer exposures and,

**Figure 3.29 A telescope set up for "piggy-back" astrophotography; the cable release on the camera, and the illuminated reticle eyepiece seen in the telescope's focuser, are almost essential for sharp, guided photos.**

**Figure 3.30 An old SLR film camera attached with an adapter directly to a telescope's focuser, for prime-focus astrophotography.**

since any movement of the field of view will cause blurring, this means that almost perfect guiding is necessary. Such photographs require careful polar alignment and special ways of guiding the telescope that are beyond the scope of this beginner's guide.

Figure 3.31 A small digital astrocamera, made to photograph bright objects like the Moon and planets. It requires a computer at the telescope.

There are also new digital **astrocameras** specifically designed for use as prime-focus instruments. They use digital chips originally designed for "web cams," and are relatively inexpensive. They are meant for taking images of the Sun, Moon, and planets; and can be purchased for less than $100. However, to use them, it is necessary to have a computer at the scope.

Newer-model digital SLRs (or **DSLR**s) are also capable of taking any of the three types of astrophotos just described. (Older models cannot take long exposures because of excessive electronic "noise.") These cameras are still quite expensive, but if you already own one, it can certainly be used. In fact, they have many features that make them especially well suited to astrophotography; being able to view the results immediately is one of the most important. However, the very sensitive electronic sensors used in these cameras (instead of film) are equipped with filters that block most of the red light seen in certain deep-sky objects. In order to photograph these objects well, DSLRs must be professionally modified. The cost of this modification is also quite high.

## "Finally, what should I buy?"

There is no easy answer to this question. There are too many choices, and too many variables amongst those choices. Most importantly, there is personal preference.

If you want to see the largest number of objects possible, and you *know* you will not mind the extra effort, then you should opt for the most aperture for the money: a Newtonian reflector on a Dobsonian mount. A good 203-mm (8-inch ) f/6 (1200-mm focal length) model can easily be had for under $500; even a 254-mm (10-inch ) f/5 (1250-mm focal length) is possible. Such a large aperture will certainly show you the brightest views, and thus also the most objects. Many experts think this is the ideal choice for the beginner.

If you are primarily viewing from a city or town where light pollution will block dimmer objects, and/or are primarily interested in the Moon and planets, and/or you really need a compact package, then a Mak might be for you. A 102-mm (4-inch) model is available on a GEM within our price range.

If you want versatility and ease of use, and you can handle a somewhat larger outfit, or you plan to use your scope for group viewing or astrophotography, then an equatorally mounted refractor may be best. I have seen both a 102-mm (4-inch) f/10 (1000-mm focal length), and a 120-mm (4.7-inch) f/8.3 (1000-mm focal length) scope package for under $500.

Yet, of course, a reflector can be put on an equatorial mount. Short-tube refractors and reflectors are compact, too. Many combinations are possible. The choice is yours, and you should now know enough at least to ask the right questions when choosing.

Still, if you cannot make up your mind, then here is my advice:

Get a refractor; get the largest aperture you can afford, with a focal length from 600–1200 mm; and, if possible, a 2-inch focuser and a piggy-back adapter. Get it on the sturdiest German equatorial mount possible (many scopes come on mounts too flimsy for them); a mount that is upgradable for use with a clock drive. Get two Plossl eyepieces, a good 2× Barlow, and a diagonal. With this set-up, you will be ready, and then some, to begin exploring the Universe.

Others would surely disagree with me, and recommend a Dobsonian reflector: to get the "most bang for the buck," because there is very little to set up, and because aiming is as simple as pushing and pulling the tube in the right directions. They may be right. But Dobs have their own problems (size, collimation, balancing, near-constant re-centering at higher magnifications), which I believe could be very frustrating for someone absolutely new to telescope use. Once someone gets over the initial learning curve of using a refractor on an equatorial mount – and it *is* confusing at first – they are simple, versatile, and worry-free to use. I think this is most important.

## In a nutshell

Below is a table comparing telescope types available to the beginner. It would not mean much without the explanation provided on the preceding

**Comparing beginners' telescopes**

| Telescope type | Advantages | Disadvantages | Largest aperture available for less than $500 in a new, complete set-up |
|---|---|---|---|
| **Refractor** | No (or low) maintenance<br>Easy to aim<br>Fewer tube-current problems<br>Variety of focal lengths | Chromatic aberration (false color)<br>More expensive (per unit of aperture) than reflector | 120mm (4.7 inch) f/8.3 (1000-mm focal length) on equatorial mount<br>102mm (4 inch) f/10 (1000-mm focal length) on EQ mount with single-axis drive |
| **Newtonian reflector** | Least expensive (per unit of aperture)<br>Shorter cool-down time than comparable Maks<br>No false color | Requires maintenance: frequent collimation and eventual re-coating of mirrors<br>Awkward to aim<br>Coma<br>Larger mirrors need long cool-down<br>Some tube currents | 254mm (10 inch) f/5 (1250-mm focal length) with Dobsonian mount<br>152mm (6 inch) f/5 (750-mm focal length) on EQ mount with single-axis drive<br>130mm (5.1 inch) f/6.9 (900-mm focal length) on EQ mount<br>130mm (5.1 inch) f/5 (650mm-focal length) on go-to mount |
| **Maksutov–Cassegrain** | Well-corrected for aberrations<br>Small tube size (allows smaller mount, portability)<br>Easy to aim | Long cool-down time<br>High f/ratios make wide-field viewing difficult<br>Most expensive beginner type (per unit of aperture) | 102mm (4 inch) f/12.7 (1300-mm focal length) on equatorial mount with single-axis drive |

pages, but with an understanding of the terms used, this table may help you to decide what is best for you.

## Resources

### Telescope dealers

I strongly recommend that you buy your astronomy equipment from a store that specializes in it. Many camera stores sell telescopes, but most (though certainly not all) have a very limited selection, especially in good equipment for beginners; and, correspondingly, the staff may have little knowledge about what they sell.

However, if there are no telescope shops nearby, which is likely for most people, there are some advantages to shopping at least at a store that sells telescopes as a side line, even if you end up buying your scope through a mail-order company. It is always helpful actually to see and touch something before buying it. If you have done research and have a good idea about what you would like to get, check any local stores that might carry it. The odds are high that you will not find just what you are looking for, but even something similar will help you in making a final decision. If the total package is even somewhat close to what you have been considering, is it portable enough for your storage and transportation needs? Can it be used comfortably? How smooth is the focuser when you turn it? Is the mount solid and stable enough for the telescope included with it? (Give the end of the telescope a good tap with one hand while resting the other hand on the mount. Does the telescope bounce back and forth? How long does it take for the whole thing to stop vibrating? Vibrations should be slight, and should not take more than a few seconds to stop.) Answers to these questions are nice to have *before* making a purchase, rather than upon assembling a package that has been paid for and shipped to you.

Hundreds of companies specialize in selling telescope equipment. They range in size from large mail-order houses that publish glossy catalogs to small, one- or two-person operations. I have had great experiences at both ends of this spectrum. Get a current copy of *Astronomy* or *Sky & Telescope*, and page through the advertisements. You should find a company that can meet your needs. Or, you could do a web search. The internet and amateur astronomy are very cozy companions. Try a search for "telescope dealers." I just did, and got 2580 results. One of the first sites listed was a page with links to more than a hundred companies around the world that sell astronomy equipment.

*Customer service* is an important factor when choosing a dealer. Astronomy equipment is delicate, and subject to defects. Unfortunately, as in every field, there are unscrupulous astronomy dealers. You may want to contact other astronomers to find out what they know. Try calling the astronomy department at a nearby college, or your local **astronomy club**. There are also telescope-review websites. They can not only help you decide what equipment to buy; some of them have reviews of dealers, too.

*Upgrades:* If you think that a telescope package you are considering comes on too flimsy a mount (they often do), or if the eyepieces included are not what you want, or if other aspects are not quite what you are looking for, you might want to ask a dealer if they would give a discount for an upgrade to another level. There is no guarantee that they will, but if two or more dealers are offering similar packages, it is worth shopping around to find out.

*Mixing and matching:* It is also worth thinking about making your own package by shopping around for different components. Many dealers sell mounts, eyepieces, and **optical tube assemblies (OTAs)** separately. One might have a great deal on an OTA (a telescope without a mount), while another has a sale on mounts, and you might see a great deal on eyepieces on eBay. Pre-packaged scopes are always less work to find, and they are usually the least expensive way of buying components from one place; but shopping around can often save money, and is more likely to provide exactly the set-up you are trying to assemble.

*Used equipment or "scratch and dents:"* If you are willing to use less than pristine equipment, you might consider buying it used, or slightly damaged. Many dealers sell such equipment, often with warranties. Look on their websites, or ask them on the phone if they have any used equipment, or "seconds." A scratch in the paint on the tube could save you so much money that you might be able to afford something outside the scope of this guide. And you cannot see scratches in the dark!

You might also consider buying equipment through online *classified advertisements*. There are several such websites that I know of. Astromart.com is very popular, and takes great care to prevent fraud, and many reputable telescope dealers also sell items at this site. But it is mostly individual amateurs who use it, and it is always risky to send money to people you do not know. Still, if you are looking for something very specific, or want to save money, try Astromart.com, or do a web search for "astronomy classifieds" or "telescope classifieds."

## Telescope reviews

If you have a specific telescope, or mount, or eyepiece, in mind, or you cannot decide which among several you should choose, try looking for a review of it online. New equipment comes onto the market so often and so quickly that this may be the best way to find out about it. Try searching for "telescope reviews." There are several excellent websites that review specific scopes, binoculars, mounts, eyepieces, Barlows, etc. Some of them even have categories devoted specifically to beginner equipment. Cloudy Nights Telescope Reviews at cloudynights.com is a site I particularly enjoy.

## Star parties

Both professional observatories and local or regional astronomy clubs have occasional get-togethers called **star parties**. The public is invited to come and share the night sky through their own, or others', equipment. Some of these parties are huge, require reservations, attract people from everywhere, and last for days. These are usually annual events, and their dates are often published in *Astronomy* or *Sky & Telescope*. Other star parties are smaller and last just one evening. These are often given on a scheduled basis by clubs, planetariums, or observatories: every other week, or once a month, and so on.

Star parties are excellent places to "test drive" equipment before you buy, or to try equipment that you could never afford yourself. They are also places to meet other, like-minded people. But *beware:* you could catch *aperture fever.* You look through the eyepiece of a monster scope and see things you did not know were possible; and, besides, the scope itself is so cool! You get very excited, and start trying to figure out how to find the money to get one yourself. This really happens. Believe me.

### Astronomy clubs

You may want to try to find out if there is a local club in your area. They often have access to dark-sky sites, and many clubs even have their own observatories. Membership in such groups gives you the chance to meet others who share your interest, and you can often get discounts on various astronomy goodies. Try another web search, for "astronomy clubs," and include your town, state, province, or country in the search. Or, in the USA, look up the *Astronomical League* at astroleague.org. They are a federation of local and regional groups, and can point you toward a club in your area.

### Books

Here are a couple of books that go into more complex and complete detail than I have about astronomy equipment. They cover a wider range of options as well, since they are not devoted to the absolute beginner. If you think you want to know more, or simply have more money to spend, you may want to check them out. The first one is primarily devoted to equipment; the second is more general, and covers a wide range of topics. Both are excellent. Try to get the latest editions available.

1. *Star Ware: The Amateur Astronomer's Ultimate Guide to Choosing, Buying, and Using Telescopes and Accessories* (4th edition), by Phillip S. Harrington. John Wiley & Sons, 2007.
2. *The Backyard Astronomer's Guide*, by Terence Dickinson and Alan Dyer. Firefly Books, 2002.

There are also books specifically written for people interested in astrophotography. Here are four that include information that could be useful for beginners.

1. *Astrophotography: An Introduction*, by H.J.P. Arnold. Sky Publishing, 1999.
2. *Astrophotography for the Amateur* (2nd Edition), by Michael A. Covington. Cambridge University Press, 1999.

3. *Digital SLR Astrophotography*, by Michael A. Covington. Cambridge University Press, 2007. (This book is written in a personable, very clear, step-by-step way, and should be of great help to anyone new to using DSLRs for astro-imaging.)

4. *Wide-Field Astrophotography*, by Robert Reeves. Willmann-Bell, Incorporated, 2000.

Figure 4A A wide-angle astrophoto, taken on August 12, 2007, shows several items described in Part II. (1) A Perseid meteor; (2) the Double Cluster, an open star cluster; (3) the constellation Cassiopeia, the Queen; (4) the Andromeda Galaxy; (5) the Hyades, an open star cluster; (6) the planet Mars; and (7) the Pleiades, or Seven Sisters, another open star cluster. All of these – except that specific meteor – are visible to the naked eye under dark skies, and most are greatly enhanced with either binoculars or a telescope.

# Part II
# What's up there?

*To infinity...or close enough!*

Once you have the equipment needed – or better yet, before you do – you will want to get right out there to start exploring the night sky. Part II attempts to answer the question its title asks in as basic a way as possible. It will not give you all the information you need to find the objects in the sky; for that you will need a field guide, or star charts. It is merely meant to be a very brief tour, to give the novice some understanding of what the various objects are, as well as an idea of what they might look like through a beginner scope. The trip will start nearby, and move outward.

However, before getting to specific object descriptions, I should start with a *warning:* you will only be disappointed if, when looking through your telescope, you expect to see things as they appear in books or magazines, or on television or the internet. The brightness and, especially, the color in those images cannot be seen through the eyepiece of an amateur scope. Such things are only possible to record through **time-exposure** astrophotos, and such photos are often taken through giant telescopes at professional observatories, or even by artificial satellites and other spacecraft. Amateurs have taken and do take great astrophotos, and even a beginner's equipment is enough for shooting some targets. All the photos in this book were taken, by me, with modest amateur equipment. But the best amateur equipment available cannot show such detail or color directly to the eye: the objects are just too dim.

On the other hand, the human eye is *more* sensitive than photography in at least one way: it can see detail in images across a very wide range of brightness *at the same time*. Through the eyepiece, it is possible to see, for example, fine detail in the wisps of **hydrogen** on the edges of the **Great Orion Nebula**, while also seeing the individual bright stars in the center of it. A photograph would either wash out the brightness, or miss the faint detail. It is true that computer software can now be used to combine photos to show more range in detail, but the eye does it literally "live."

So, what can you expect to see? Within the Solar System, even with the minimum 80-mm aperture, you will be able to see close-up views of mountains, and "seas," and hundreds of craters on the Moon. With a proper solar filter, you can see features like sunspots on our very own star, the Sun. Seven other planets, and several of their moons, can be seen. If you know when to look, you can see meteor showers (no telescope necessary). If you know where to look, you can view **asteroids** and comets, smaller objects that also orbit the Sun.

Beyond the Solar System, you will, of course, be able to see countless stars: single stars, **double stars**, **multiple stars**, and star clusters, both open and **globular**. Stars are one type of object often bright enough to show their color in telescopes: red, orange, yellow, white, and blue are the most common, but some people even report seeing green or turquoise.

Nebulae (of several kinds) and galaxies (including our own) are the other targets of the deep sky. Some are actually visible to the unaided eye, but the apertures of binoculars and telescopes make it possible to view hundreds of these fantastic, varied objects.

Every night, the sky is filled with potential celestial targets. They are beautiful things to behold, simply as lights and patterns in the sky, and can be greatly admired and appreciated as such. However, knowing what they are, how far away, how large, how old, how they came to be, and what might one day happen to them – all of this knowledge *adds* to the impact of seeing them "live and in person" through the eyepiece. From small rocks spinning and hurtling around the Sun, to entire "island universes" containing hundreds of billions of suns: when you know what it is you are looking at, and you try to get your mind around the vastness and complexity of it all, it is easy to get truly lost in awe. And guess what? You can do it again and again, night after night, and never come close to seeing it all. The information in Part II might help you get *started* on the road to that awe-inspiring knowledge. I hope your own experiences at the eyepiece make you hungry for more!

**More info**

Here's something else: stargazing takes practice, like any other skill. The more you do it, the better you get at it. Just like playing the piano or typing, where your fingers eventually seem to know what to do, with stargazing it is your eyes that seem to change, gaining in the ability to see faint, fine detail. Amateurs with years of experience can truly see things, through the same telescope and eyepiece, which less experienced observers cannot.

# The Solar System

# 4

Our own star is called **Sol**. It, and all that orbit it, are part of the Solar System. This includes, at this time, eight planets (or maybe nine; or possibly twelve or more: the scientific definition of "planet" is under debate), a large and growing number of known moons that orbit these planets, and millions of smaller asteroids and comets, which range in size from several hundred kilometers in diameter to microscopic.

Most of these objects are not visible to even the largest telescopes, since the tiny greatly outnumber the large. But many of them are excellent targets for even the smallest scope. The Sun and the Moon are the only two bodies in the heavens that clearly appear as disks to the unaided eye, rather than as points of light. Seven planets show disks when viewed through telescopes. Some of their moons are also visible.

However, what makes these objects into targets worth visiting often is that they *change*. The **phases** of the Moon, Mercury, and Venus; the appearance and passage of spots across the surface of the Sun; the polar caps of Mars; the dance of the moons of Jupiter and Saturn; all of these and more are there to see, even through beginner scopes.

In addition to these regular denizens of the solar neighborhood, there are periodic visitations from lesser-known objects. These include asteroids, comets, and meteors.

This chapter will give a brief description of all of these objects. The sections on the Moon, the Sun, and the planets will start with some statistical information; one term needs definition: **angular size** is the *apparent* diameter of an object, when viewed in the Earth's sky. It is given in degrees (one 360th of a circle,) minutes (one sixtieth of a degree,) or seconds (one sixtieth of a minute.) The Sun and the Moon, for example, each have an angular size of about 30 minutes (written 30′) or half a degree. This is an amazing coincidence, and is the reason that **total solar eclipses** are so spectacular. But the angular sizes of these two bodies have nothing to do with their *actual* sizes: the actual diameter of the Moon is 3476 kilometers (2140 miles); the diameter of the Sun is about 1,390,000 kilometers (862,000 miles). The Sun is about

**More info**

Many "deep-sky" objects, strange as it may seem, appear much larger in our sky than anything in the Solar System. The Andromeda Galaxy, for example, spreads across at least three degrees; more than six times the diameter of the Moon. Of course, it is also very dim, especially at the edges; though under even medium-dark skies, it can be seen with unaided vision. It *appears* so large because it is large: it measures more than one hundred thousand **light years** across!

400 times wider than the Moon. Now here's the amazing coincidence: the Sun is also about 400 times farther away from the Earth! That is why the two appear to be the same size when seen in our sky.

All other objects in the Solar System appear so small that their measurements must be given in seconds of arc. This means that, even at very high magnification, they will not fill much of the field of view in an eyepiece. Still, with good optics, you will be able to see **surface features** on other worlds, hundreds of millions of kilometers away!

**Measurements**

**The Moon**

| | |
|---|---|
| **Angular size:** | 30′ |
| **Diameter:** | 3476 kilometers (2140 miles) |
| **Average distance from Earth:** | 380,000 kilometers (238,000 miles) |

## Our nearest neighbor

### The Moon

The closest heavenly body regularly seen from the Earth is, of course, the Moon. It is a dead ball of rock, with no life, and no atmosphere. It is covered with craters caused by the bombardment of thousands of objects early in the history of the Solar System, a few billions of years ago, and less frequently, since then.

The Moon orbits the Earth in about a month (the word "month" is derived from "moon"). The force of the Earth's **gravity** has caused one side of the Moon to become "locked" toward us; this means that it turns once on its axis during one orbit of the Earth. So, though only one side ever faces us, both sides of the Moon get exposed to the Sun (and that means there is no "dark side of the Moon;" but there is a "far side"). This may be hard to visualize at first.

*Moon phases*

The Moon orbits us, and we orbit the Sun. As Figure 4.2 shows, when the Moon is on the opposite side of the Earth from the Sun, the side of the Moon that faces us is completely lit up by the Sun. This is the **full Moon**.

When the Moon is close to the Sun in the sky, the other side of it is lit, and the side facing us is dark. This is called the **new Moon**.

A week after the new Moon comes the **first quarter Moon**. This may seem confusing, since *half* of the Moon's face is lit up, but it is actually a way of saying "the first quarter of the month."

**Figure 4.1 No object in the night sky shows as much detail to us as the Moon; and the detail changes from night to night on Earth, as the line between night and day on the Moon slowly crosses its face.**

A week after the full moon comes the **last quarter Moon**.

The Moon is a **crescent** when it is less than half lit up, but not new, and **gibbous** when more than half lit, but not full. It is said to be **waxing** as it is "growing" between new and full; and **waning** as it "shrinks" between full and new.

As a month passes (not a calendar month, but an orbit of the Moon around the Earth), the Moon rises later each night. When it is new, it is very

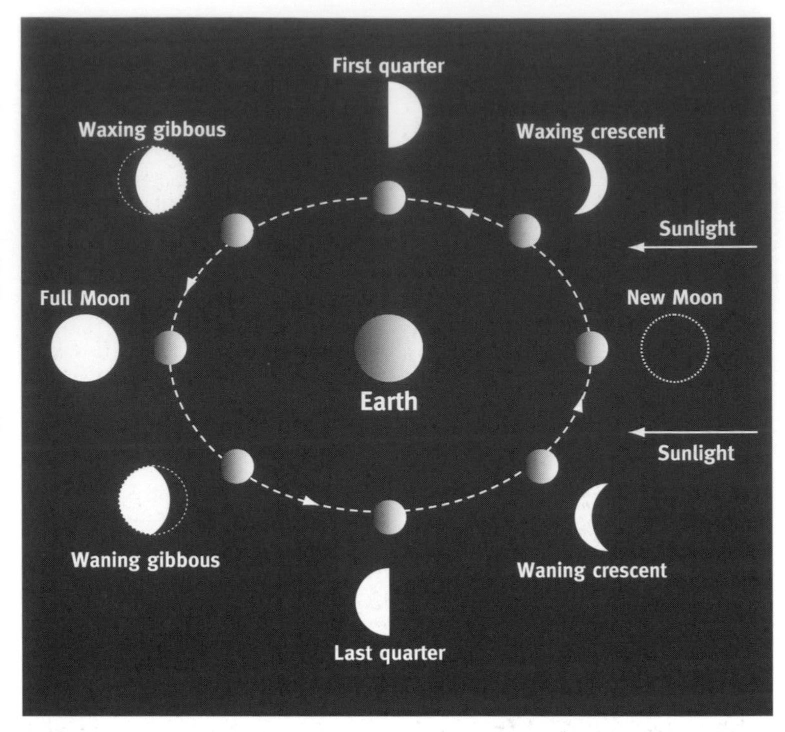


Figure 4.2 The changing face of the Moon as it orbits the Earth each month. The inner circle of small images shows the Moon's movement around the Earth. The outer circle shows the phases of the Moon as they appear to viewers in the northern hemisphere on Earth. Southern-hemisphere viewers see an inverted view (and can just turn the book upside-down), since they themselves are "upside-down" relative to people in the north.

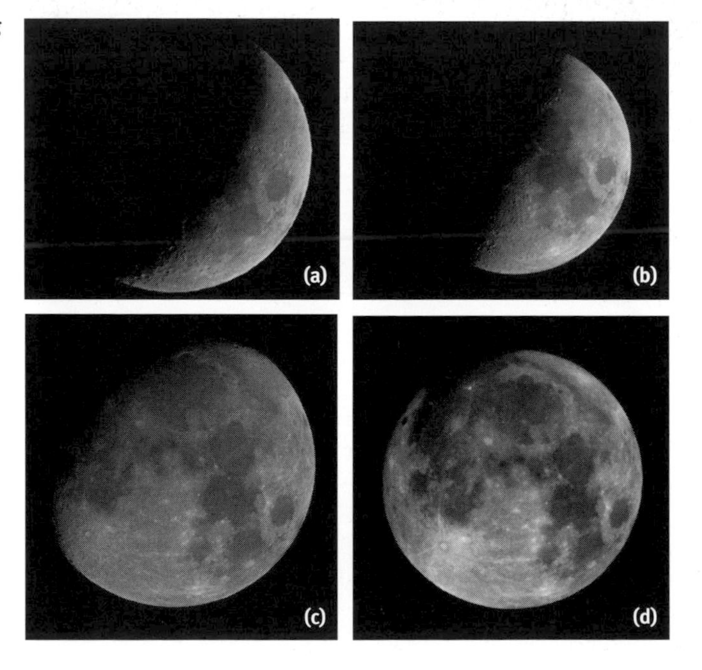


Figure 4.3 The waxing (growing) Moon: (a) crescent; (b) first quarter; (c) gibbous; and (d) full.

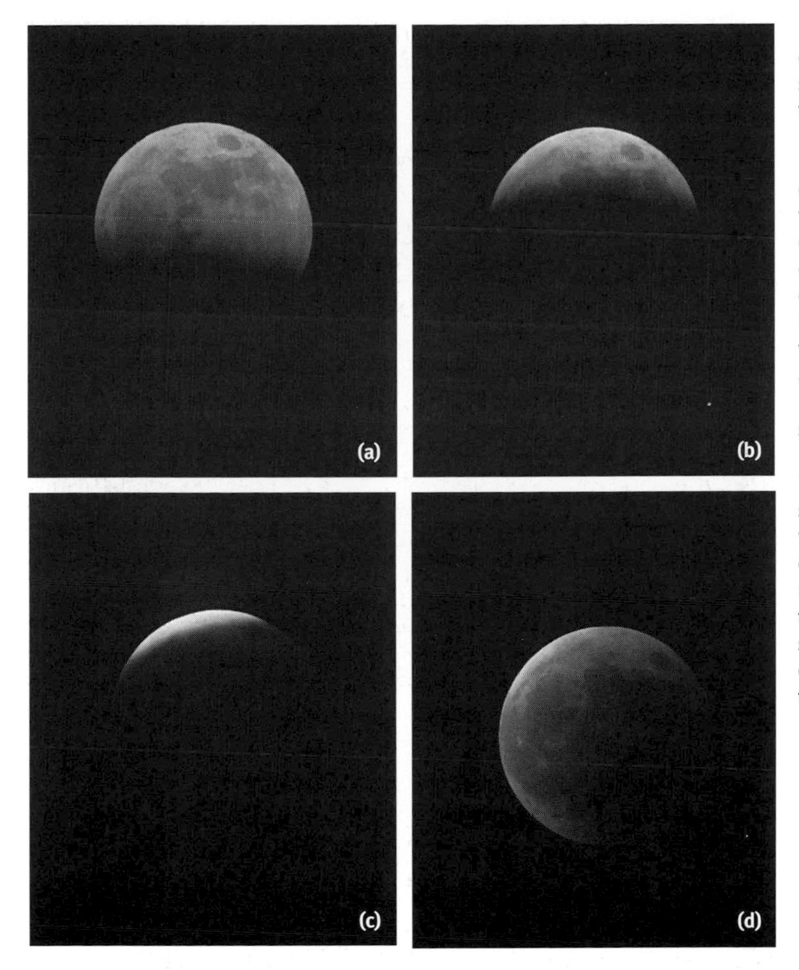


Figure 4.4 A lunar eclipse. The Earth's shadow slowly covers the face of the full Moon in (a), (b), and (c); but it completely covers it in (d), which was a much longer exposure, showing an orange Moon in the original slide, darkly lit by sunlight shining through the Earth's atmosphere (see the back cover). Note the star visible to the Moon's right in (d). It is only possible to see stars this close to the full Moon during an eclipse; they are usually washed out by the light of this, the second brightest object in the sky, after the Sun.

close to the Sun in the sky, so it rises and sets at about the same time as the Sun. At first quarter, the Moon rises around midday, and sets around midnight. The full Moon rises near sunset, and stays in the sky all night. The last quarter Moon rises around midnight, and sets around noon.

*Observing the Moon*

About the *worst* time to view the Moon through binoculars or a telescope is when it is full. The Moon's surface is covered with craters, mountains, and valleys. During a full Moon, the Sun comes closest to shining straight down on all of these, making any shadows cast by them very short. To see detail on the surface, it is best to view where the shadows are longest: along the line between the lit and unlit parts of the surface. This line is called the **terminator**, and it is slowly moving – at about 10 miles (16 kilometers) per hour – all the time.

A great time to watch the Moon with your eyes, binoculars, or at low telescope magnification is during a **lunar eclipse**. Lunar eclipses happen when the Earth gets directly between the Moon and the Sun (and so, of course, only when the Moon is full). Because the Earth has an atmosphere, its shadow as it crosses the Moon varies in darkness. Sometimes, the Moon passes through the very dark center of the Earth's shadow; at other times, it skirts along the edge. So some lunar eclipses are very dark, while others are less so. But all seem, to me, both beautiful and sort of eerie.

Lunar eclipses are sedate and slow, compared to solar eclipses, because the Earth's shadow is so much larger than the Moon. It can cover the Moon completely for well over an hour.

No other object in the sky has as much detail visible through the telescope as the Moon: it is our closest neighbor; it is quite large (compared to other moons); and it has one more thing going for it, at least from an astronomer's point of view: it has no atmosphere to get in the way of viewing. There are many hundreds of named features on its surface, and many books are devoted entirely to observing it.

## Our own star

### Sol (the Sun)

Next to the Moon, there is nothing else in the sky that is easier to see and study than the Sun. Surface details are visible without any magnification. Sunspots larger than the Earth are often visible as they crawl across the solar surface while it slowly spins on its axis about once every 27 days.

The Sun is a star, and a fairly average one. Stars are huge balls of gas (mostly hydrogen, usually). They are held together by their own gravity, and they have so much **mass** that gravity squeezes them tightly enough to cause **nuclear fusion**: hydrogen atoms are forced together to form **helium** atoms, a huge amount of energy, and some leftovers. This nuclear energy is what causes stars to shine. It also keeps them stable. Stars have so much mass that, without the outward pressure of the heat from the nuclear reactions, the inward pressure of gravity would (and sometimes does) cause them to collapse into much smaller, denser bodies. The Sun has been steadily burning for more than four billion years, and should continue to do so for billions more.

**Measurements**

| **Sol (the Sun)** | |
|---|---|
| **Angular size:** | 30′ |
| **Diameter:** | 1,390,000 kilometers (862,000 miles) |
| **Average distance from Earth:** | 150,000,000 kilometers (93,000,000 miles) |

*Observing the Sun*

The first thing to be said on this subject is that *it is never safe to view the Sun without a solar filter approved for the purpose!* In fact, even some of those should not be used. I have read that some telescopes may still be sold with solar filters that are meant to be screwed onto an eyepiece. *These filters are not safe and should be thrown away!* The objective lens or primary mirror of a telescope will focus the intense energy of the Sun on such a filter, which could crack it and cause immediate, serious damage to the eye of anyone looking through it. Solar filters should fit over the object end of the telescope, thus filtering the Sun's energy before it ever reaches a lens or mirror. Some filters are made of a plastic material called **mylar**, which is inexpensive, but can turn the image of the Sun blue. Other plastic filters are available that give a more "natural" orange appearance. Glass filters are also available, but cost much more.

Indirect observation of the Sun is also possible. After you cover or remove the finder scope for safety, you can aim your unfiltered scope at the Sun by watching the telescope's shadow on the ground as you adjust its position (obviously, you do not want to look through the telescope or its finder at the Sun). When the shadow of the tube becomes a circle, you are centered on the Sun. You can then hold a stiff piece of white paper several inches behind the eyepiece. Move it in and out until a clear image of the Sun appears. Tripod-mounted binoculars also work well for this purpose, as long as you cover one objective lens.

**Figure 4.5 A solar filter in its proper place: in front of the objective lens or mirror of a telescope. This one is made of mylar.**

**Figure 4.6 A photo of the Sun, taken in March, 2008. The Sun follows an approximately 11-year cycle of activity. A new cycle had just begun a few months before this photo was taken, but solar cycles overlap slightly, and these sunspots were actually one of the last groups produced by the one just ending. This photo was taken through the "planetary scope" shown in Figure 3.8 on page 36, equipped with a safe yet inexpensive homemade solar filter (total cost was less than $10). This was also the last photo taken for the book, and the only one taken with a DSLR.**

Sunspots are clearly visible. They appear to be small, dark splotches on the face of the Sun. But they are actually huge (remember, the Sun is nearly a million miles across!), and they are only dark in comparison to the extremely bright regions around them.

Though relatively few people are lucky enough to see many of them, total solar eclipses are one of the most impressive spectacles in nature. When the Moon is the right distance from the Earth (and it can vary), and it passes directly between us and the Sun, a total solar eclipse is the result. These are quite rare for a given spot on the Earth – the next over the United States will not occur until 2017 – and **totality** never lasts longer than several minutes. **Annular eclipses** occur when the Moon is farther from the Earth, and does not completely mask the Sun. **Partial eclipses**, when the Moon covers only a portion of the Sun's disk, are much more common, and seeing a "crescent" Sun is very impressive in itself.

Even more rare than solar eclipses are **transits** of the planets Mercury and Venus, though they are visible over a larger geographical area. Transits occur when the orbits of Earth and these two planets line up such that they pass between us and the face of the Sun. Because it is much closer to the Sun, transits of Mercury are more common, and they occur thirteen or fourteen times per century, though times between transits vary, from 3.5 to 13 years. Partially because their timing involves the orbit of the

Earth (which, by definition, lasts exactly one year), transits always happen around the same dates in our calendar. Mercury transits are in either May or November. Since 1993, there have been four Mercury transits, but we are now in the midst of a long gap. Here are the dates for the next few transits of Mercury across the face of the Sun: May 9, 2016; November 11, 2019; and November 13, 2032.

Transits of Venus are far more rare. They occur in pairs, eight years apart, almost to the day. But between those paired transits more than a century passes. We are currently in the middle of one of the pairings. A transit occurred on June 8, 2004; and the next will happen on June 6, 2012. Watch this event – through a proper solar filter! – and know that you may not have another chance to do so in your lifetime, since the next transit of Venus will not occur until December 11, 2117.

## The planets

Even a moderate beginner's scope will allow you to see all the planets in the Solar System, except for tiny, distant **Pluto** (currently downgraded to the status of "dwarf planet"). However, for different reasons, only three of the seven show much detail, even to the largest Earth-bound observatories.

Finding the planets in the sky is more challenging than finding deep-sky objects, at least in one respect. Since both the Earth and the other planets orbit the Sun, each at a different distance and speed, their positions across the background of the sky change. The change is slow for those planets farther from the Sun, but it can be fairly quick for the inner planets.

Luckily for the observer, none of them move randomly across the whole sky. All of the planets (except, again, for lonely, virtually invisible, controversial Pluto) orbit the Sun in a fairly flat disk, which means that they seem to stay near a single, specific line that encircles our night sky. This line is called the **ecliptic**. It has this name because it is along this line that the Sun, like the planets, seems to travel around the sky each year (see Figure 4.7.) So it is also the line along which solar eclipses occur. The ecliptic passes through twelve famous constellations: the signs of the **zodiac**. Several of these constellations are fairly dim, but they are famous because of **astrology**. The ecliptic passes through a few other, less famous constellations as well, including **Cetus** (the Whale) and **Ophiuchus** (the Snake Holder.) The positions of the planets are plotted each month in *Sky & Telescope* and *Astronomy*, as well as on various websites.

**More info**

Pluto has become controversial. In 2006, the International Astronomical Union, a group charged with the authority to give astronomical definitions, downgraded its status to "dwarf planet." This decision has met with protest from some quarters. Still, large telescopes have found other bodies in the outer reaches of the Solar System that may be nearly as large as Pluto, and at least one that is larger. Is Pluto a planet, or isn't it? And if it is, what of the others out there?

Figure 4.7 Four naked-eye planets together in the sky. From left to right, starting with the triangle just left of center: Saturn, Mars, and very bright Venus. Using the triangle as an arrow to find it, with Venus as the tip, Mercury can be seen just above the center tree's top. The fifth and last of the naked-eye planets, Jupiter, was not far away to the upper left but does not appear in the photo. All of them are near the ecliptic, a path around the sky that corresponds to the flat disk of the Solar System.

## Mercury

Mercury is the closest planet to the Sun; so close, in fact, that it is never visible above the horizon when the sky is truly dark. Much of the time, it is so close to the Sun in our sky that it is not viewable.

Even when it is apparent in our sky, Mercury is so small (only Pluto is smaller, and a couple of moons are bigger), and so far away, and so washed out by the glare of the Sun, that no features are visible on its surface. All that can be seen of it through telescopes are its phases. It has phases, like the Moon, because it is closer to the Sun than we are; so, like the Moon, it can get between us and the Sun, and show us its night side.

**More info**

When the Sun, traveling along its path on the ecliptic, is high in the sky during summer days, most of the planets, when they are up during summer evenings, tend to be low in the sky. This is because they are then located on the part of the ecliptic where the Sun will be in winter. Conversely, when the Sun is low in the winter, evening planets are usually much higher in the sky, making them easier to view, since there is much less of Earth's atmosphere to look through.

**Measurements**

| **Mercury** | |
|---|---|
| **Angular size:** | 5–13″ |
| **Diameter:** | 4878 kilometers (3031 miles) |
| **Average distance from Sun:** | 58,000,000 kilometers (36,000,000 miles) |

## Venus

Venus is the second planet from the Sun. It also shows phases, since it too is closer to the Sun than we are. It appears smallest when it is "full" because at that time it is on the opposite side of the Sun from the Earth. When it is a very thin crescent, it appears far larger, because it can then be as close to us as about 40 million kilometers (25 million miles), closer than any other planet. Because it gets so close, it is often the third brightest object in the sky, after the Sun and Moon.

Venus is called the Morning or Evening "Star" because it never strays too far from the Sun, again like Mercury. However, because it orbits further out, Venus can be seen much higher above the horizon than Mercury, and it can stay up much longer than Mercury does after the Sun sets, or rise quite a while before dawn.

Though Venus gets closer to us than any other planet, it is as featureless as Mercury through a beginner's scope. This is because of its thick, impenetrable atmosphere, 100 times thicker than our own. (This thick, highly reflective atmosphere is another reason that Venus is so bright.) Scientists knew almost nothing about its surface until radar was bounced off it a few decades ago. They could not even be sure how long it took to spin on its axis, or if it did at all. The phases are easily visible, but that is about all.

**Measurements**

**Venus**

| | |
|---|---|
| **Angular size:** | 10–64″ |
| **Diameter:** | 12,104 kilometers (7521 miles) |
| **Average distance from Sun:** | 108,200,000 kilometers (67,200,000 miles) |

**More info**

The thick atmosphere of Venus makes it, on average, the hottest planet in the Solar System; even hotter than Mercury, since the Sun's heat is trapped and spread by the poisonous gases in the Venusian sky. Though closest to Earth in both distance and size, there is possibly no planet in the Solar System less hospitable to life than Venus.

### Mars

Mars is the fourth planet from the Sun, and can be the second closest to the Earth. It has two small moons, called **Phobos** and **Deimos**, which are too small and close to the planet to be seen in beginners' scopes.

Mars is much smaller than the Earth, but the two planets have much in common. Mars has an atmosphere, though it is only one percent as dense as ours. Temperatures there can get above the freezing point of water (though not usually by much, and not very often). Satellites orbiting Mars have photographed areas that seem to have been carved by running water, and recent satellite surveys seem to prove that there is indeed much water

**Figure 4.8 Venus and the Moon often appear near each other in the sky; sometimes the Moon actually passes in front of Venus. Because they are both so bright, photographing them is not difficult. Here, the photo has been over-exposed to show Earthshine on the part of the Moon not directly lit up by the sun. It is caused by sunlight reflecting off of the Earth onto the Moon, and then back to the Earth again.**

**Figure 4.9 A crescent Venus, bright but featureless, as it might be seen through a beginner's scope. The thick atmosphere of Venus never shows much detail and, as with a crescent Moon, this crescent Venus was quite close to the Sun as viewed from Earth. The Sun itself was just below the horizon, so Venus was very low in the sky. Objects near the horizon have much more atmosphere blocking them than things higher up, so the lack of detail seen here is no surprise.**

just below the surface. All of this makes Mars one of the most likely places other than Earth to harbor life or, at least, life's remains. This in itself makes Mars an interesting place, and gives you thoughts to ponder while looking through the eyepiece.

Because it has a thin atmosphere, and because at its closest approach it can come to within 56 million kilometers (35 million miles) of the Earth, Mars is the only other planet in the Solar System whose true surface features can be seen through Earth-bound telescopes, even beginners' scopes (the "surface features" of any other planets are actually cloud-tops). And, because Mars has an atmosphere, those surface features can change; which is one more reason to return to Mars again and again.

*Observing Mars*

Mars has detail worth seeing, so it is desirable to get as close a view as possible. Mars takes 687 days, or almost two years, to travel around the Sun. The Earth, of course, takes one year. Because both planets are traveling in the same direction, it takes just over two years to go from one close approach of Mars to the next. At such times, Mars is said to be at **opposition**. This term is used because Mars (or any other body at opposition) is on the exact opposite side of the Earth from the Sun. Because both the Earth and Mars are traveling in orbits with the Sun at or near their centers, it makes sense that the shortest line between the planets happens around opposition (see Figure 4.11).

Through smaller telescopes, the most visible feature on Mars will be the polar ice caps, made of both water and frozen carbon dioxide ("dry ice"). It is possible, though difficult, to see other features: lighter and darker areas on the planet's ruddy surface. But it takes patience and practice, since the disk is never large.

**Measurements**

| **Mars** | |
|---|---|
| **Angular size:** | 4–25″ |
| **Diameter:** | 6787 kilometers (4222 miles) |
| **Average distance from Sun:** | 228,000,000 kilometers (142,000,000 miles) |

**Figure 4.10 A photo of Mars, taken with an inexpensive "web cam," during its close approach to the Earth in 2005, through a 120 mm (4.7-inch) refractor and a 3× Barlow. To see this amount of detail through the eyepiece at a beginner's scope would be very difficult. Patience and practice are needed to tease things out.**

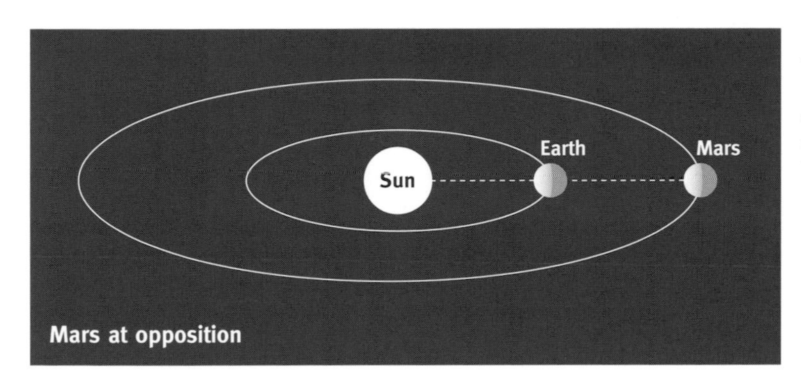


**Figure 4.11 Mars and the other "outer" planets are typically closest to Earth when on the opposite side of it from the Sun.**

The best time to view Mars is for a few months around opposition. It is a small planet, and when farther from the Earth it is not a very rewarding target. However, during a close approach, it can rival Jupiter in brightness. Below is a table showing the dates for oppositions of Mars through to the year 2020, and the maximum angular size it will reach. Notice that it is much larger during some oppositions than others. This is because Mars has an **eccentric orbit**. Its distance from the Sun can vary by tens of millions of kilometers. Every fifteen to seventeen years, Mars comes closest to Earth around the same time that it is closest to the Sun. Such oppositions are called **perihelic** (closest to the Sun). In 2003, it came closer to us than it has in recorded history (though not by much). The next close opposition will be in 2018. These are clearly the best times to view Mars from the Earth, and astronomers eagerly await them.

How do you find Mars in the sky? Well, at or near these times of opposition, look for the brightest reddish "star" in the sky! By definition, an object at opposition reaches its highest point in the sky around midnight.

**Oppositions of Mars**

| Date | Angular Size |
|---|---|
| January 29, 2010 | 14.0″ |
| March 3, 2012 | 14.0″ |
| April 8, 2014 | 15.1″ |
| May 22, 2016 | 18.4″ |
| July 27, 2018 | 24.1″ |
| October 13, 2020 | 22.6″ |

**Measurements**

**Jupiter**

| | |
|---|---|
| **Angular size:** | 31–48″ |
| **Diameter:** | 142,800 kilometers (88,700 miles) |
| **Average distance from Sun:** | 778,000,000 kilometers (484,000,000 miles) |

For more detailed information, or to find its location at other times, check out *Sky & Telescope* or *Astronomy*, or go online.

## Jupiter

By far the largest planet in the Solar System, Jupiter is also one of the most dynamic, interesting targets in the night sky, for several reasons.

First, it is the biggest, not only in reality, but through the eyepiece of a telescope, most of the time. (Venus can appear larger at times, when it is a crescent near the Sun in the sky.) Any time Jupiter is visible in the night sky, it is worth looking at.

Second, even small scopes will show some surface detail on the planet: a couple of prominent, dark **belts**, and the paler **zones** between them. Larger apertures will show more detail, including the **Great Red Spot**, a giant storm that has been raging for at least several hundred years. And since Jupiter spins once on its axis in about ten hours, it is possible to see a large part of its surface in a single, patient night of observing.

**Figure 4.12 A close-up of Jupiter and one of its four large moons.**

**Jupiter's travels**

| Year | Months | Constellation |
|---|---|---|
| 2008 | | Sagittarius |
| 2009 | January | Sagittarius |
| | February–December | **Capricornus** |
| 2010 | January | Capricornus |
| | February–May | **Aquarius** |
| | June–October | **Pisces** |
| | November–December | Aquarius |
| 2011 | January–February | Pisces |
| | March | Cetus |
| | April–June | Pisces |
| | July–December | **Aries** |
| 2012 | January | Pisces |
| | February–May | Aries |
| | June–December | Taurus |
| 2013 | January–June | Taurus |
| | July–December | **Gemini** |
| 2014 | January–June | Gemini |
| | July–September | **Cancer** |
| | October–December | Leo |
| 2015 | January | Leo |
| | February–May | Cancer |
| | June–December | Leo |
| 2016 | January–July | Leo |
| | August–December | **Virgo** |
| 2017 | January–October | Virgo |
| | November–December | **Libra** |
| 2018 | January–November | Libra |
| | December | Ophiuchus |
| 2019 | January–November | Ophiuchus |
| | December | Sagittarius |
| 2020 | | Sagittarius |

**Figure 4.13 Jupiter and three of its four large moons, looking much as they would through a beginner's scope.**

Finally, Jupiter has four moons that are easily visible, even through steadily held or mounted binoculars (it has many others too small to see). Discovered by Galileo around 1610, they are called, in order of distance from the planet, **Io**, **Europa**, **Ganymede**, and **Callisto**. They are so large and bright that, were it not for the glare of the giant they circle, each would be visible to the naked eye under dark skies. They dance around Jupiter at different distances and speeds, and often cross in front of it in a transit. At such times the moon casts a black shadow on the bright cloud tops of Jupiter. The moons also disappear behind Jupiter or its shadow (an eclipse). All of these happenings are quite predictable, and information on times is available in the magazines, or on various websites.

*Observing Jupiter*

Since Jupiter is a worthy target whenever it is in the sky, knowing when it is at opposition is not necessary (though it *is* best at that time). Knowing where it is in the sky is another matter. Jupiter takes 12 years to travel through a complete circle of the constellations along the ecliptic. Opposite is a table that shows what constellations Jupiter will be in through to the year 2020. You can use this table and a planisphere to figure out when and where Jupiter will be in your sky, and the best time of year to see it. For example, from mid 2012 to mid 2013, Jupiter will be traveling through **Taurus**. Taurus is highest in the sky during December and January, a fact your planisphere will show you; so Jupiter will be as well. Jupiter will almost certainly be the brightest object in or around whatever constellation it's in, unless Venus is nearby.

**Measurements**

**Saturn**

| | |
|---|---|
| **Angular size:** | 15–21″ (not including rings) |
| **Diameter:** | 120,660 kilometers (74,600 miles) |
| **Average distance from Sun:** | 1,427,000,000 kilometers (886,700,000 miles) |

## Saturn

Many people consider Saturn the most striking object in the sky. Though not as large or bright through the eyepiece as Jupiter, the amazing and beautiful ring system that surrounds it gives Saturn a three-dimensional reality that,

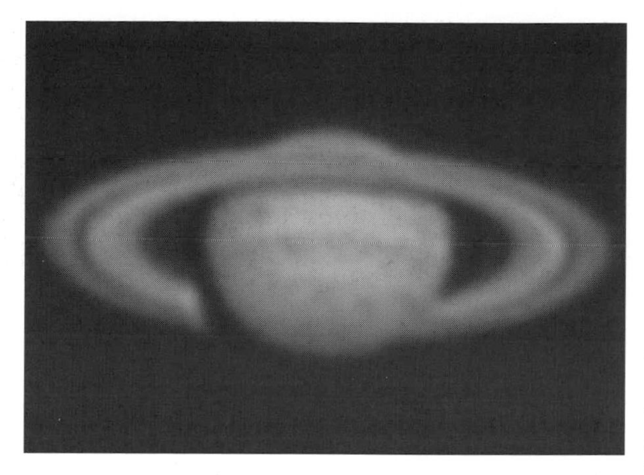

**Figure 4.14 Saturn, with its rings nearly wide open. Though this photo captures a fair amount of detail (good "seeing" – or a larger aperture – would have made it sharper), it does not convey the amazing reality that one sees when viewing through the eyepiece. Viewed "live" through a telescope, Saturn seems clearly to be what it actually is: a brilliant, fully three-dimensional globe and rings hanging in the blackness of space.**

I believe, nothing else can match. Even a small scope will show the rings distinctly, and good optics should allow you to see a break within them called **Cassini's Division** (named for the astronomer **Giovanni Cassini** who first described it in the seventeenth century). Saturn is a fantastic sight, and photographs just do not seem to catch the apparent depth and magic of the real thing.

Saturn takes nearly 30 years to orbit the Sun. Like the Earth, its axis of rotation is tilted in relation to its orbit. This is lucky for us. If it were not tilted, the rings would be very difficult to see since they would appear edge-on, or nearly so. As it is, there are still times during its 30-year cycle when the rings are difficult to see. They slowly open up and then close again over a period of 14 to 15 years (half the planet's orbital period). As I write this, in 2008, they are several years past a peak in their expansion. By 2009, they will be nearly invisible. Then the rings will start to slowly open up, until they again reach their widest in 2016.

In addition to the rings, Saturn also has many moons, at least one of which should be visible in beginner's scopes. This is **Titan**. Titan is the only moon in the Solar System known to have a significant atmosphere. It can always be seen a few ring widths away from Saturn. Since Saturn is tilted, so are the orbits of its moons. This means they do not usually cross the planet's face or hide behind it, but seem to travel around it in a flattened oval.

### *Observing Saturn*

Like Jupiter, but unlike Mars or Venus, Saturn does not vary much in apparent size during the year. This is because it is *always* very far away: its relative distance from us does not vary as much, so neither does its size through the eyepiece. So, again like Jupiter, Saturn is a good target whenever it is in the nighttime sky.

Below is a table showing what constellation Saturn can be found in through the year 2020. As you can see, it moves much more sedately across the sky than Jupiter.

**Saturn's stroll**

| Year | Months | Constellation |
|---|---|---|
| 2008 | | Leo |
| 2009 | January–September | Leo |
| | October–December | Virgo |
| 2010 | | Virgo |
| 2011 | | Virgo |
| 2012 | | Virgo |
| 2013 | January–May | Libra |
| | June–August | Virgo |
| | September–December | Libra |
| 2014 | | Libra |
| 2015 | January–May | **Scorpius** |
| | June–September | Libra |
| | October–November | Scorpius |
| | December | Ophiuchus |
| 2016 | | Ophiuchus |
| 2017 | January–February | Ophiuchus |
| | March–May | Sagittarius |
| | June–November | Ophiuchus |
| | December | Sagittarius |
| 2018 | | Sagittarius |
| 2019 | | Sagittarius |
| 2020 | January–March | Sagittarius |
| | April–June | Capricornus |
| | July–December | Sagittarius |

Saturn does not always dominate the section of sky it is in the way Jupiter does. But it is usually quite bright, and shines with a pale yellow hue. If you can locate the constellation it is in, finding it should not be very difficult, since it will not appear on a planisphere or star chart, so it will be the extra "star" you see.

### Uranus

Uranus is the seventh planet from the Sun, and the first to be discovered through a telescope. It is actually visible to the keen-eyed without magnification, and is fairly easy to spot through binoculars. The problem is, you might look right at it and not know whether it was a planet or star. Since it is so far away, it will seem like just another point of light. It takes a telescope aimed in the right direction to know for certain. At higher magnifications, Uranus will appear as a small, but distinct, disk; stars are *always* points.

For the beginner, just finding Uranus is the thrill; it is too distant for amateur scopes to show anything but a bluish dot. Larger scopes do allow some of its moons to be seen.

| **Measurements** | |
|---|---|
| **Uranus** | |
| **Angular size:** | 3–4″ |
| **Diameter:** | 51,000 kilometers (32,600 miles) |
| **Average distance from Sun:** | 2,871,000,000 kilometers (1,784,000,000 miles) |

Uranus takes 84 years to circle the Sun, so no tables are necessary to show its location: it moved into Aquarius in 2003, and will stay there pretty much through to 2010, when it will move in for a long stay in Pisces. Finding it takes patience, and an accurate star chart. *Sky & Telescope* publishes such a chart in an issue once a year. Locating a finder chart on the Internet should also be easy.

### Neptune

**Neptune** is like Uranus, only more so. It can be seen in amateur scopes, but that is about all; and because it has an even smaller angular size, finding it is more of a challenge. It takes twice as long as Uranus to orbit the Sun: 165 years! It will be in Capricorn until 2010, when it will move into Aquarius. Charts showing its location appear in the same issue of *Sky & Telescope* mentioned above.

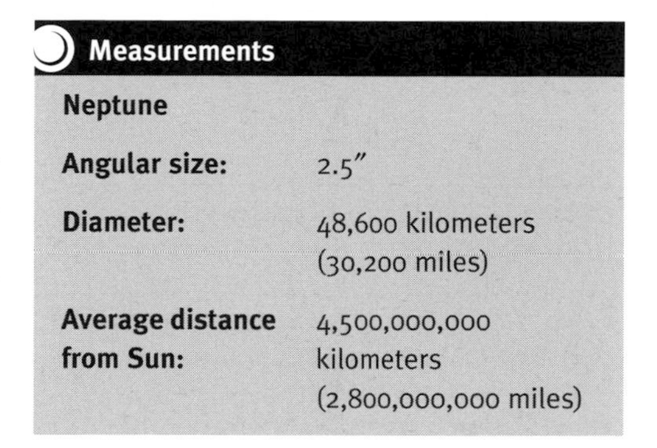

| **Measurements** | |
|---|---|
| **Neptune** | |
| **Angular size:** | 2.5″ |
| **Diameter:** | 48,600 kilometers (30,200 miles) |
| **Average distance from Sun:** | 4,500,000,000 kilometers (2,800,000,000 miles) |

## Other objects in the Solar System

The Solar System contains countless other objects, but most of them are too dim for beginners' equipment, or their appearance is not predictable. Following are brief descriptions of several types of object which you might encounter.

### Meteors

These are small to tiny pieces of material that happen to be traveling through space when they suddenly run into the Earth's atmosphere, about 80 kilometers (50 miles) up. Though the atmosphere is thin there, these objects are moving so quickly that slamming into the atmosphere causes

**Some well-known meteor showers**

| Meteor Shower | Location of radiant | Date of shower peak |
|---|---|---|
| Quadrantid* | **Draco** | January 4 |
| Lyrid | **Lyra** | April 22 |
| Eta Aquarid | Aquarius | May 6 |
| Delta Aquarid | Aquarius | July 28 |
| Perseid | Perseus | August 12 |
| Orionid | Orion | October 21 |
| Leonid | Leo | November 17 |
| Geminid | Gemini | December 14 |

*Named for a defunct constellation that used to exist in the part of the sky where we now see Draco.

them to burn up in a brilliant flash. Most meteors are no larger than a pebble, though a very few are much larger (but they are certainly never "shooting stars"!). Bright meteors sometimes leave behind what is called a **meteor train**; a dim cloud of hot gas that can last several minutes.

On any given night under dark skies, you may see a few meteors; but during certain times, the numbers rise dramatically. These times are when meteor showers occur, and they happen when the Earth travels through the trails of old comets. Perhaps the most famous shower is the **Perseid**, so called because an imaginary line from the actual paths of the meteors lead back to, or **radiate** from, a spot – aptly called the **radiant –** in the constellation **Perseus**. The Perseid shower is popular because it is fairly reliable in its numbers, and because it peaks on the night of August 12, a time of year when stargazing weather can be quite pleasant (in the northern hemisphere, at least). But there are many other showers that occur each year. The table above lists several of the best.

*Observing meteor showers*

Viewing meteors is one astronomical pursuit that is actually hindered by the use of binoculars or a telescope. This is because their locations are unpredictable; most flash into and out of existence in a second or less; and they cross a large arc of sky. Observing them requires the "wide-eye" technique mentioned earlier in the guide. It is also best to be in a reclining position when viewing, so the most sky can be comfortably viewed simultaneously.

The number and brightness of meteors varies greatly between the different showers, since each is caused by different comet trails. Also, though some showers are fairly consistent in their numbers from year to year, others are quite variable.

Figure 4.15 A Leonid meteor and the Big Dipper. The constellation Leo is "below" the Big Dipper in the sky, and this meteor appears to be coming "up" from it.

Sky darkness is very important for viewing the most possible meteors, since many are quite dim. Moonlight and light pollution wash out the light of many. Once you know that the Moon will not be in the sky on the night you will be observing, it is best to get to the darkest site you can.

Typically, meteor showers peak after midnight, since they happen as the leading edge of the Earth plows through debris from a comet. Before midnight, an observer is on the trailing edge of our planet. Still, many showers take place over several nights, so evening viewing is still worthwhile.

## Comets

At least every few decades, a bright comet appears in our skies. The last one visible in the northern hemisphere was Hale-Bopp, in 1997. Comet McNaught, in 2007, was actually brighter, but was visible primarily to people in the southern hemisphere.

Comets have been called "dirty snowballs." They are mostly made of frozen gases, but they have some solid material in them as well. Usually just a few kilometers across when frozen solid, comets are ancient objects, mostly unchanged since the time the Solar System formed. Most are believed to come from the outer fringes of the area influenced by the Sun. Something disturbs them in their far, cold orbits, and they begin to fall slowly inward. As they near the Sun, they speed up; and when they get quite close, they begin to "melt". It is then that they can become brilliant. The small, original **nucleus** becomes surrounded by a huge ball of gas (often

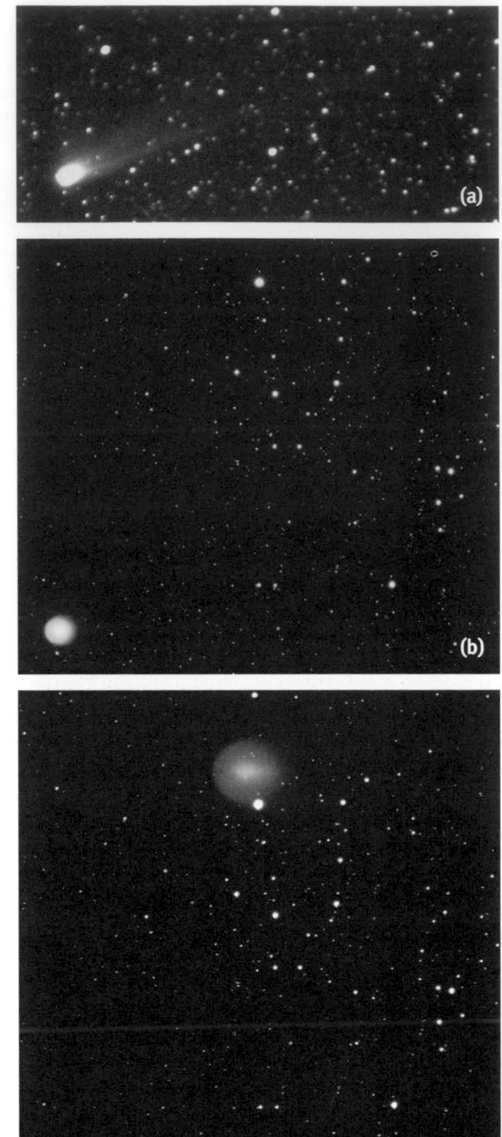


**Figure 4.16** Two very different comets. (a) A typical comet that passed through the Earth's evening sky in 2004, never getting bright enough to see without binoculars. (b) The very strange comet 17P/Holmes, on November 3, 2007. (The "17P" signifies that this was the seventeenth periodic comet discovered; Holmes is the name of the man who discovered it, in 1892.) On October 24, 2007, this comet, which never comes closer to the Sun than Mars, and which is usually much too dim to see in even large amateur scopes, suddenly brightened by nearly a million times. (c) Comet Holmes on November 20, 2007. Over the weeks following the outburst, the shell of gas and dust surrounding the nucleus expanded until it actually became larger than the Sun! Note the motion of the comet over the seventeen days between photos: the brightest star in both photos is Mirfak (seen just below the comet in (c)), the brightest star in the constellation Perseus. The comet was still visible in binoculars, and much larger yet, as late as early 2008.

much larger than the Earth) called the coma. Tails of gas and dust millions of miles long can trail across long stretches of the night sky.

Some comets sweep into the inner Solar System, whip around the Sun, and fly out again on courses that will take them back out of history. Others, periodic comets, are caught, and will return someday. The most famous of these is Comet Halley, which will return in 2061.

What most people do not know is that comets are fairly common visitors. The brilliant headline makers are fairly rare; but dimmer, less flashy comets, often only visible through binoculars or telescopes, occur in most years. There is something ethereal and almost magical about looking at these silent, ghostlike travelers through an eyepiece. You know they are moving at incredible speeds, but they seem to be just quietly drifting through the night; and they might be visible for several nights, weeks, or even months. Since it is not possible to predict when the next new comet will be spotted, and because they get little media attention unless they are spectacular, few people other than astronomers notice when most of these visitors travel through.

## Asteroids

There is really no *true* difference between an asteroid and a rocky planet, except size. (The current definition of a "planet" is under debate.) Both are solid bodies orbiting the Sun. There are now more than 100,000 known asteroids. The larger ones, and some smaller ones that come close to Earth, are often visible through amateur scopes. *Sky & Telescope* often publishes information on finding them. Through the eyepiece they will appear as dim stars, and the only way to be sure you've spotted one is to watch it move over time. The closer they are to us, the more quickly they appear to move.

# The deep sky 5

Targets for the amateur astronomer that are outside the Solar System are both easier and more difficult to find. Easier, because they stay in pretty much the same spot on the celestial sphere for centuries or longer, so planispheres, field guides and star charts that show their positions can be used at any time of year, and for many years. More difficult, because they are dimmer, usually *much* dimmer, than the planets, though they often cover a much larger area of the sky. This makes many deep-sky objects prime targets for binoculars rather than telescopes, since the higher magnification of telescopes makes it impossible to fit many of these objects entirely within view at one time.

There are several types of deep-sky objects that can be viewed by the amateur: stars, star clusters, galaxies, and nebulae (or nebulas); and all of these categories contain their own types. Each is described below. There are other types of objects out there, like quasars and black holes, but they are not visible in beginners' scopes (or at all!), so they will be left alone.

## Stars

Stars are immense balls of compressed gas, so squeezed by their own gravity that nuclear reactions take place within them, and they begin to shine. They can be smaller than the Earth in diameter, or larger than the diameter of the orbit of Mars. They can be red, orange, yellow, white, or blue; and to the human eye at the telescope, other colors are also seen. Yet no matter how large in reality, or what color, stars are so far away that they always appear as points of light. No amount of magnification will change this.

### Star brightness

Because of their varying distance from Earth, and because of actual physical differences between them, some stars shine brighter in our sky than others.

**Figure 5.1 A wide view of some constellations seen in the evening skies in December. Orion, the Hunter, on the lower left, uses his shield to fend off Taurus, the Bull, above. The Pleiades, or Seven Sisters, are in the upper right.**

Charts use various methods to show a star's **magnitude**, or level of brightness. However, there is a standardized system of measurement. The lower the magnitude number, the brighter the star. A star of magnitude 1 is about 2.5 times brighter than a star of magnitude 2, which is itself about 2.5 times brighter than a star of magnitude 3, and so on. A difference of five magnitudes is a difference of exactly 100 times in brightness. For example, the brightest star in the sky, **Sirius**, actually has a negative magnitude number; it shines at –1.46. Polaris, the North Star, has a magnitude of about 2.1.

## Star names

There are two primary methods used to name individually the brighter stars. First, several hundred of the brightest have proper names, most of them very old, and many derived from Arabic. However, the names often "evolved" as

> **More info**
>
> To add a little complexity, many stars go up and down in brightness over time. The period of time can vary greatly, and so can the change in brightness. These are called **variable stars**, and there are several reasons for the variability. Some actually change in brightness because of physical changes in a single star (Polaris is one of these). Others change because the "star" is actually two or more stars that circle so close to each other that we cannot separate them visually, and when one sometimes passes in front of (eclipses) the other, overall brightness changes. These, naturally enough, are called **eclipsing variable stars**. The most famous of these is the star **Algol** in the constellation Perseus. Every three days or so, its brightness drops by more than half for about 10 hours.

> **More info**
>
> Other deep-sky objects are also assigned magnitude numbers, but they are somewhat misleading: the number indicates the object's total brightness, but since the object is spread over a large area, it will appear dimmer than its number would suggest.

they were translated from one language to another. For example, Orion's left shoulder is the star **Betelgeuse**, which some sources claim comes from an Arabic word for "shoulder" or "armpit." Others claim it comes from words that originally meant "hand of the center one." Sirius, the brightest star, is said to get its name from a Greek word meaning "scorching."

The other way brighter stars have been named is through a system created several centuries ago. Greek letters are used to designate the brightest stars in a given constellation, starting with alpha ($\alpha$), the first letter, for the brightest star; and continuing through the Greek alphabet (see Appendix 1 on page 133) as stars decrease in brightness. So Sirius, the brightest star in the constellation **Canis Major** (the Greater Dog) is called $\alpha$ (alpha) Canis Majoris. This system is a good one, but can be a bit confusing, since mistakes were made with it when stars were first named. Betelgeuse is called $\alpha$ (alpha) Orionis, but is actually the *second* brightest star in the constellation Orion.

Of course, this system also works only with the brightest stars. There are only 24 letters in the Greek alphabet, and only 88 constellations. This provides a total of 2112 star designations, and there are hundreds of billions of stars in our galaxy alone (though only a very small percentage of these have been cataloged.) Other systems are used for naming dimmer stars.

### Double and multiple stars

The Sun is a single star, which is probably lucky for us. Many stars come in pairs or groups, held together by gravitational attraction. They are so close together in our sky that, to the naked eye or at low magnification, they appear to be a single object. They are even listed under one name on charts. But when magnification is increased, separate stars are seen. Often there are two (this is a *double star*); sometimes three or even more (a *multiple star*). Some are so close together that they can not be separated visually, and are only known to be pairs by their effect upon each other, or through other indirect means. Others only *appear* close together from Earth, because they happen to fall along the same line-of-sight, but are actually not related at all, and are very distant from each other.

Some amateurs are keenly interested in double and multiple stars. They are challenged by the task of trying to see the separation between close doubles, or are captivated by the contrast between the colors of some groupings.

One famous double star is **Albireo**, the "head" of the swan in the constellation **Cygnus**. It is easily separable in beginners' scopes. It contains two stars that appear yellow and blue–green, and is considered by some to be the most beautiful double star in the sky.

## Star clusters

Sometimes, stars are gathered into larger groups called clusters. They are beautiful objects to view, and can cover both very large and very small parts of the sky. Clusters are divided into two forms: *open clusters*, and *globular clusters*. The two are different in both structure and size.

### *Open star clusters*

These are irregular groupings of stars that are smaller, and closer to the Earth than globulars. They can contain anywhere from several stars to many hundreds of stars. Some open star clusters are called "tight" because their stars are more tightly packed together; others are called "loose." The most famous open cluster in the sky is M45, the **Pleiades**, or Seven Sisters, in the constellation Taurus. Easily visible to the naked eye as a group of about six stars (the keen-eyed report more), this cluster is stunning through binoculars, and dozens of stars become visible. The **Hyades**, a much looser open star cluster, is not far away in the sky (see Figure 5.2). The **Double Cluster**, in the constellation Perseus, is, as its name states, two clusters which appear close together in our sky, and which make for great viewing through a pair of binoculars.

### *Globular star clusters*

Much larger than open star clusters, globulars usually appear smaller because they are far more distant. Globular star clusters are also much more uniformly structured. Almost all are spherical, and the stars grow ever more tightly packed together toward the center. Some are so closely packed that the smaller apertures of beginner scopes will only show the cluster as a fuzzy patch, almost like an unfocused star.

The most famous globular cluster visible to northern observers is **M13** in the constellation **Hercules**. But the brightest of them all is ω **(omega) Centauri**, in the southern constellation **Centaurus**. It is visible to the naked eye even under somewhat light-polluted skies as a blurry star, but telescopes show it to be what it really is: a huge, dense collection of millions of stars.

**Figure 5.2 Two open clusters and a wandering visitor. The Pleiades, or Seven Sisters, is possibly the most well-known open cluster in the sky, and is prominent on the lower right in this photo. Less easily seen to the left is the larger, more open, V-shaped cluster called the Hyades. Both are in the constellation Taurus, the Bull. Between them is the bright planet Mars, which passed through Taurus when this photo was taken, in August, 2007.**

## Galaxies

Galaxies have sometimes been called "island universes." They are almost unimaginable in their size: up to perhaps a trillion stars collected together in a group by their gravitational attraction. They are divided into three main types, based on their shapes: **elliptical, spiral**, and **irregular**.

Our own spiral galaxy, the Milky Way, is visible at dark sites as a long, dim cloud that spans the entire sky (see Figure 5.3.) Certain areas of it are brighter than others, and when viewed through binoculars or telescopes at low magnifications, these cloudy patches reveal themselves as dense patches of countless stars. The view toward the center of the Milky Way is located in the constellation Sagittarius, but it is blocked by huge clouds of interstellar dust. Still, the area around Sagittarius and neighboring constellations is an amazing place to explore at low powers.

The closest large galaxy to our own is M31, the Andromeda Galaxy. Though it is about two and a half *million* light years away, it still spreads across more than three degrees of sky: six times the width of the full Moon! Visible to the unaided eye under dark skies as a dim, slightly elongated smudge, it is the furthest naked-eye object visible to most people. Good binoculars show more detail, including dust lanes in the arms of our spiral neighbor.

Two other galaxies are easily visible to southern viewers: the **Small** and **Large Magellanic Clouds**. These are smaller, irregular galaxies, and are much closer to us than M31. Though small by galactic standards, they still contain billions of stars between them.

**Figure 5.3 In the brightest part of this photo, in the constellation Sagittarius, you can see the center of our own galaxy, the Milky Way. The bright areas going from bottom to top are clouds of stars and glowing gas; the dark areas are caused by obscuring dust. Our galaxy is a flattened disk, and since we are inside of it, the disk forms a great circle around our sky.**

All other galaxies are too small and dim in our sky to be seen with unaided vision for any but the keenest-eyed under the darkest skies. But many are visible through binoculars, and far more through telescopes. Some of the brightest of these include **M51**, known as the **Whirlpool Galaxy**, and the pair of **M81** and **M82**, all near the Big Dipper. A bright galaxy for southern viewers is **M83**, a spiral in the constellation **Hydra**, the Water Snake. It should be noted that none of these will appear as much more than "faint fuzzies" through a beginner's scope, but just think about what they are!

**More info**

**Charles Messier** was an eighteenth century comet hunter. He compiled a list of more than 100 "nebulous objects" that could be mistaken for comets, so that he and others could avoid them when searching (though it stretches the imagination to see how some of them could ever be mistaken for comets). This **Messier Catalog** has become the most famous list of deep-sky objects among amateurs. It includes clusters, galaxies, and nebulae; all of them visible in basic equipment, comparable to – or better than – what Messier had available over 200 years ago. Messier objects are designated by numbers preceded by the letter "M", as in M31, M45, M81, etc.

## Nebulae

The word "nebula" (plural "nebulae") is Latin for *mist* or *cloud*, and before the advent of modern, large telescopes, it was used to describe nearly all objects in the night sky other than stars and planets. Today, it is still used for several types of objects: **emission nebulae**, **reflection nebulae**, **dark nebulae**, **planetary nebulae**, and **supernova remnants**.

### Emission nebulae

In regions of space where stars are being formed, large clouds of gas have their atoms "excited" by the light of nearby giant stars. This causes the gas (primarily hydrogen) to glow. In long exposure photographs, they appear red, but through the eyepieces of beginners' scopes, they will appear pale gray or green.

There are many well-known emission nebulae, but perhaps the most famous is the Great Orion Nebula, M42. It is part of Orion's "sword," and easily visible to the naked eye from nearly anywhere on Earth. Binoculars and telescopes show that it is a complex structure with wisps and strands of ghostly material. Photographs do not usually capture all these details, since some parts of the nebula must be over-exposed in order to bring out the dimmer areas (see Figure 5.4.)

There are many emission nebulae to be seen in or around the constellation Sagittarius. These include **M8** (the **Lagoon Nebula**), **M20** (the **Trifid Nebula**), and **M16** (the **Eagle Nebula**). Perhaps the largest known emission nebula is the **Tarantula Nebula**, located in the Large Magellanic Cloud, so not viewable to those in the north. Since it is in another galaxy, it is *much* farther away than other observable nebulae, so its size in our sky is smaller than one might suppose.

### Reflection nebulae

These are different from emission nebulae in that they are starlight reflected from great clouds of dust. They usually appear blue in photographs because they are reflecting the light of bright, hot, blue stars. They are typically dimmer than emission nebulae because they do not actually shine on their own. The Pleiades are embedded in a reflection nebula, and larger apertures will barely show some of it; this nebula shows up much better in long-exposure astrophotos. Another famous example is the Trifid Nebula, M20. It contains areas of both emission and reflection, indistinguishable at the eyepiece.

**Figure 5.4 The Great Orion Nebula.**

## Dark nebulae

Also composed of interstellar dust, dark nebulae simply block the light of whatever is behind them, whether star fields in the Milky Way, or the light of emission or reflection nebulae. Perhaps the most famous is the **Horsehead Nebula** in Orion, though this one is too small and dim to see though a beginner's equipment. The Trifid Nebula has dark lanes of dust running through it that divide it into the three main sections that give it its name.

## Planetary nebulae

These are completely different, at least in their origin. Planetary nebulae got their name because they are typically small and round and appear about as large in the eyepiece as one of the planets. They are shells of gas that have been blown off stars as they go through changes when they age. They are much smaller than most emission nebulae, though they glow for the same reason: excited particles of glowing gas.

Two bright examples of planetary nebulae are the **Ring Nebula, M57**, in the constellation Lyra; and the **Dumbbell Nebula, M27** in **Vulpecula**. There are hundreds of others in the sky.

### Supernova remnants

These are the gaseous remains of exploded stars. Stars end their lives in a variety of ways, and supernovas are the most spectacular. Only large, bright, relatively short-lived stars go out with the bang of a supernova, so they are fairly rare, especially at distances close enough to Earth to leave behind debris visible in small telescopes. (It has been centuries since the last close supernova was seen.) Supernova remnants are typically quite dim, and often spread over a wide swath of sky. The **Crab Nebula**, **M1**, is one of the few remnants that can be seen through a small scope. It resulted from a supernova seen in the year 1054.

## Resources

### Books

The field guides already mentioned go into more detail than I have about what's out there, but there are many excellent books written specifically to describe the origins, workings, and ultimate fate of what you will be seeing. Some deal specifically with the Solar System; others describe what's beyond it. Many contain beautiful photographs. Here is a warning, however: the technology that is currently (or soon to be) available to astronomers has revolutionized the field. If you are interested in current thinking, get the latest publications you can find. Knowledge advances so quickly that, by the time a book reaches the shelf, its information may well be outdated. This is why I will not make specific recommendations here about books dealing with the *science* of astronomy; and it is one more reason to check out the magazines: they are always publishing the latest thoughts, theories, and images.

There are, however, two more books I would like to recommend. Each contains more detailed information on specific objects to observe than I have, and using either (or both) of them will provide many, many hours of stargazing opportunities.

1. *Turn Left at Orion: A Hundred Night Sky Objects to See in a Small Telescope – and How to Find Them*, by Guy Consolmagno and Dan M. Davis. Cambridge University Press, 2000.
2. *NightWatch: A Practical Guide to Viewing the Universe* (Fourth edition) by Terrence Dickinson. Firefly Books, 2006.

## Star charts

A field guide (or even a planisphere) is all that is needed to get started on your search of the cosmos; but once you have been at it for a while, you may want to graduate to the next step. Star charts show a larger area of sky, on a larger scale, and, typically, in more detail, than a field guide. Here is one of the standards:

> *Sky Atlas 2000.0*, by Wil Tirion and Roger W. Sinnott. Sky Publishing and Cambridge University Press, 1998.

The *Sky Atlas* charts are available in three different formats, but all contain 26 maps of the entire sky, plus seven close-up charts of areas of special interest. Tens of thousands of stars, and 2,700 deep-sky objects are accurately plotted. These are the charts I own, and it will be a long time before I need any others.

## Software

You can get much more detail than is possible with paper charts by using a computer. There are dozens of different programs available that all fit under the heading of *planetarium software*. Some have only very basic capabilities and show relatively few objects; while others may be able to show you millions of different stars, many thousands of deep-sky objects, and the past, current, and future positions of the Sun, Moon, planets, and thousands of asteroids and comets. Many of the most detailed programs must be purchased, and can be quite expensive. But there are plenty of others, many of them quite complete, which are available for free on the Internet. A web search for "free planetarium software" should lead you to some of them. Try more than one, since each is slightly different in the way it works, and in the presentation of the sky.

It is easy to lose yourself in all the information and graphics these programs offer. It may be very interesting – and loads of fun – to recreate a solar eclipse in a simulation of the sky over Rome on July 17, 709 BC. But be sure to remember to go outside sometime soon to see the *real* sky!

# Part III
# A stargazing glossary

This section of the guide is meant to be used as a reference to astronomy terminology. It contains all the words in the text that were printed in **bold** type. Some of the more obscure words will have their pronunciations given. Deep-sky objects that have Messier Numbers (for example, M42 or M57) will be found under their common names, unless they do not have a well-known common name. An Internet search using many of these terms would give much more information; more than almost anyone could use.

**aberration** A problem, a deviation; either inherent in a design, or a flaw in the manufacture of telescopes.

**achromatic refractor** A refracting telescope that uses a lens with two elements (separate pieces of glass) to help relieve chromatic aberration (false color).

**afocal** Not a part of the focusing system; separate from the telescope. In astrophotography, when a camera is placed near a telescope's eyepiece, and focused separately.

**air-spaced achromat** An achromatic lens with a very short space between the two elements.

**Albireo (al-BEER-ee-oh)** A beautiful double star that is the "head" of the swan in the constellation Cygnus.

**Algol** An eclipsing variable star in the constellation Perseus.

**alt-az mount** Short for altitude-azimuth; a type of mount that is adjustable up and down, and left and right.

**Andromeda** A constellation in the northern sky, named for the mythical maiden rescued by the hero Perseus. The constellation is best known for containing M31, the Andromeda Galaxy.

**Andromeda Galaxy (M31)** The closest spiral galaxy to our own Milky Way, but still more than two million light years away! It is so large that, even at that distance, it takes up more than three degrees of sky. It is visible to the unaided eye under dark skies, and features become discernible with binoculars.

**angular size** The *apparent* size of an object in the Earth's sky, measured in degrees of arc.

**annular eclipse** An eclipse of the Sun by the Moon that occurs when the Moon is far enough from the Earth in its orbit that it does not completely block out the entire disk of the Sun.

**aperture** Literally, an opening or hole; in a telescope, it is expressed as the diameter of the main light-gathering surface.

**apochromatic** Various lens designs that virtually eliminate chromatic aberration in refractors; the word means "without color."

**Aquarius** A constellation in the zodiac; the Water Bearer.

**area (of a circle)** A measure of the space contained within the circumference of a circle. It is measured using the formula $A = \pi r^2$; area ($A$) equals $\pi$ (pi: about 3.14) times the radius ($r$) times itself (squared).

**Aries** A constellation in the zodiac; the Ram.

**asterism** A part of an official constellation whose shape is recognized and named; like the Big Dipper, which is actually part of the constellation **Ursa Major**, the Great Bear.

**asteroid** Any small body that orbits the Sun rather than a planet, and that is not a planet itself, or a comet (comets differ in that they sometimes glow within a halo of their own melted gases). Tens of thousands are now known, at least one surpassing the size of Pluto. Sometimes they are called "minor planets."

**astrocamera** Any camera designed specifically for taking astrophotos. They can range from relatively simple and inexpensive modified "web cams" to liquid-cooled large-format digital cameras costing tens of thousands of dollars.

**astrology** A system of belief that claims to be able to predict personal fate or world events based upon the positions of the stars and planets at the time of a person's birth. It has no basis in fact.

**astronomy** The modern study of the heavens and everything contained within them. It is entirely based on experiment, observation, and theory based on these.

**astronomy club** An organized group, usually local or regional, formed in order to share the science and joy of stargazing.

**astrophoto** Any photo that is primarily of something astronomical.

**astrophotography** The art and science of taking photographs of astronomical phenomena.

**atmosphere** A layer of gas that surrounds some massive objects in space. All of the planets have significant atmospheres, except for Mercury. Pluto (the planetary status of which is now in some doubt) may have one that exists as a gas only when Pluto is at its closest to the Sun; farther out it may freeze. (There is a probe on its way to Pluto to find out.) Only one moon (Saturn's Titan) has an atmosphere.

**aurora** The Northern or Southern Lights; caused by interactions between material ejected from the Sun and the Earth's magnetic field in its upper atmosphere. Strong Solar storms can result in spectacular curtains of color that sometimes fill the entire sky.

**axis (or axis of rotation)** The imaginary line about which a spinning body rotates.

**BAK-4** A type of glass (barium crown) used in optics, particularly binocular prisms. It is typically superior for this purpose to another type: BK-7.

**Barlow lens** A special lens that, in effect, lengthens a telescope's focal length, thereby increasing magnification. It does so without decreasing eye relief.

**belts** The dark bands visible across Jupiter's surface, even in beginners' scopes.

**Betelgeuse** The bright orange star that marks the left shoulder of the constellation Orion, the Hunter. The name comes from Arabic origins.

**Big Dipper** The name used in North America for the familiar, bright grouping of seven stars in the northern sky that form the rough shape of a dipper, or ladle. In other parts of the world it is known as the Plough, or the Great Wagon, or other names. It is not actually a constellation itself, but a part of the constellation Ursa Major, the Greater Bear. Such groupings within constellations are called ***asterisms***.

**binoculars** Two small telescopes bound together to allow the use of both eyes when viewing. Special prisms bend the light paths to give correct, real-world images, while also providing for compact, lightweight housings.

**BK-7** A type of glass (borosilicate) used in optics, particularly binocular prisms. It is usually not as good for this purpose as another type: BAK-4.

**cable release** A device that is attached to a camera that allows the shutter to be tripped from a distance to avoid shaking during an exposure.

**Callisto** One of the four Galilean moons of Jupiter (named after their discoverer), and the furthest of the four from the giant planet. With a diameter of 4,800 kilometers (about 2,900 miles), it is about the same size as Mercury.

**Cancer** A constellation in the zodiac; the Crab.

**Canis Major** A constellation, the Greater Dog. It is near Orion in the sky, and contains the brightest star, as seen from Earth: Sirius.

**Capricornus** A constellation in the zodiac; the Sea Goat (?!)

**Cassini, Giovanni (1625–1712)** Among other things, this astronomer was the first to record the division in the rings of Saturn that now bears his name; he also discovered several of its moons.

**Cassini's Division** A gap between two of the rings of Saturn; visible even in some small scopes, it was first recorded by the Italian astronomer Giovanni Cassini.

**Cassiopeia** A constellation in the northern sky; the Queen.

| | |
|---|---|
| **catadioptric** | Any telescope design that incorporates both mirrors and lenses. |
| **celestial** | From a Latin word meaning "of the sky" or "of the heavens." |
| **Celestial poles** | The two points in the sky directly above the Earth's poles; the only places in the sky that do not appear to rotate as the Earth spins. |
| **Celestial Sphere** | An ancient concept about the workings of the night sky. At one time, people believed that all the stars in the heavens existed as they appear to be: immovable lights upon a huge sphere that spun in the sky above the Earth. It was thought that the planets traveled around this bowl, or that they somehow moved in spheres of their own. |
| **Celestron** | One of the two largest telescope manufacturers in the United States; the other is Meade. |
| **Centaurus** | A constellation in the southern sky; the Centaur (a mythical creature, half horse, half human). It is a constellation filled with interesting objects, including the brightest globular cluster – ω (omega) Centauri – and the closest star system to our own – α (alpha) Centauri. |
| **central obstruction** | The design of all reflecting telescopes requires a second mirror to bounce the image out of the telescope tube and into the focuser. This mirror is typically in the center of the tube, and blocks some of the incoming light. |
| **Cetus** | A constellation through which the ecliptic passes (though Cetus is not part of the zodiac); the Whale. |
| **chromatic aberration** | A problem in refractors, where the shape of the objective lens allows for magnification, but also acts as a sort of prism, breaking light into its component colors. |
| **clock drive** | An electric motor which, when attached to an equatorial mount, rotates it around its right ascension axis in order to compensate for the rotation of the Earth. When the RA axis is accurately aligned with the celestial pole, objects will stay centered in the eyepiece. |
| **coated lenses** | Good quality lenses are coated with various materials to reduce internal reflections. |

**collimation** The lining up of the centers of the mirrors in a reflecting telescope with the center of the focuser; extremely important for sharp focusing.

**coma** 1. A type of aberration common in reflecting telescopes, in which stars toward the edge of the field appear stretched, like little comets. 2. The large shell of glowing gas and dust surrounding the nucleus of a comet.

**comet** A ball of frozen gas, dust, and other solid material. Comets dwell far from the Sun, but sometimes their orbits are disturbed and they begin to fall inward. When they near the Sun, the gases begin to melt, sometimes giving spectacular views before flying off into the void again.

**computer-guided mount** Telescope mount with the electronic ability to find targets contained within a computer database.

**concave** A mirror or lens face that curves inward (like a cave). The opposite of concave is convex.

**constellation** One of the 88 recognized groupings of stars that divide up the entire sky as seen from Earth. The official names of constellations are usually expressed in Latin.

**contrast** The difference in brightness between the light and dark areas in an image.

**convex** A mirror or lens surface that curves outward. The opposite of convex is concave.

**cooling time** The time necessary for the temperature of a telescope to cool to the temperature outside, or at least close enough to make for decent viewing. Until this happens, currents of air often form inside the telescope, distorting the images seen through it.

**correct image** An image in an eyepiece that appears right-side-up and non-inverted, like the real world.

**corrector plate** The lens on the front of a catadioptric telescope that helps these designs overcome aberrations.

**Crab Nebula (M1)** A supernova remnant in the constellation Taurus; visible, but not very impressive, in beginners' scopes.

**crescent** Of the Moon (or Mercury or Venus), when less than half of the surface, as viewed from Earth, is lit up by the Sun.

| | |
|---|---|
| **cross-hairs** | Small threads that form an x-shaped pattern; contained inside eyepieces and finder scopes to help center an image. |
| **Cygnus** | A constellation, through which the Milky Way runs; the Swan. |
| **dark nebula** | Clouds of interstellar dust which block the light of objects behind them. |
| **dark site** | Any location relatively unaffected by artificial light; a diminishing resource. |
| **declination (Dec)** | Celestial latitude; the angular distance an object is from the celestial equator, at 0 degrees; the poles are at 90 degrees. |
| **deep sky** | Everything outside the Solar System. |
| **deep-sky object** | Any object outside the Solar System; the term is usually used to designate all objects other than stars. |
| **degree (of arc)** | An angular distance equal to 1/360th of the circumference of a circle. |
| **Deimos** | The smaller and outermost of the two moons of Mars. Deimos is named for one of the dogs of the god of war, and means "panic." |
| **diagonal** | A handy device that uses a prism or mirror to angle the light path of a telescope for more comfortable positioning of the eyepiece. |
| **diameter** | The width of a circle, measured through its center. |
| **diffraction spikes** | "Rays" that appear on stars in Newtonian reflectors, caused by the vanes of the spider that holds the secondary mirror. |
| **digital camera** | A camera that uses a special sensor that is electronically responsive to light, rather than photographic film. |
| **Dobson, John** | The amateur astronomer who popularized the use of an inexpensive alt-az mount for reflectors that has come to bear his name. |
| **Dobsonian mount** | A simple alt-az mount popularized by John Dobson for use with Newtonian reflectors. |
| **Double Cluster** | A pair of beautiful open star clusters, so close together in the constellation Perseus that they can be |

viewed at the same time in binoculars, or at low magnification through a telescope.

**double star** Two stars so close to each other in the sky (though not always close in reality) that they appear as one star to the unaided eye.

**Draco** A constellation, the Dragon, which lies in the far northern sky.

**DSLR** A digital single-lens reflex camera. Like a film SLR, this type of camera uses interchangeable lenses, and focuses through the lenses and a mirror and prism. But rather than using film, it uses a special computer chip to collect and store the light of the image.

**dual-axis drive** Motor drives on both the declination and right ascension axes of a telescope mount.

**Dumbbell Nebula (M27)** A large, bright, planetary nebula in the constellation Vulpecula, the Fox.

**Eagle Nebula (M16)** A star cluster imbedded in an emission nebula in the constellation Serpens, the Snake.

**Earth** The third planet from the Sun; *possibly* the only body in the Solar System to currently contain life.

**Earthshine** Sunlight which reflects off of the Earth onto the Moon and then back again, allowing the dark regions of the Moon (those unlit at the time by the Sun) to be dimly seen.

**eccentric orbit** Orbits can be almost exactly circular, or very elongated ovals; the more elongated, the more *eccentric*.

**eclipse** The blocking of the view of one celestial body (the Sun, the Moon, a planet, or other moon) as seen from Earth, by the passing of another in front of it.

**eclipsing variable star** A "star" that varies in brightness because it is actually two or more stars orbiting very close together. When one passes in front of (eclipses) the other, the overall brightness of the "star" changes.

**ecliptic** The line around the entire sky along which the Sun seems to travel as the Earth orbits it. The planets (except Pluto) all stay very close to this line as they cross our sky, since they and the Earth orbit the Sun in a flat disk.

**elliptical galaxy** An oval shaped galaxy that is more or less evenly lit across its surface.

**emission nebula** A nebula whose light is produced by the glow of particles being excited by hot, bright, nearby stars.

**equatorial mount** A telescope mount that has one axis of rotation that is set to match the motion of the wheeling sky.

**Europa** One of the four Galilean moons of Jupiter (called so after their discoverer); it is the smallest of the four, with a diameter of 3,100 kilometer (about 1,900 miles), a little smaller than our own Moon. Current theory says there may be a vast water ocean under a relatively thin layer of ice on Europa, a possible abode of life.

**exit pupil** The diameter of the circle of light that leaves the eyepiece.

**eye lens** The lens in an eyepiece that comes closest to the eye during use.

**eye relief** The farthest distance, usually expressed in millimeters, at which you can hold your eye from an eyepiece and still see a clear image. This is an important factor to consider, especially if you wear eyeglasses.

**eyepiece** The lens (or, usually, group of lenses) on a telescope to which you hold your eye for viewing.

**false color** The usually blue halo seen around brighter objects viewed through achromatic refractors, which suffer from chromatic aberration.

**fast film** Photographic film is sold in different "speeds," which are indicated by special numbers (the ISO number). Slower film, up to ISO 100, is usually very sharp, but the shutter of the camera must stay open longer to produce an image. Fast film, ISO 400 or above, can be "grainier" than slow film, but images form much more quickly.

**field of view** The circular area of sky visible through the eyepiece.

**filter** An optically flat piece of glass (or sometimes plastic for solar filters) that is coated in some way to change the color or brightness of an image viewed through it.

Filters can only remove some of the incoming light; they do not add to it.

**finder** A small, low-power telescope or other device attached to the main scope, used to assist in initial pointing.

**first quarter Moon** When a waxing (growing) Moon has half its Earth-facing side lit by the Sun; it really means "first quarter of the month."

**focal length** The distance from a lens or mirror to its focal point, or more accurately, focal plane. Longer focal length in a telescope means *higher* magnification; longer focal length in an eyepiece means *lower* magnification. This is because magnification is equal to the focal length of the telescope divided by the focal length of the eyepiece.

**focal point** The place at which light rays converge after being reflected or refracted by a lens or curved mirror; in reality this is not a point, but a small spot.

**focal ratio** This is equal to focal length divided by aperture.

**fork mount** A mount that has two tines that attach to either side of a telescope; really only appropriate for designs with short tubes, like Maks or Schmidt – Cassegrains.

**full Moon** When the entire Earth-facing side of the Moon is lit by the Sun.

**fully coated** A telescope, pair of binoculars, or eyepiece that has been coated on all exposed lens surfaces.

**fully multi-coated (FMC)** A telescope, pair of binoculars, or eyepiece that has multiple coatings on all exposed lens surfaces.

**galaxy** A vast collection of stars held together by gravitational attraction; may contain millions to hundreds of billions of stars. Our own galaxy is called the Milky Way.

**Galileo Galilei (1564–1642)** The Italian scientist who first turned a telescope skyward to discover mountains and "seas" on the Moon, moons around Jupiter, "handles" on Saturn, and countless stars in the Milky Way.

**Ganymede** One of the four Galilean satellites of Jupiter; third out from the giant planet. With a diameter of 5,200 kilometers (about 3,200 miles) Ganymede is

| | |
|---|---|
| | the largest moon in the Solar System, and is even bigger than either Pluto or Mercury. |
| **Gemini** | A constellation in the zodiac; the Twins. |
| **gibbous** | Of the Moon or the planets, when more than half, yet less than all of the Earth-facing side is lit by the Sun. |
| **globular star cluster** | A typically spherical, tightly packed group of many thousands of stars. |
| **go-to mount** | A computer-guided telescope mount. |
| **gravity** | One of the basic forces of the universe; all matter exerts this attractive force, but it is so weak that only large bodies show its effects. |
| **Great Orion Nebula (M42)** | The brightest emission nebula that is visible throughout most of the world; easily visible to the naked eye, even in suburban areas, and beautiful when viewed at any magnification. |
| **Great Red Spot (GRS)** | A giant storm visible in the atmosphere of Jupiter, first recorded by Cassini in 1665. In recent years, it has grown smaller, and is more beige than red, making it more difficult to spot in smaller scopes. In 2006, a smaller white spot also became reddish. Some are calling it Red Spot Junior. |
| **guide** | To use a star centered in an eyepiece (preferably one with cross-hairs) to steer a telescope while taking a long-exposure photograph. |
| **hand controller** | A small box with electric switches, used to steer a telescope with a motor drive. More advanced hand controllers are used with computer-guided scopes. |
| **helium** | The second simplest and second most common element (after hydrogen) in the Universe. |
| **Hercules** | A constellation named for the mythical Greek hero; it contains the brightest globular cluster visible to most northern hemisphere viewers: M13. |
| **Herschel, William (1738–1822)** | The German-born English astronomer who discovered Uranus in 1781. |
| **high eyepoint** | A term for binoculars with long eye relief, suitable for those who wear glasses. |

**Horsehead Nebula** A dark nebula with the shape of a horse's head in the constellation Orion. Not visible in beginners' scopes, but often photographed in long exposures.

**Huygens** A very old and, by modern standards, not very useful eyepiece design.

**Hyades** (HI-uh-deez) A loose open star cluster in the constellation Taurus.

**Hydra** A long, southern constellation; the Water Snake.

**hydrogen** The simplest and most abundant element; most stars are made primarily of hydrogen.

**illuminated reticle eyepiece** A special eyepiece that is used to help guide a telescope during a long photographic exposure. The eyepiece has cross-hairs (the reticle) that meet in the middle, which are lit up (illuminated) by a dim red LED so they can be seen in the dark.

**International Dark-Sky Association** An organization dedicated to the reduction of light pollution; visit them at www.darksky.org.

**Io** (EYE– oh) One of the four Galilean moons of Jupiter, and the one closest to the planet; so close, in fact, that tidal forces caused by Jupiter's gravity make Io the most volcanically active body known. It has a diameter of 3,600 kilometers (2,200 miles), slightly larger than the Moon.

**iris** In the eye, the colored muscles that close around the pupil as light gets brighter; a mechanical device with a similar purpose, and the same name, is used in camera lenses.

**irregular galaxy** Any galaxy with a shape that is not spiral or elliptical.

**Jupiter** The fifth planet from the Sun, and the largest in the Solar System.

**Kellner eyepiece** An eyepiece design no longer held in favor with many amateur astronomers; sometimes included with beginner scopes.

**Lagoon Nebula (M8)** A bright emission nebula in the constellation Sagittarius; it can be seen with the unaided eye under relatively dark skies. It is less than two degrees away from another bright nebula: M20, the Trifid Nebula.

**Large Magellanic Cloud** A "small," irregular galaxy that is a satellite of our own Milky Way. Not visible in the northern hemisphere, it is named for the Portuguese explorer, Ferdinand Magellan (1480[?]–1521), who commanded the first circumnavigation of the world. It lies in the constellations Dorado, the Dolphin, and Mensa, the Table Mountain. It contains billions of stars.

**last quarter Moon** When exactly half of the Earth-facing side of a waning Moon is lit by the Sun, short for "last quarter of the month".

**latitude** The angular distance north or south of the equator, measured in degrees.

**LED** Light emitting diode; a type of electric light source that uses very little power, yet is bright enough to allow the reading of star charts. A flashlight that uses red LEDs is ideal for use in astronomy.

**lens** Any transparent material with one or both surfaces curved in order to focus light.

**lens elements** Nearly all modern eyepieces and camera lenses are actually made up of several separate pieces of glass working together. Each of these pieces of glass is one of the lenses' *elements*.

**Leo** A constellation in the zodiac; the Lion.

**Libra** A constellation in the zodiac; the Scales.

**light pollution** Artificial light that is inadvertently aimed outward and/or upward to no useful purpose. It is one source of pollution for which there are easy and economically beneficial solutions.

**light year** Not a measure of time, but of distance: how far light travels in a year; more than nine trillion kilometers (about six trillion miles).

**longitude** The angular distance east or west, measured in degrees, from the Prime Meridian, which (arbitrarily) runs from the south pole, through Greenwich, England, and on to the north pole.

**lunar eclipse** An event that occurs when the Earth is directly between the full Moon and the Sun. The Earth's shadow slowly covers the Moon, and because Earth is much larger than the Moon, the eclipse can last much longer than a Solar one. The Earth's shadow can completely cover the Moon for over an hour; solar eclipses never last longer than several minutes. The Earth's large size also makes lunar eclipses more common than solar eclipses.

| | |
|---|---|
| **Lyra** | A small, northern constellation, the Lyre, or Harp, that contains the planetary nebula, M57, the Ring Nebula. |
| **M13** | A globular cluster in the constellation Hercules; the brightest visible to many northern hemisphere viewers. |
| **M81** | A beautiful, almost perfect spiral galaxy in the constellation Ursa Major, the Great Bear. Visible in small scopes as an oval smudge (but visible!), it is about half a degree distant from M82. |
| **M82** | An irregular galaxy in the constellation Ursa Major, the Great Bear. Visible in small scopes, it is about half a degree distant from M81. |
| **M83** | A bright spiral galaxy in the constellation Hydra, the Water Snake. |
| **magnification** | Through optical means, the enlarging of an image of an object. |
| **magnitude** | A measure of a celestial object's brightness. By definition, a star of magnitude 1 is 100 times brighter than a star of magnitude 6 (a difference of five magnitudes). There is an approximate increase in brightness of 2.5 times between each magnitude level, and brightness increases as magnitude levels decrease. *Apparent magnitude* is the magnitude of a star as seen in the Earth's sky, and does not take into account the varying distances of stars; *absolute magnitude* is a star's true brightness, when measured from a standard distance. |
| **Mak** | Short for Maksutov; there are two common telescope designs that were originated by this optical designer in the first half of the twentieth century: the Maksutov–Cassegrain, and the Maksutov–Newtonian. |
| **Maksutov, Dmitri (1896–1964)** | The Russian optical designer who originated two popular types of telescope that bear his name: the Maksutov–Cassegrain and the Maksutov–Newtonian. |

**Maksutov–Cassegrain** A telescope design that uses spherical curves on its mirrors and lenses to deliver well-corrected views in an inexpensive, compact, rugged package.

**Maksutov–Newtonian** A telescope design by Dmitri Maksutov that combines features of the Newtonian reflector and the Maksutov–Cassegrain.

**Mars** The fourth planet from the Sun, and the only planet other than Earth that has an atmosphere that is transparent enough to show true surface detail. It comes closer to Earth than any other planet except Venus, and its proximity and surface detail have tantalized astronomers for centuries.

**mass** Basically, the amount of matter an object contains. The air inside a balloon may have the same mass as a paper clip, though it is much less densely packed. A human who weighs 60 kilograms (132 pounds) on the Earth may weigh only 10 kilograms (22 pounds) on the Moon, because the Moon's gravitational pull is less; but that person's mass would be the same in both places.

**Meade** One of the two largest telescope manufacturers in the United States; the other is Celestron.

**meniscus lens** A crescent-shaped lens that curves inward on one side, and outward on the other.

**Mercury** The closest planet to the Sun, and the second smallest, after Pluto (or the smallest, according to some, since Pluto has been demoted).

**Messier Catalog** A catalog of deep-sky objects compiled by French astronomer Charles Messier.

**Messier, Charles (1730–1817)** The French astronomer and comet hunter who compiled a list of deep-sky objects that now bear his name, or at least his initial. The commonly accepted story is that he compiled the list, and accurately marked each object's celestial position, so that these objects could be avoided when searching for comets. The telescopes he used would only be comparable to a very moderate modern beginner scope, so all the 109 objects in his Catalog, which includes galaxies, nebulae, and star clusters, are easily accessible with a beginner's equipment under moderately dark skies.

**meteor** A small piece of interplanetary debris, usually no larger than a pebble or a grain of sand, that strikes the Earth's upper atmosphere at incredible speed. Friction usually causes it to burn up entirely in no more than a second or two. Before hitting the atmosphere, these objects are called *meteoroids*; if they are large enough that part of them actually reaches the ground, that remnant is called a *meteorite*.

**meteor shower** Periodic occurrences during which the number of meteors greatly increases because the Earth passes through the debris path left by a comet. The meteors seem to radiate from a specific spot in the sky, which is aptly called the *radiant*. Regularly occurring showers are given names based on the constellation that contains the radiant. For example, the Perseids radiate from Perseus, and peak on August 12; the Leonids radiate from Leo, and peak on November 17–18.

**meteor train** A dim cloud of hot gas, high in the atmosphere, that can last up to several minutes; caused by the interaction of larger meteors with atmospheric particles.

**Milky Way** Our own galaxy, which under dark skies can be seen to span the entire sky.

**minus-violet filter** An eyepiece filter designed specifically to help remove the false color often seen in inexpensive refractors using air-spaced achromatic lenses.

**minute of arc** A unit of measurement equal to 1/60th of a degree.

**Mirfak** The brightest star in the constellation Perseus. Mirfak (sometimes spelled "Mirphak") comes from an Arabic word meaning "elbow." Another name sometimes used for this star is Algenib, "the side."

**modified achromat eyepiece** An eyepiece of inexpensive and mediocre design that is often included in beginner's scope packages.

**Moon** Our nearest celestial neighbor. It is so large in comparison to its host planet (Earth) that some have suggested we be called a double planet. Its diameter, 3476 km (2140 miles) is more than one quarter of the Earth's.

**motorized tracking** A system that uses electric motors to cause a mount to compensate for the spinning of the Earth, thus keeping objects centered in the eyepiece.

**mount** The device upon which a telescope is placed which allows it to be smoothly and precisely moved, either between targets, or to follow a target as the Earth turns; not to be confused with a tripod, which is simply one type of stand for the telescope and mount.

**multi-coated** Several layers of chemical film that are applied to a lens to reduce internal reflections.

**multiple star** A group of up to several stars that appear as a single star until magnified.

**mylar** A type of nearly opaque plastic film that can be used as a solar filter.

**nebula** (NEB-you-la, plural: nebulae, NEB-you-lee); from the Latin for *mist* or *cloud*. Before large modern telescopes, any deep-sky object that was not a star appeared as a cloudy smudge when viewed through the eyepiece, and was given the designation *nebula*. Now, the term applies to actual clouds of interstellar gas and dust which make up *emission, reflection, dark*, and *planetary nebulae*, as well as *supernova remnants*.

**Neptune** The eighth planet from the Sun, and the second to be discovered through a telescope. Because it is a large planet, it is visible (barely) through small scopes, even at the distance of 4.5 billion km (2.8 billion miles). Because of the strange path of Pluto's orbit, Neptune was, in the last two decades of the twentieth century, the farthest planet from the Sun. It is still the most distant Solar-System object visible in most beginners' scopes.

**new Moon** When the Moon is closest to the Sun in the Earth's sky, and has virtually none of its Earth-facing side lit.

**Newton, Sir Isaac (1646–1723)** An English scientist who, among other things, first described gravity and invented the telescope design that bears his name.

**Newtonian reflector** A simple telescope design invented by Isaac Newton that uses a single curved surface, the primary or main mirror, to reflect a magnified image off a smaller flat mirror, which is placed at an angle to bounce the

| | |
|---|---|
| | image into the focuser on the side of the tube; the least expensive telescope design, per unit of aperture. |
| **nocturnal** | Active primarily at night; the dream of many amateur astronomers. |
| **nuclear fusion** | The combining of atoms of one element to form atoms of another, a great amount of energy, and some leftovers. |
| **nucleus** | The small core of a comet. |
| **objective lens** | The lens in a refracting telescope or binoculars that is aimed at an object; the main light-gathering surface. |
| **ω (Omega) Centauri** | The brightest globular cluster in the sky, in the southern constellation Centaurus, the Centaur. |
| **open star cluster** | A collection of several to hundreds of loosely gravitationally attracted stars. |
| **Ophiuchus** | (oaf-ee-YOO-cuss), a constellation through which the ecliptic passes (though it is not part of the zodiac); the Snake Holder. |
| **opposition** | When a celestial object is on the opposite side of the Earth from the Sun, which means it will reach its highest point in the sky around midnight. The outer planets are at about their closest approach to Earth at these times. |
| **optical tube assembly (OTA)** | A telescope by itself, without a mount; and usually without eyepieces or other accessories. |
| **optics** | The lenses or mirrors used in the gathering of light for observation. Also, the name of the branch of physics that deals with light and its properties. |
| **orbit** | The path that a body follows around another more massive body to which it is gravitationally attached. The shapes of orbits, as Johannes Kepler realized some four hundred years ago, are elliptical (oval), though many are very nearly circles. |
| **Orion** | The brightest constellation in the sky, and the location of some of the sky's most sought-after targets; the Horsehead Nebula and The Great Orion Nebula (M42) are just two of them. |
| **partial eclipse** | When part of one celestial body partially covers the view of another as seen by an observer. |

**perihelic** (pear-uh-HEE-lick) Literally, "near the Sun."

**perpendicular** At a right angle to; separated by 90 degrees of arc.

**Perseid (PER-see-id) meteor shower** A meteor shower that radiates from the constellation Perseus, peaking on August 12. Often known simply as Perseids.

**Perseus (PER-see-us)** A constellation, named after the mythical Greek hero who destroyed the Medusa.

**phase** The name for the various states of illumination of the Moon (or Mercury or Venus): first quarter, gibbous, full, etc.

**Phobos** The larger and innermost of the two moons of Mars. Phobos is named for one of the dogs of the god of war, and means "fear."

**pi (π)** A letter in the Greek alphabet, and a mathematical symbol representing the number equal to the length of any circle's circumference divided by its diameter: equal to somewhere in the neighborhood of 3.141592653589793238462643383279502884197 1693993751.

**piggy-back** An astrophotography set-up in which a camera is mounted (piggy-backed) to a telescope. While the camera and its lens expose the film, the telescope is used to track the motion of the sky.

**Pisces** A constellation in the zodiac; the Fishes.

**planet** The definition of this word is currently under debate. A planet is a body that orbits a star; but when does it become so small that it is considered only an asteroid? And when does it become large and "hot" enough to be considered a star? Discoveries of huge "planets" orbiting other stars, and huge asteroids far out in our own Solar System, have raised questions about what has, until recently, been an easy definition. A planet *used to be* one of the nine bodies we learned about in school; now even that assumption has become controversial: is Pluto a planet, or merely a "dwarf planet?"

**planetary nebula** A huge shell of glowing gas thrown off by a star in later life. These nebulae are called "planetary" because, when first seen through more primitive telescopes centuries ago, they appeared round, and about the same size as the planets, though much dimmer.

**planetary scope** A telescope with a "slow" focal ratio: a long focal length for the size of its aperture; making it most useful for viewing bright objects at higher magnifications, which is just right for the planets.

**planisphere** A simple device made of two round wheels of cardboard or plastic attached together in the center, and able to spin freely. The front wheel has an oval hole in it that allows one to see part of the map of the entire sky (or, that part of it visible from a band of given latitude on Earth) that is printed on the back wheel. By matching an edge containing all the dates of the year on one wheel with an adjacent edge on the second wheel that charts the hours from midnight through midnight, one can see what section of sky will be visible at any time on any date of the year.

**Pleiades (M45) (PLEE-uh-deez)** The brightest and most famous open star cluster, located in the constellation Taurus, the Bull. Though called the Seven Sisters (after the daughters of mythical Atlas), six stars in a tight dipper shape are visible to most people without aid. Binoculars and telescopes show at least dozens more.

**Plossl eyepiece** The eyepiece design favored as the best all-around choice for the budget-minded amateur.

**Pluto** Until recently, the ninth, and probably the last, planet out from the Sun. It was also the smallest planet; with a diameter of 2320 kilometers (1,440 miles) it is smaller even than several moons, including our own. It was discovered in 1930 by American astronomer Clyde Tombaugh (1906–1997), who found it by searching photographic plates. It has three moons of its own, one of which has a diameter that is more than half of Pluto's. Discovered in 1978, this larger moon is called Charon (KARE-on). The other two were only recently discovered; they are much smaller, and little is known about them. More will be learned when a fly-by spacecraft mission reaches Pluto in 2015.

**polar alignment** Lining up the right ascension axis of an equatorial mount with the celestial pole so that it matches the axis of the Earth. This allows the tracking of targets with adjustments in only one direction.

**polar alignment scope** — A small, wide-field telescope that usually is placed in the center of the shaft of the right ascension axis of an equatorial mount, used to help in gaining accurate polar alignment.

**Polaris** — The Pole Star, or North Star; in the constellation **Ursa Minor**, the Lesser Bear, which is better known as the Little Dipper. Polaris is the star at the outer end of the Dipper's "handle." It is the closest bright star to the north celestial pole, at a distance of about one degree. Contrary to a commonly held belief, it is *not* the brightest star in the sky (Sirius is); it is not even in the top 20!

**porro prism** — The most common type of prism used in binoculars, and the most economical for use with the large apertures necessary for astronomy.

**primary mirror** — The main light-gathering source in any type of reflecting or catadioptric telescope.

**prime focus** — The focusing point of the main mirror or objective lens of a telescope. A *prime-focus astrophoto* is one in which a camera is attached directly to a telescope, without an intervening eyepiece or camera lens; the telescope is used, in effect, as a very long camera lens.

**prism** — A piece of glass that is cut in such a way that it changes the direction or orientation of a light beam's path, and/or splits it into its component colors.

**pupil** — The small hole in the eye through which light enters.

**radiant** — The point from which meteors of a given meteor shower radiate.

**radiate** — To extend in straight lines from a central point.

**radius** — The distance from the center of a circle to its edge; half the diameter.

**Ramsden eyepiece** — A cheap, almost useless eyepiece design.

**red-dot finder** — A finder that does not magnify an image, but rather projects a small red dot against a piece of clear glass, showing the point in the sky at which the telescope is aimed.

| | |
|---|---|
| **reflection nebula** | A nebula whose light is formed by reflection of the light of nearby bright stars off interstellar dust. |
| **reflector** | Any telescope that uses a mirror as its main light gathering surface, without the addition of intervening lenses. |
| **refractor** | Any telescope that uses a lens as its main light-gathering surface. |
| **resolution** | The fineness of detail that can be viewed through a telescope; also called resolving power. |
| **retina** | The area in the back of the eye that collects light and transmits visual information to the brain. |
| **right ascension (RA)** | Celestial longitude, measured in a circle of 24 hours (each divided into minutes and seconds), with the zero hour being the vernal equinox (the point at which the Sun, traveling along the ecliptic in the spring, crosses the celestial equator); located in the constellation Pisces, the Fishes. |
| **Ring Nebula (M57)** | A bright, donut-shaped planetary nebula in the constellation Lyra, the Lyre. |
| **roof prism** | The less common type of prism used in binoculars. While allowing for a lighter, more compact package, good roof prisms are expensive to manufacture, and are not usually available in sizes useful or affordable to amateur astronomers. |
| **Sagittarius** | The Archer; a constellation in, possibly, the most interesting part of the sky. The center of our galaxy lies behind clouds of obscuring dust in Sagittarius; and no less than fifteen Messier objects are charted within it (more than any other constellation). |
| **satellite** | any body, even an artificial one, which orbits a larger one. |
| **Saturn** | The sixth planet from the Sun; the second largest; and by almost anyone's standard, the most beautiful because of its bright rings. |
| **Schmidt–Cassegrain telescope (SCT)** | A popular, portable catadioptric telescope design that is relatively inexpensive to mass produce compared to some other designs, but still outside the price range of this book. Both Meade and Celestron have sold many thousands of them. |

| | |
|---|---|
| **Scorpius** | A constellation in the zodiac; the Scorpion. |
| **secondary mirror** | In a reflector or catadioptric telescope, the smaller mirror that reflects the image from the main mirror into the focuser. |
| **second of arc** | A unit of measure equal to 1/60th of a minute of arc, or 1/3600th of a degree. |
| **seeing** | A term used by astronomers to describe the steadiness of the atmosphere. Excellent seeing means that stars appear as stable points through the eyepiece; when seeing is poor, stars (and everything else in the sky) roil and shimmy. |
| **single-axis drive** | A single motor attached to the right ascension (RA) axis of an equatorial mount, used to rotate the mount to follow the spinning of the Earth. |
| **single-lens reflex (SLR)** | a type of camera design that uses interchangeable lenses, and which is focused through the lens with a mirror and prism. |
| **Sirius** | The brightest star in the sky, at magnitude –1.46; in the constellation Canis Major, the Greater Dog. Sirius is said to come from a Greek word meaning "scorching." |
| **Small Magellanic Cloud** | A "small" irregular galaxy that is gravitationally attached to our own Milky Way. Visible only to southern hemisphere viewers, it was named for the Portuguese explorer, Ferdinand Magellan (1480[?]–1521), who commanded the first circumnavigation of the world. It lies in the constellation Tucana, the Toucan. |
| **Sol** | The name of our own star, from which we get "Solar System." It is better known simply as the Sun. |
| **solar filter** | An acceptable solar filter is one that fits over the object end of a telescope, and that blocks almost all of the incoming light. |
| **Solar System** | The region of space that is under the influence of our local star, Sol (the Sun). |
| **spherical** | A curve that is shaped as a section of a sphere. |
| **spider** | The vaned mechanism that holds a Newtonian reflector's secondary mirror in place. |
| **spiral galaxy** | Any galaxy with arms that spiral out from the center. |
| **squaring** | Multiplying a number by itself. |

| | |
|---|---|
| **star** | A ball of gas, usually made up mostly of hydrogen, so massive that its own gravity forces it to collapse until nuclear fusion takes place within it, causing it to shine. |
| **star charts** | Maps of the Heavens, either in a field guide, or in larger formats. Some show only brighter stars; more detailed, comprehensive charts show stars and objects that have higher magnitudes, which are both dimmer and more numerous. |
| **star party** | A scheduled meeting of amateur and/or professional astronomers for the purpose of sharing their equipment, their time, and the night sky. |
| **star trails** | Curved lines seen in the simplest of astrophotos, left by stars during long exposures as the Earth spins on its axis. |
| **Sun** | Our own star; an apparently average one, in the middle of life. |
| **sunspots** | Cooler areas on the surface of the Sun that only appear to be dark when contrasted against the neighboring brightness. Sunspots occur in places where magnetic fields burst through the surface. |
| **supernova remnant** | The highly charged shell of gaseous remains left after a large star collapses and then explodes at the end of its life. |
| **surface features** | The actual features visible on the Sun, Moon, or planets; sometimes actual details of the solid surface, but in some instances, the tops of their atmospheres. |
| **Tarantula Nebula** | A huge emission nebula in the Large Magellanic Cloud. In actual size, it dwarfs any other known nebula anywhere near our galaxy. |
| **Taurus** | A constellation in the zodiac; the Bull. It contains two famous open star clusters: the Pleiades (M45); and the Hyades. |
| **telephoto lens** | Any camera lens that magnifies an image. |
| **telescope** | Any instrument that uses mirrors and/or lenses to give magnified images of objects toward which it is aimed. |
| **terminator** | The line between the sunlit and dark sides of a celestial body. |

**time-exposure photo** — A photograph in which the film is exposed for an extended period of time.

**Titan** — The largest moon of the planet Saturn, and the only moon in the Solar System to possess an atmosphere. It is the second largest moon in the Solar System, just barely smaller than Jupiter's Ganymede.

**total solar eclipse** — By an amazing coincidence, the Moon and the Sun both usually appear about the same size in the Earth's sky. Occasionally, the Moon crosses in front of the Sun and completely covers its disk for a short time (up to several minutes, but often much shorter), providing lucky observers beneath with one of nature's greatest shows.

**totality** — The time during which an entire celestial body is eclipsed by another.

**tracking** — Following a celestial object with a telescope as it wheels across the sky; it can be done either manually or electronically.

**transit** — When a smaller celestial object crosses the face of a larger one as viewed by an observer.

**Trifid Nebula (M20)** — A fairly bright nebula in the constellation Sagittarius, the Archer. It has dark portions, emission portions, and reflection portions (though the last two can only be distinguished by their color in long-exposure photographs). It lies less than two degrees from the brighter Lagoon Nebula (M8).

**tripod** — A three-legged stand used to hold cameras, or telescopes and their mounts.

**tube currents** — Waves of air that form in the tubes of telescopes when the temperature inside the tube is different from the temperature outside, often causing distorted views.

**turbulence** — Disturbed currents of air that cause star images to flutter and roil.

**Uranus** — The seventh planet from the Sun, and the first to be discovered through a telescope, by William Herschel in 1781. Visible even without aid, if you know exactly where to look, it is best to try to view Uranus through a telescope. When this is done, the giant planet can clearly be seen as a disk. Since no star ever shows as anything but a point, you can then be certain you have actually viewed the seventh planet.

**Ursa Major** A constellation, the Greater Bear, located near the north celestial pole; the Big Dipper is contained within it.

**Ursa Minor** A constellation, the Lesser Bear; better known as the Little Dipper (in North America, at least.)

**variable star** A star that varies in brightness over time. Some vary by a little, others by a large amount. Some cycles of variability are very regular; others are erratic. There are several types of variable star.

**Venus** The second planet from the Sun, and the one that approaches closest to the Earth. Its thick atmosphere and nearness to the Sun make it the hottest planet known.

**Virgo** A constellation in the zodiac; the Virgin.

**Vulpecula** A constellation; the Little Fox. It contains the large, bright planetary nebula, M27, the Dumbbell Nebula.

**waning** Shrinking.

**waxing** Growing.

**Whirlpool Galaxy (M51)** One of the brightest face-on spiral galaxies, visible as a blur in small scopes. It is just below the handle of the Big Dipper.

**wide field** Basically, low magnification. There are times when a wide field of view is desirable: when trying to aim a telescope or mount at a target, or when trying to view a celestial object that takes up a large section of sky.

**zodiac** The twelve ancient constellations that lie along the ecliptic, and that are so important to believers in astrology. As a group, their importance to amateur astronomers is limited to the help they can give one who is searching for planets.

**zones** The lighter areas between the dark bands on the surface of Jupiter.

**zoom lens** A type of lens design that allows a change in magnification through the internal movement of lens elements. They are not typically useful in astronomy.

# Appendix 1: the Greek alphabet

| | | | | | |
|---|---|---|---|---|---|
| α | alpha | ι | iota | ρ | rho |
| β | beta | κ | kappa | σ | sigma |
| γ | gamma | λ | lambda | τ | tau |
| δ | delta | μ | mu | υ | upsilon |
| ε | epsilon | ν | nu | φ or ϕ | phi |
| ζ | zeta | ξ | xi | χ | chi |
| η | eta | ο | omicron | ψ | psi |
| θ or ϑ | theta | π | pi | ω | omega |

# Appendix 2: the constellations

Below is a table showing all 88 of the constellations recognized by the scientific community of the world today. The table also shows how to pronounce the names, and what the constellations are often called in English. They are not always direct translations; Cassiopeia is the name of a particular mythical queen, not just a word meaning "queen."

The entire sky is divided into 88 jigsaw-like pieces, and these are the names that go with those pieces. Most of the constellations are very ancient, and refer to mythological creatures, gods, and heroes; often very strange to us now (what, for instance, is a sea goat?) But many of the constellations were created much later, by some of the first European astronomers to travel far enough south to see new stars (new to them, at least) and shapes made by those stars. The shapes they saw and named were then put on charts and accepted by astronomers at the time. They have since come down to us and are now official. Oddly, many of these shapes are also named for things obscure to us now, especially those named (rather unimaginatively) after scientific instruments of the time. (What, exactly, is an octant, and how do you use it?)

I should also say that some of the constellations are very dim, with not a single bright star. Many of these (for example: Lynx, Camelopardalis, and Antlia) were created in later years just to fill in the gaps between the brighter spots in the sky. A few, like Cancer and Equuleus, are ancient.

| Constellation name | Pronunciation | English translation or description |
|---|---|---|
| Andromeda | *an-DRAH-mih-duh* | The Chained Lady |
| Antlia | *ANT-lee-uh* | The Air Pump |
| Apus | *APE-us* | The Bird of Paradise |
| Aquarius | *uh-KWAIR-ee-us* | The Water Bearer |
| Aquila | *ACK-will-uh* or *uh-QUILL-uh* | The Eagle |

## Appendix 2: the constellations

| | | |
|---|---|---|
| Ara | *AIR-uh* or *AR-uh* | The Altar |
| Aries | *AIR-eez* | The Ram |
| Auriga | *or-EYE-guh* | The Charioteer |
| Boötes | *bo-OH-teez* | The Herdsman |
| Caelum | *SEE-lum* | The Chisel |
| Camelopardalis | *cuh-MEL-oh-PAR-duh-liss* | The Giraffe |
| Cancer | *CAN-ser* | The Crab |
| Canes Venatici | *CANE-eez* (or *CAN-eez*) *ve-NAT-iss-eye* | The Hunting Dogs |
| Canis Major | *CANE-iss* (or *CAN-iss*) *MAY-jer* | The Greater Dog |
| Canis Minor | *CANE-iss* (or *CAN-iss*) *MY-ner* | The Lesser Dog |
| Capricornus | *CAP-rih-CORN-us* | The Sea Goat |
| Carina | *car-EYE-nuh* or *car-EE-nuh* | The Keel |
| Cassiopeia | *CASS-ee-uh-PEE-uh* | The Queen |
| Centaurus | *sen-TOR-us* | The Centaur |
| Cepheus | *SEE-fyoos* or *SEE-fee-us* or *SEF-ee-us* | The King |
| Cetus | *SEE-tus* | The Whale (or Sea Monster) |
| Chamaeleon | *cuh-MEAL-ee-un* | The Chameleon |
| Circinus | *SIR-sin-us* | The Surveying Compass |
| Columba | *cuh-LUM-buh* | The Dove |
| Coma Berenices | *COE-muh BARE-uh-NICE-eez* | Berenice's Hair |
| Corona Australis | *cuh-ROE-nuh aw-STRAL-iss* | The Southern Crown |
| Corona Borealis | *cuh-ROE-nuh bor-ee-AL-iss* | The Northern Crown |
| Corvus | *COR-vus* | The Crow |
| Crater | *CRAY-ter* | The Cup |
| Crux | *CRUCKS* or *CROOKS* | The Southern Cross |
| Cygnus | *SIG-nus* | The Swan |
| Delphinus | *del-FINE-us* or *del-FIN-us* | The Dolphin |
| Dorado | *doh-RAH-do* | The Dolphin (fish; not mammal) |
| Draco | *DRAY-co* | The Dragon |
| Equuleus | *eh-KWOO-lee-us* | The Little Horse |
| Eridanus | *eh-RID-un-us* | The River |
| Fornax | *FOR-naks* | The Furnace |
| Gemini | *JEM-uh-nye* or *JEM-uh-nee* | The Twins |
| Grus | *GRUSS* or *GROOS* | The Crane |

| Hercules | *HER-kyuh-leez* | Hercules |
|---|---|---|
| Horologium | *hor-uh-LOE-jee-um* | The Clock |
| Hydra | *HIGH-druh* | The Water Snake (or Sea Monster) |
| Hydrus | *HIGH-drus* | The Male Water Snake |
| Indus | *IN-dus* | The Indian |
| Lacerta | *luh-SER-tuh* | The Lizard |
| Leo | *LEE-oh* | The Lion |
| Leo Minor | *LEE-oh MY-ner* | The Lesser Lion |
| Lepus | *LEEP-us* or *LEP-us* | The Hare |
| Libra | *LEE-bruh* or *LYE-bruh* | The Scales |
| Lupus | *LOOP-us* | The Wolf |
| Lynx | *LINKS* | The Lynx |
| Lyra | *LYE-ruh* | The Lyre (or Harp) |
| Mensa | *MEN-suh* | Table Mountain |
| Microscopium | *my-cruh-SCOPE-ee-um* | The Microscope |
| Monoceros | *muh-NAH-ser-us* | The Unicorn |
| Musca | *MUSS-cuh* | The Fly |
| Norma | *NOR-muh* | The Carpenter's Square |
| Octans | *OCK-tenz* | The Octant |
| Ophiuchus | *OFF-ee-YOO-kus* or *OAF-ee-YOO-kus* | The Serpent Bearer |
| Orion | *oh-RYE-un* | The Hunter |
| Pavo | *PAY-vo* | The Peacock |
| Pegasus | *PEG-us-us* | The Winged Horse |
| Perseus | *PER-see-us* or *PER-syoos* | The Hero |
| Phoenix | *FEE-nix* | The Phoenix |
| Pictor | *PICK-ter* | The Painter's Easel |
| Pisces | *PICE-eez* or *PISS-eez* | The Fishes |
| Piscis Austrinus | *PICE-iss* (or *PISS-iss*) *aw-STRY-nus* | The Southern Fish |
| Puppis | *PUP-iss* | The Stern |
| Pyxis | *PIX-iss* | The Magnetic Compass |
| Reticulum | *rih-TICK-yuh-lum* | The Reticle |
| Sagitta | *suh-JIT-uh* | The Arrow |
| Sagittarius | *SAJ-ih-TARE-ee-us* | The Archer |
| Scorpius | *SCOR-pee-us* | The Scorpion |
| Sculptor | *SCULP-ter* | The Sculptor |

| | | |
|---|---|---|
| Scutum | *SCOOT-um* or *SCYOOT-um* | The Shield |
| Serpens | *SER-punz* | The Snake |
| Sextans | *SEX-tunz* | The Sextant |
| Taurus | *TOR-us* | The Bull |
| Telescopium | *tel-ih-SCOPE-ee-um* | The Telescope |
| Triangulum | *try-ANG-gyuh-lum* | The Triangle |
| Triangulum Australe | *try-ANG-gyuh-lum aw-STRAL-ee* | The Southern Triangle |
| Tucana | *too-KAY-nuh* or *too-KAH-nuh* | The Toucan |
| Ursa Major | *ER-suh MAY-jur* | The Great Bear |
| Ursa Minor | *ER-suh MY-ner* | The Lesser Bear |
| Vela | *VEE-luh* or *VAY-luh* | The Sails |
| Virgo | *VER-go* | The Maiden |
| Volans | *VOH-lanz* | The Flying Fish |
| Vulpecula | *vul-PECK-yuh-luh* | The Little Fox |

# Some simple star charts

Here are eight maps of the night sky; one for each of the seasons, for both the northern and southern hemispheres. They were created, in part, with the planetarium software program *SkyChart III*, (available at www.southernstars.com/skychart/). Because of the scale used for this guide, the charts show little detail, but I have included them as a starting point. Under the starry sky, they should help you to find the brighter constellations; and the line marking the ecliptic might help you find planets. The lines used to show the constellation figures are those designed by H.A. Rey, for the book *The Stars: A New Way to See Them* (see page 15).

To use them outside, simply face in one of the directions marked around the edge of the chart that corresponds closely to the time and date at which you are using it. Then lift the book up, turning it so that the direction you are facing is at the bottom. You should now be able to match the chart with the sky. Start with a shape you are familiar with (the Big Dipper in the north, and the Southern Cross in the south are visible at virtually any time of year), and use it and the chart to move on to new constellations.

I used the latitudes 45 degrees north and 30 degrees south because these seemed the best for the most people around the world. If you live north or south of these latitudes, the sky will not match the charts; but, unless you are quite distant, you should still get some use from them.

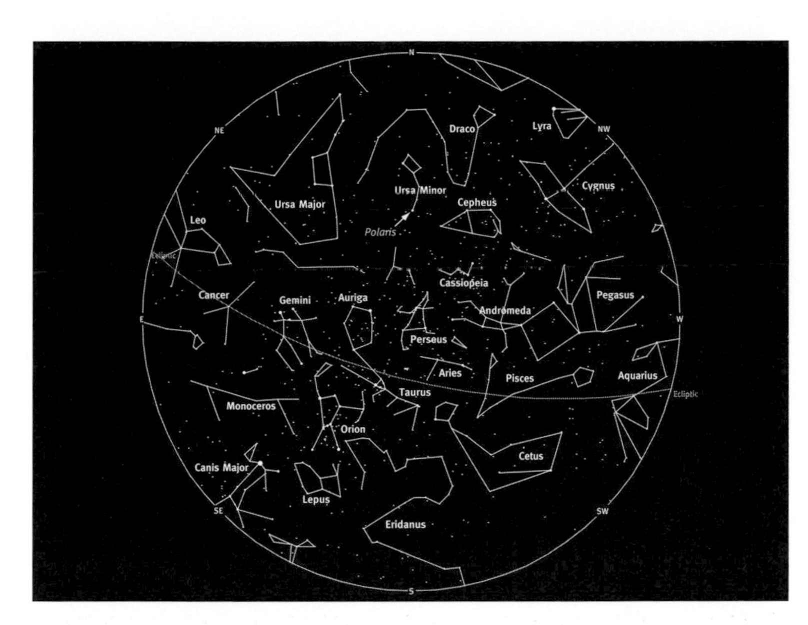


**1 Northern evening sky, winter (latitude 45 degrees). The sky as it appears at these times on these dates:**

**12:00 am**
**November 15**
**11:00 pm**
**November 30**
**10:00 pm**
**December 15**
**9:00 pm**
**December 30**
**8:00 pm**
**January 15**
**7:00 pm**
**January 30**

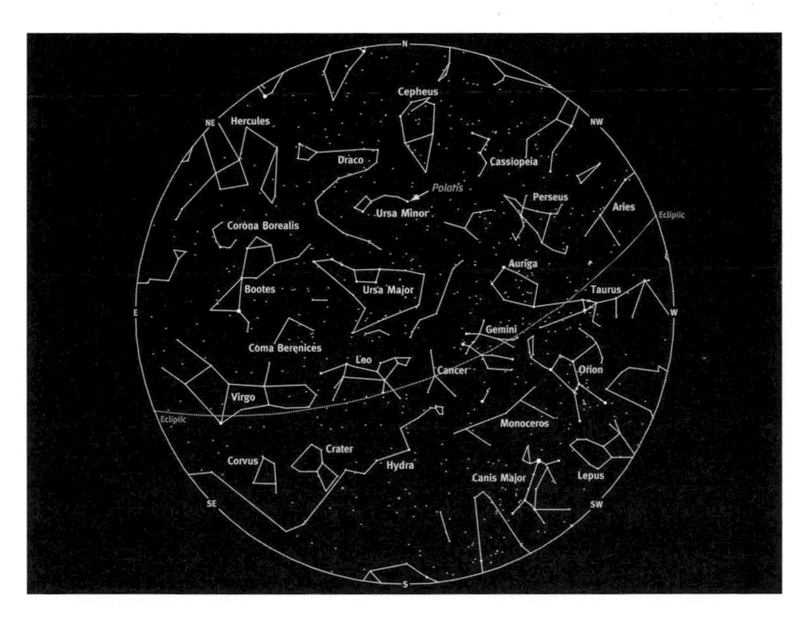


**2 Northern evening sky, spring. The sky as it appears at these times on these dates:**

**12:00 am**
**February 15**
**11:00 pm**
**February 30**
**10:00 pm**
**March 15**
**9:00 pm**
**March 30**
**8:00 pm**
**April 15**

**3 Northern evening sky, summer. The sky as it appears at these times on these dates:**

**12:00 am**
**May 15**
**11:00 pm**
**May 30**
**10:00 pm**
**June 15**
**9:00 pm**
**June 30**

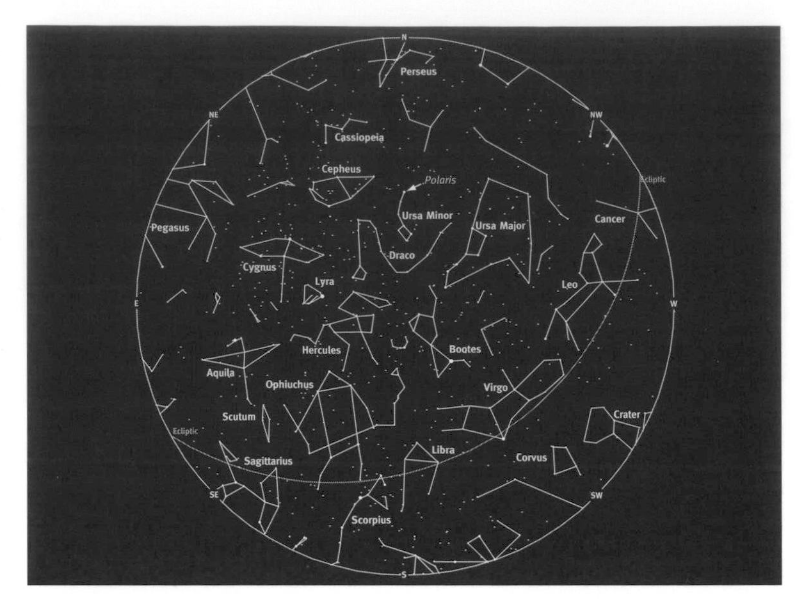


**4 Northern evening sky, autumn. The sky as it appears at these times on these dates:**

**12:00 am**
**August 15**
**11:00 pm**
**August 30**
**10:00 pm**
**September 15**
**9:00 pm**
**September 30**
**8:00 pm**
**October 15**
**7:00 pm**
**October 30**

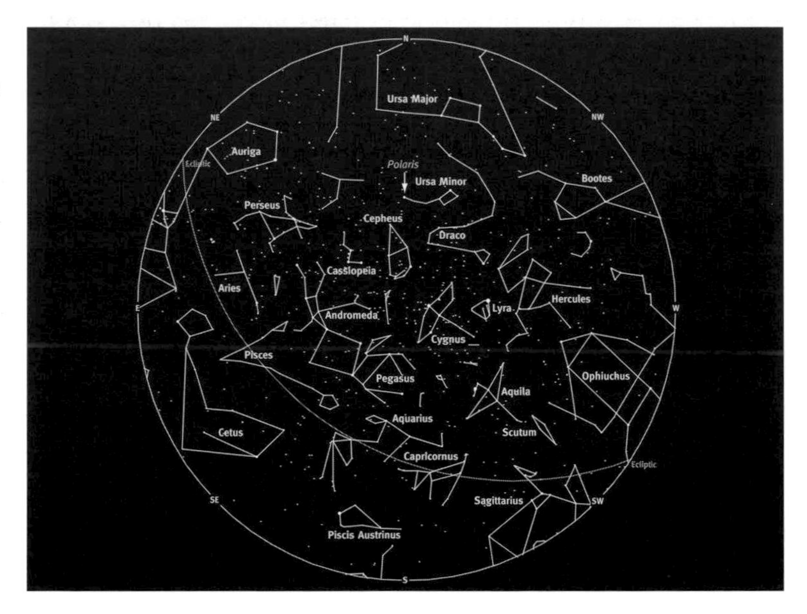

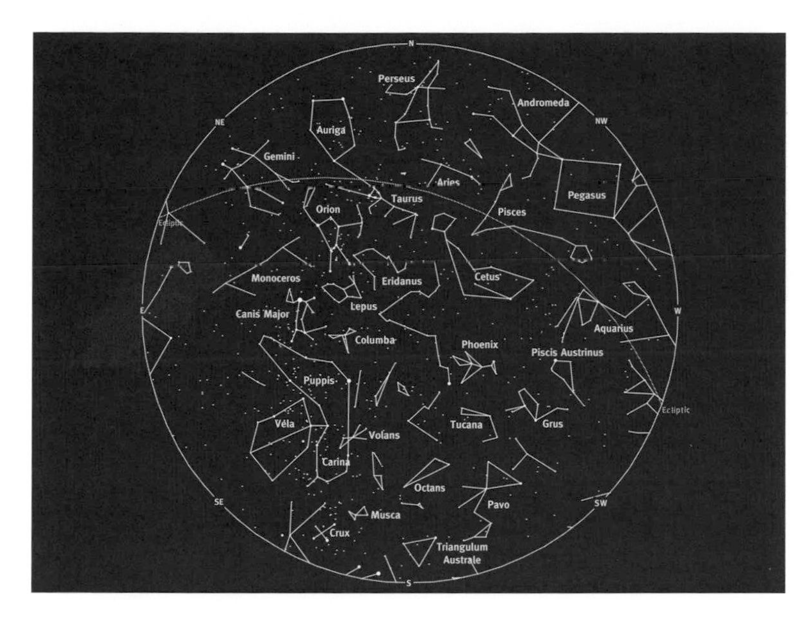


**5 Southern evening sky, summer (30 degrees latitude). The sky as it appears at these times on these dates:**

**12:00 am November 15**
**11:00 pm November 30**
**10:00 pm December 15**
**9:00 pm December 30**

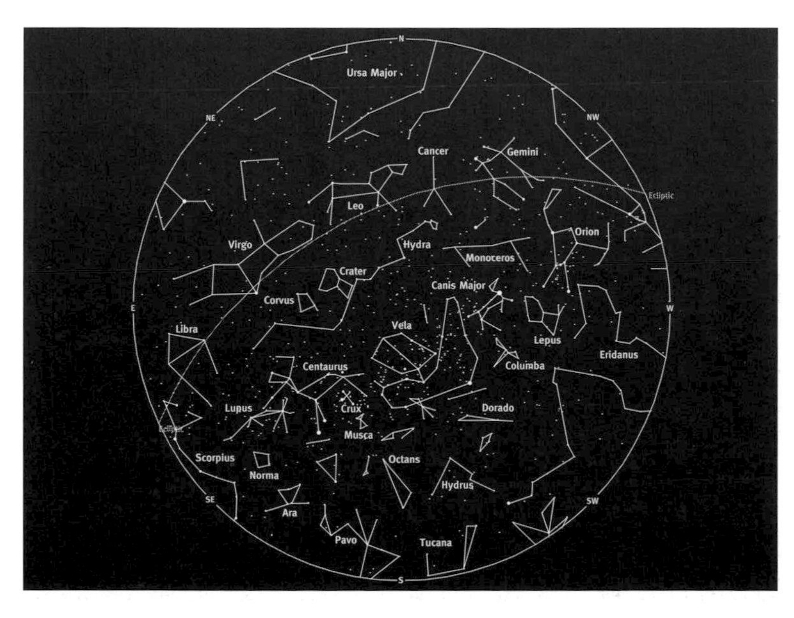


**6 Southern evening sky, autumn. The sky as it appears at these times on these dates:**

**12:00 am February 15**
**11:00 pm February 30**
**10:00 pm March 15**
**9:00 pm March 30**
**8:00 pm April 15**

**7 Southern evening sky, winter. The sky as it appears at these times on these dates:**

**12:00 am**
**May 15**
**11:00 pm**
**May 30**
**10:00 pm**
**June 15**
**9:00 pm**
**June 30**
**8:00 pm**
**July 15**

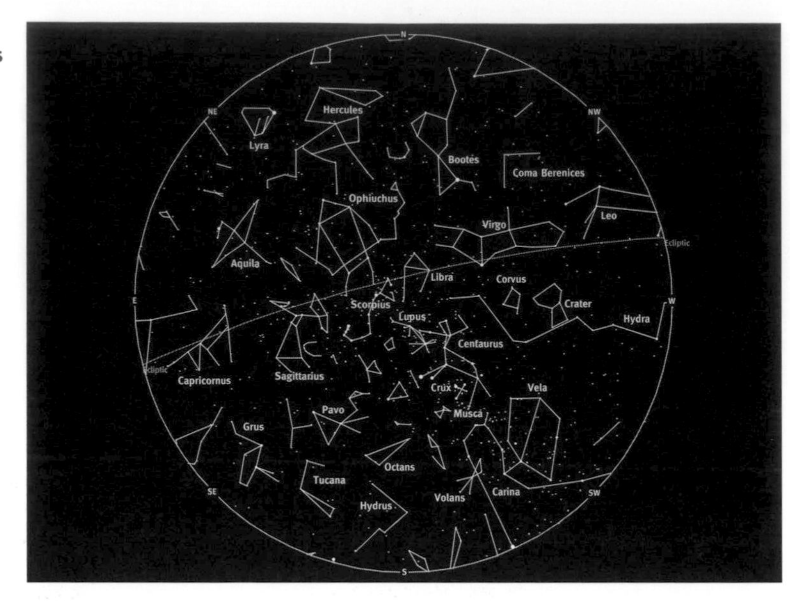


**8 Southern evening sky, spring. The sky as it appears at these times on these dates:**

**12:00 am**
**August 15**
**11:00 pm**
**August 30**
**10:00 pm**
**September 15**
**9:00 pm**
**September 30**
**8:00 pm**
**October 15**

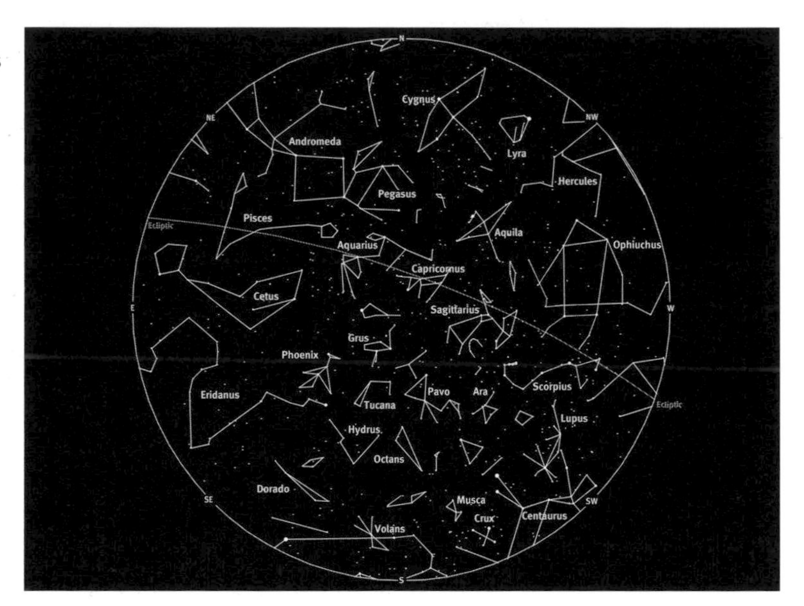

# Index

(*Italics* indicate an illustration; **bold** print indicates an entry in the glossary.)

## Index

ANGELA PROBERT

CARROLL & BROWN LIMITED

*For my Issie, my reason, my inspiration.*

First published in 2013 in the United Kingdom by
Carroll & Brown Publishers Limited
Winchester House
259-269 Old Marylebone Road
London NW1 5RA



A CIP catalogue record for this book is available
from the British Library.

ISBN 978-1-907952-41-8

10 9 8 7 6 5 4 3 2 1





## PART 1
## PREGNANCY CONCERNS



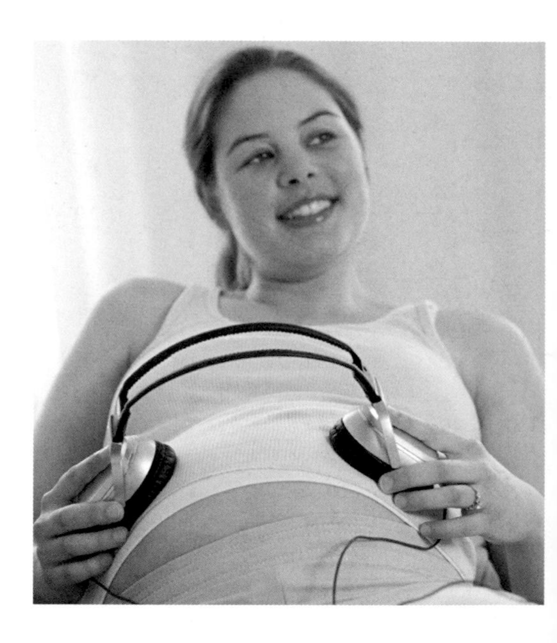

# Contents

# Introduction

During my experiences of being a parent I have frequently turned to books, friends, parents, magazines, the internet, (as well, I'm sure, many other sources) to guide me in making the best decisions I could for my daughter. Occasionally I struggled my way through the masses of information that was so often conflicting. When I started planning how to wean my daughter I had an added concern because coeliacs disease runs in the family, as do asthma and eczema. I wanted to know exactly what was the best time to introduce things like gluten into my daughter's diet, should there be a way of preventing her developing the illness. Finding little information relevant to this somewhat unique question in parenting books I turned to actual research to provide me with information that was totally objective. The factual content of the research allowed me to amass information that was trustworthy and not based on opinion or unique experience. This gave me the confidence to make the best choice I could for my little girl.

Whilst I was researching this topic and others, it occurred to me that perhaps others could also benefit from the objectivity of research in making some of the many decisions of parenthood. And so began my goal of summarising the often prolific number of research articles to provide accessible relevant information for making all those decisions that come with the job of parenthood, easier.

I have benefited hugely from the information I have gathered in this book. The topics have been chosen because the majority have been applicable to my family, as well as others. It has equipped me in making the best decisions I can for my little girl and it is my hope that it will also benefit you.

**How research works**

There exists a tension between the scientists and the non-scientists of this world. Scientists have an appreciation and even 'a patience' for the scientific method, non-scientists, justifiably, tend to be more easily frustrated by it. Scientific research, particularly clinical research, does not always yield conclusive proof. One well designed study can suggest something that totally opposes the findings of another well designed study. For some this is a frustrating concept and one that tends to decrease the credibility of all scientific research. I have the greatest sympathy for public health policy makers who have to take into account all of the conflicting and ever-updated research and make a decision that will affect millions of lives.

An example of how policy has changed dramatically was the age at which women were advised to wean their babies. Not long ago it was set at four months, it is now six months. This change in policy has caused many parents to disregard the new advice, particularly those who have a frustration or even a lack of understanding of how scientific research functions. Policy changes because new data is found. The science of medical research is more like an onion than an orange. Onions have layers, we peel back a layer with each new discovery and policy gets more and more accurate with each new peeling. Other scientific research, however, is more like an orange, you do the research and you get a definite answer, there's only one layer. Unfortunately, most of the relevant research for parents and babies is of the onion-kind and patience is required.

In this book are occasional examples of where the research is either incomplete or simply not in agreement. It is easy to become frustrated with the lack of absoluteness but hopefully a better understanding of this scientific process will encourage patience and a hope that the evidence will become absolute in the not too distant future.

**My research**

Most scientists would agree that the majority of research is flawed to an extent, some more than others. Anyone reading research or indeed a media report on a piece of research, needs to add a degree of caution. Many journals will require a 'peer review' of articles before they publish them, during which an expert will make a judgement on the quality of the research, point out any errors and ultimately help make the decision whether or not it is accurate enough to be published. Note the phrase "accurate enough", few articles are without errors but that doesn't mean they cannot shed light on issues and aid decision making.

Where possible, in doing my research, I deliberately chose to read reviews on relevant topics, rather than just individual studies. Reviews take into account all the papers written on a subject, summarise the main findings, and report consensus, if consensus was found. They often also make a judgement on the quality of the research that has been conducted; indeed some reviews only include quality research. Using reviews has allowed me to place a greater trust on what the research has uncovered

### How to conduct your own research – the wonderful world of Pubmed

If you have access to the internet then you can easily conduct your own research from the comfort of your own home, or during baby's nap time. Not all topics of interest to you will be covered in this book but that doesn't mean you can't still make a well informed decision. I encourage you to seek your own data when important decisions must be made.

Pubmed is an internet interface for searching research papers. It includes a database called Medline where many medical articles are indexed. Simply going to the website http://www.ncbi.nlm.nih.gov/PubMed/ and typing in your area of interest will likely yield hundreds of papers on that topic. You can access most of the abstracts (summaries) and a number of full papers also. Although not necessary for a casual user, you can also hone your research methods by learning from a librarian or by reading the book 'How to Read a Paper' by Trisha Greenhalgh.

Another useful link is the Cochrane Library (http://www.thecochranelibrary.com) which is a series of review type articles whereby all included papers must meet certain criteria, ensuring that the evidence given is high quality and trustworthy. The reviews also include an introduction to the subject that in my opinion is often very well written and easy to understand. You can also skip to the conclusion at the end should you want an immediate answer!

and to a certain extent, eliminate the effect of errors, so often present in published research.

For those that enjoy reading research it may come as no surprise that there are a number of extremely interesting individual studies written on the subjects discussed in this book. Some have even been reported in the media. Occasionally I have tried to include summaries of these papers because if I, as a mother, found them interesting, chances are you will too! But caution is required when applying these individual findings and predominantly the results of these studies are left out of the final summaries in each section.

All in all, I hope that the information summarised in this book will enable you to make informed decisions for the benefit of you and your baby. The lack of foolproof advice may be frustrating at times but that is, unfortunately, how the scientific method operates. Also, try not to throw the baby out with the bathwater, per se, there are plenty of examples where the evidence provides clear, reliable information for your decision making.

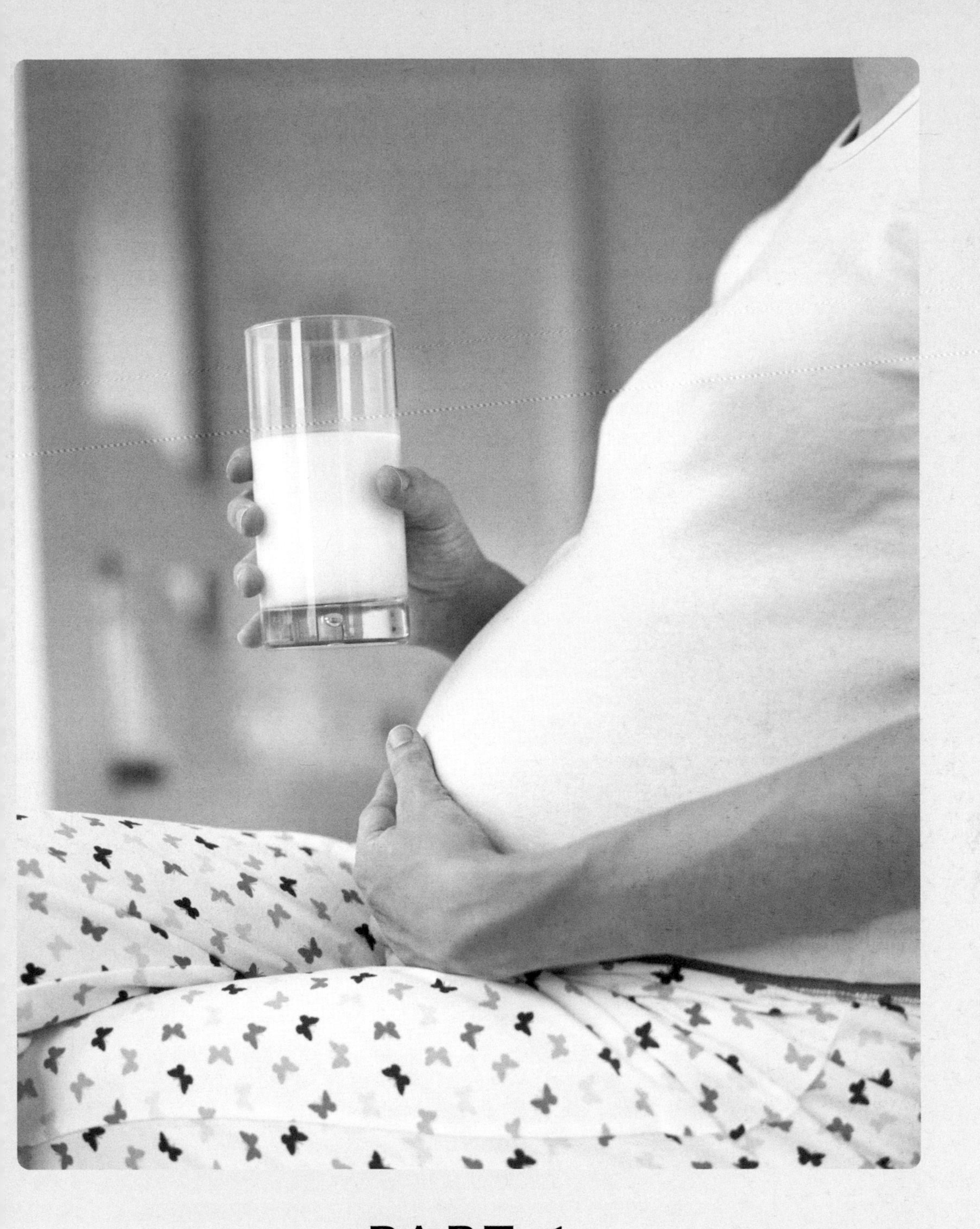

# PART 1
# PREGNANCY CONCERNS

# Nutritional supplements

The decisions about her diet that a pregnant woman makes (or indeed even when trying to conceive) can have lasting consequences, not just for this generation but for those to come. What she chooses to eat during pregnancy can dictate the health and well-being of her baby and her baby's babies. Every ounce of nutrient passed to a baby must come via the mother; the baby is totally dependent on her. Is eating a healthy diet sufficient to ensure a baby gets everything he needs? Are supplements necessary, a bonus or totally superfluous?

## CURRENT RECOMMENDATIONS

The NHS currently recommends supplements of folic acid and vitamin D only but is likely in future to recommend taking them as part of a multivitamin, which also contains vitamin C. If, during her pregnancy, a mother becomes anaemic, then an iron supplement will be prescribed.

## WHAT THE RESEARCH SAYS

### Calcium

A very important nutrient, calcium is vital for muscle contractions, enzyme and hormone building and is a key ingredient in bones and teeth. A pregnant or breast-feeding mother needs up to 600mg extra calcium per day. If a pregnant mum does not get enough calcium in her diet, she will be at risk from pre-eclampsia (a serious pregnancy condition involving high blood pressure and swelling of the feet and legs), muscle cramps and, later on, osteoporosis. Deficiency in the mum has also been associated with a baby being smaller than it should and with weaker bones although there is not enough research to suggest that the former is true.

Calcium supplementation during pregnancy has, fairly conclusively, been found to decrease the risk of pre-eclampsia. There is also some evidence that the mineral density of a baby's bones are increased with calcium supplementation but the effects of this in later life are unknown as are the chances of him developing osteoporosis.

### Folic acid

In the UK, 400 micrograms is the recommended dose for women trying to conceive and whilst pregnant until, at least, the 12th week of pregnancy. This amount of folic acid has been shown to decrease the risk of neural tube defects such as spina bifida by 41%. Although a considerable amount of research supports the recommendation, many women still do not

| Common measures | | |
|---|---|---|
| Unit | Measurement | What it means |
| mg | milligrams | 1/1000 of a gram |
| µg or mcg | micrograms | 1/1000 of a milligram |

get enough folic acid during the most crucial point of their pregnancy so there is a call to fortify bread and other products to encourage an increase in levels in the population (Lane IR, 2011).

Despite taking folic acid, there is still a risk of neural tube defects but the risk can be further reduced by taking a vitamin $B_{12}$ supplement as well whilst trying to conceive and for at least the first trimester (Molloy, AM, et al, 2008).Some research also says that taking folic acid supplements before pregnancy and possibly during pregnancy can protect against the development of childhood brain tumours (Milne, E, et al, 2012) and reduce a child's risk of delayed language development (Roth, C, et al, 2011).

A link has also been made between late folic acid supplementation (i.e. during the third trimester) and a child developing asthma. However, the research is not conclusive as some researchers have found no connection (Sharland E, et al, 2011).

**Iodine**

Pregnant women are recommended to take about 220mcg of Iodine per day, and breastfeeding women to take 290mcg daily. The demand for iodine to synthesise thyroid hormones increases by 50% during pregnancy. If this demand is not met or a woman becomes iodine deficient, miscarriage, slower growth of the baby, detrimental effects on the child's IQ and possible development of ADHD (attention deficit hyperactivity disorder), amongst others, are possible.

Since the 1920s, British cattle have been given iodine-enriched feed, which boosts the amount of iodine in milk and other dairy products. Iodised salt can also counteract deficiency but this is not readily available in UK supermarkets. However, S.C. Bath et al (2012) found that fewer dairy products are being consumed today and those produced by organic farming contain 42% less iodine than conventional milk. M.P. Vanderpump et al (2011) determined that the population is suffering from iodine deficiency, particularly young women.

Currently, iodine supplementation is not recommended. If you choose to artificially boost your iodine levels, it's safest to take a multivitamin tablet containing iodine; avoid iodine supplements on their own as these are often made from kelp, which can cause you to overdose. Some common multivitamins do not contain iodine, so check the label before buying.

**Iron**

Iron is needed for transporting oxygen around both a mum's and her baby's body so a pregnant mum needs more iron than usual. The extra iron goes, not just to her

## Supplements for pregnancy

| Nutrient | RDA during pregnancy | Shown to help with.... |
|---|---|---|
| Calcium | 700mg | Pre-eclampsia and mineral density of baby's bones. |
| Folic Acid | 400mcg | Preventing neural tube defects. It protects against childhood brain tumours and prevents language delay. |
| Iodine | 220mcg | Making thyroid hormones. |
| Iron | 27mg | Normal development of baby and allowing oxygen to reach the baby. |
| Magnesium | Around 350mg | Decreases the chances of preterm delivery and low birth weight. |
| Selenium | 60mcg | Reducing oxidative stress. |
| Zinc | 11mg | Reducing preterm birth. |
| Vitamin A | 700mcg | The development of vision and healthy skin and teeth. |
| Vitamin B | $B^6$ 2mg/$B^{12}$ 1mg | Pregnancy sickness and tooth decay. |
| Vitamin C | 60mg | The absorption of iron and preventing anaemia |
| Vitamin D | 10mcg | Absorption of calcium and regulating blood glucose levels. |
| Vitamin E | 10mg | Preventing pre-eclampsia. |
| Omega-3 | 250mg | Reducing the risk of preterm birth, preventing allergies in children, increasing birth weight and better cognitive skills. |
| Prebiotics and Probiotics | | Reduces the chances of preterm births, supports the health of a baby's digestive system and reduces the risk of dermatitis. |

baby and the placenta, but also to her own increased blood volume.

Iron deficiency during pregnancy can cause low birth weight, early delivery, or in severe cases, death. It can also cause a mum to feel tired and weak. If a mum is mildly iron deficient during her pregnancy, her baby may also develop reduced stores, which can contribute to cognitive and behavioural problems later on.

In the UK, iron supplementation is only recommended after a blood test shows low iron levels, as supplements can cause constipation and can interfere with the absorption of other nutrients. In the USA, however, the CDC (Centers for Disease Control and Prevention) recommend 30 mg iron supplements be given daily to all pregnant women. This low dose is designed to avoid the complications of too much iron.

| Deficiency during pregnancy may lead to ….. | Chances of Deficiency |
|---|---|
| Higher risk of pre-eclampsia, maternal osteoporosis and muscle cramps, weaker bones in baby, | Low (depending on diet) |
| Conditions associated with neural tube defects such as spina bifida or stillbirth. | High |
| Pregnancy loss, slower growth of the baby, detrimental effects on a child's IQ and possible development of ADHD. | Mid-low |
| Low birth weight, early delivery, fatigue for mum. Increased chance of baby/toddler also becoming anaemic. | Mid-low |
| Possible increased chance of pre-eclampsia. | Low |
| Miscarriage, pre-eclampsia and restricted fetal growth. | Low |
| Prolonged labour and low birth weight. | Low |
| Preterm delivery and maternal anaemia. | Very low |
| Pre-eclampsia, carbohydrate intolerance, neurologic disease, reduced fetal growth. | Low |
| Pre-eclampsia and restricted fetal growth. | Low |
| Pre-eclampsia and gestational diabetes for the mum; poor development of baby's bones and teeth, restricted growth. Weaker immune system later on. | High |
| Preterm birth and low birth weight. | Low |
| Poorer cognitive skills for baby and increased chances of allergy. | Low (depending on diet) |

Research is being conducted into whether an iron supplement given 2–3 times a week, as opposed to every day, will have a better outcome as it allows the iron to be absorbed without affecting the absorption of other nutrients. So far, the research suggests that iron is absorbed just as well when taken sporadically as compared to everyday.

Taking lots of iron can be detrimental to women who are not anaemic. Women at high risk of developing gestational diabetes (diabetes as a result of pregnancy) – those who are obese, have had the condition before or have a family history of it – are more likely to developing it if they take iron supplements in early pregnancy when they're not needed (Helin, A, et al, 2012).

**Magnesium**

The recommended amount of this mineral is generally 350mg per day, but more for women below 18 years old (400mg) and slightly more for women aged above 31 years (360mg). It is easily found in the diet in a variety of foods so deficiency is very rare in developed countries.

Magnesium is needed in rather large amounts to regulate body temperature, repair tissues and to relax muscles (calcium does the opposite; it stimulates contraction). A deficiency during pregnancy is thought to increase the chances of developing pre-eclampsia and a lower-than-expected-birth-weight baby.

Investigations looking at the advantages of taking a magnesium supplement found fewer instances of preterm birth and fewer babies with a lower-than-average birth weight. However, supplements had no effect on pre-eclampsia.

**Selenium**

Playing an important role in defending the body from free radicals (tiny particles with unpaired electrons that are highly chemically reactive), selenium is an antioxidant that works to prevent oxidative stress (a condition in which components of cells are damaged such as with pre-eclampsia), preterm birth, miscarriage and gestational diabetes. Selenium is commonly found in food but content depends on the mineral's concentration in the soil.

Selenium levels in a pregnant mum decrease throughout her pregnancy and whilst she is unlikely to become deficient, it may be that a supplement could improve her antioxidant levels and benefit her baby although as yet research is not conclusive.

Though very rare, a selenium deficiency is associated with recurrent miscarriages, pre-eclampsia and growth restriction.

**Zinc**

Being deficient in zinc has been associated with prolonged labour, preterm birth (before 37 weeks) and low-birth-weight babies. Taking zinc during pregnancy helps to slightly reduce preterm birth, but has no effect on birth weight even for those women at risk of having low-birth-weight infants. This is thought to be because taking zinc in isolation is far less useful than taking it in a multivitamin. In any event, taking a zinc supplement during pregnancy is not thought to be particularly beneficial.

**Vitamin A**

Essential for healthy skin and teeth, and for good vision in low light, vitamin A is necessary in larger than normal amounts for healthy pregnant or breastfeeding mums. However, the extra needed is thought to be already stored in the liver so no supplementation is necessary. Women who breastfeed for extended periods may benefit from a supplement if their reserves start to run low.

M.S. Radhikaet al (2002) found that women who are deficient in this vitamin have a greater risk of preterm delivery and anaemia ( although deficiency in this country is very rare and usually occurs where there is a pre-existing condition such as coeliac disease or poor absorption of fat.

The limit for a pregnant mum is 700mcg per day because too much vitamin A can be toxic to a developing baby; overdosing has been associated with birth defects and miscarriage. Animal liver is very high in radicals and antioxidants (which 'mop' them up), a condition known as oxidative stress arises and this can cause problems in pregnancy, such as pre-eclampsia and restricted fetal growth. Vitamin C also helps to absorb iron that comes from plants so it is important for preventing anaemia in those that don't eat meat.

Vitamin C crosses the placenta into the developing baby and pregnant women need to have an increased allowance and sufficient amounts are easily absorbed through a healthy diet. Good sources include peppers, oranges, broccoli and sweet potatoes. Vitamin C can't be stored by the body, so daily sources are needed.Doses reaching 2000mg per day, however, can be dangerous for mum and baby and so can taking a vitamin C supplement during pregnancy. This may result in a higher risk of preterm birth (before 37 weeks) although the only one investigation (Steyn, PS, et al, 2003) has been done.

**Vitamin D**

Since 2008, NICE (National Institute for Health and Care Excellence) has recommended that pregnant and breastfeeding women take a daily 10mcg vitamin D supplement although they note that further research into its benefits should be carried out. Vitamin D is needed by the body to absorb and retain calcium for bones (a deficiency is the main cause of rickets (soft, malformed bones) and it plays a role in maintaining normal blood glucose levels and regulating the release of insulin. An imbalance between glucose levels and insulin (if there is insufficient to 'put the glucose into storage') is associated with diabetes.

In the last 5 years, it has become apparent that many pregnant women and babies are deficient in vitamin D. For the former, pre-eclampsia and gestational diabetes are risks; for the latter, there may be poor development of the skeleton, teeth and general growth (lower birth weights) and their fetal immune system may have been affected so that they become more susceptible to respiratory illnesses and autoimmune diseases (e.g. diabetes or lupus). A deficient child may also suffer from wheezing and an increased BMI (body mass index). Later on, if still vitamin D deficient, a child may suffer from rickets.

Only 10% of the recommended vitamin D allowance is attained through diet, so the rest must either be synthesised by the body from sunlight or be taken as supplements. Vitamin D made by the skin cells in sunlight is the best and most absorbable form; a person cannot 'overdose' via sunlight as she potentially could by taking tablets. In the UK, however, there is not usually sufficient sunshine to generate the much-needed levels, especially in the winter. Furthermore, sun blocks, which prevent access to the UVB light frequencies that are needed to generate the vitamin, are widely used.

Current guidelines for vitamin D supplementation are based on the above knowledge rather than on sound research. According to the Cochrane review (a systematic look at primary research in human health care and health policy) of the data, the need for a vitamin D supplement is inconclusive. However, others (Wagner, CL, et al, 2012) conclude that since vitamin D deficiency is not good for a baby nor his mum, it remains prudent to recommend a supplement and whilst research is ongoing.

**Vitamin E**

A deficiency in this vitamin is rare in healthy adults but too much iron or too many polyunsaturated fats can make one more likely. Deficiency is associated with preterm birth and low birth weights.

Vitamin E acts as an antioxidant and is recycled by vitamin C, which is possibly why individuals are rarely deficient, as vitamin C replenishes it. Oxidative stress (see page 10) in pregnancy (treated with antioxidants like vitamin E) can be a cause of pre-eclampsia and a smaller than expected birth weight. Unfortunately no trials, to date, have been of sufficient quality to allow a firm conclusion on vitamin E supplements although one small but well conducted trial (Chappell, LC, et al, 1999), found that women at increased risk of developing pre-eclampsia, were less likely to develop it as a result of taking vitamin E and C supplements. The research did not hold up when extended to the full population so more research is needed particularly in the light of the effects of taking too many antioxidants.

**Omega-3 supplements**

Omega-3 fatty acids, such as DHA and EPA, are long chain unsaturated fatty acids, commonly found in fish, which are needed for brain and eye development. A deficiency during pregnancy could affect a baby's cognitive skills and increase his chances of developing an allergy. It has been suggested that supplements during pregnancy could reduce the risk of a preterm birth, increase birth weight and a baby's IQ, and prevent a baby developing allergies.

Relatively sound research exists for a reduced risk of preterm birth and several studies, but not all, have shown better cognitive skills (the ability to integrate new information), neurodevelopment and vision in young children, when DHA supplements are taken during pregnancy. Other research suggests supplements during pregnancy could increase birth weight but some similar trials have found no effect. The role of supplements in preventing allergic disease in children, whilst currently inconclusive, is promising.

**Probiotics and Prebiotics**

Probiotics are 'friendly' bacteria that can drive down the number of harmful bacteria, whilst prebiotics are components of the food we eat, but cannot digest, that stimulate the growth of beneficial bacteria, such as probiotics, in the colon. Probiotics are thought to reduce the chances of preterm birth as up to half of these are associated with the mother developing an infection of the urinary tract or vagina, causing an immune response, which triggers early labour. Probiotics have been shown to kill harmful bacteria and adjust the immune response so that the triggers that cause early labour are suppressed. It's also thought that taking probiotic supplements during pregnancy could limit the chances of infection occurring in the first place because they can restore normal flora in the vagina.

Probiotics such as Lactobacillus rhamnosus, taken as a supplement in late pregnancy, have been shown to encourage the development of a healthy digestive system in a baby and, if the mother takes it during pregnancy and/or whilst breast-feeding, can prevent dermatitis in children (Pelucchi, C, et al, 2012).

It also looks as if probiotic supplementation by mum during breastfeeding can help protect vitamin A, so pregnant women are advised to avoid eating it.

Women taking multivitamins with vitamin A before and during early pregnancy have been found to be more likely to have multiple births (e.g. twins). As exciting as this may sound, it unfortunately, significantly increases the chance of having a high-risk pregnancy.

### Vitamin $B_6$ and $B_{12}$

Needed by the body to help absorb proteins and carbohydrates from food, a deficiency of $B_6$ during pregnancy is associated with pre-eclampsia, carbohydrate intolerance during pregnancy, severe nausea ('morning sickness') and neurologic disease in the baby. Supplements during pregnancy may help to alleviate nausea and may also protect against tooth decay in mums.

A deficiency of vitamin $B_{12}$ in pregnancy may result in restricted fetal growth. This vitamin is found almost exclusively in animal products so vegetarians are at greatest risk of having a deficiency.

Only small amounts are needed and for those that eat meat and dairy products, deficiencies are very rare in developed countries. However, if a mum fails to take in enough of this vitamin during her pregnancy, her baby will be deficient in it.

### Vitamin C

Needed for the growth and repair of body tissues, vitamin C along with vitamin E, works as an antioxidant, blocking some of the damage caused by free radicals. If there is an imbalance between the number of free

***Eat yourself healthy***

*A varied diet, filled with fresh fruit and vegetables is the best way to ensure you benefit from all the vital nutrients.*

## Recommendations from the research

Folic acid supplements are vital – particularly early on in a pregnancy or whilst trying to conceive. There's also a good case for taking a vitamin D supplement, as many women are deficient and you might be, too. Taking a probiotic like Lactobacillus rhamnosus might be a good idea especially if there's a family history of allergy or to prevent infections that could trigger an early labour.

Supplements of calcium, iodine, magnesium, vitamin $B_6$ and omega-3 have all been shown to be beneficial but are not likely to be needed. There is less conclusive evidence about selenium. Zinc supplementation has greater benefits if it is included as part of a multivitamin. Getting the balance right with iron is crucial, as too much or too little can be damaging. Consider vitamin $B_{12}$ supplements if you are a vegetarian or cannot eat dairy foods. Eating a balanced and varied diet is the best way to avoid deficiency in most nutrients; many, such as vitamin C, are best taken in food. If you cannot stomach a varied diet, a multivitamin (which includes minerals) could be a good idea whilst trying to get pregnant and during the first trimester. They are pricey and may not be necessary but you might find it comforting to know that anything you are missing in your diet is compensated for by a simple tablet each day.

against allergies but this finding needs confirming.

A link has been made between probiotics and preventing obesity in children but there has been little research conducted to date (Ly, NP et al 2011)

**Multivitamins**

Taking a multivitamin (which includes common minerals such as iron and zinc) whilst trying to get pregnant may increase the chances of success. R. Agrawal et al (2012) who conducted a study on women taking a multivitamin to encourage ovulation in order to become pregnant, found that these women were much more likely to be successful compared with those taking folic acid alone. Research also suggests that women who take multivitamins before getting pregnant and during the first trimester are less likely to give birth preterm or to have babies who are smaller than they should be.

Unfortunately, controversy surrounds the taking of multivitamins because many nutrients prevent the absorption of other nutrients. For example, iron can impair zinc absorption, whilst zinc can impair iron and copper absorption. Also, taking too much of any one vitamin can cause serious medical problems, which is more likely to happen if you're taking a multivitamin.

# Wine, beer and other alcoholic drinks

A cocktail or gin and tonic at the end of a long and stressful day or a glass of wine with dinner are pleasures that few can repudiate. Unfortunately, pregnant women are advised to do just that. The safety of alcohol consumption is hotly debated and for some women (who are, by no means, excessive in their drinking), 'going dry' can be a large sacrifice. Most women on becoming pregnant, or even before, will ask "Do I really need to give up all alcohol?" No one can deny that all pregnant women want the best for their unborn babies and are willing to do their utmost to ensure they are healthy and have everything they need. But still the question remains, "If I have an occasional drink will it really be harmful for my baby?"

## CURRENT RECOMMENDATIONS

The Department of Health backed up by the Royal College of Obstetricians and Gynaecologists, The National Institute for Health and Care Excellence (NICE) and the World Health Organisation (WHO) currently recommend that women who are pregnant or are trying to conceive should avoid taking alcohol altogether.

### Units of alcohol

In the UK, a unit is approximately 8g of pure alcohol, a definition that doesn't shed much light on how much a person normally drinks. Since people don't really understand what is meant by a unit of alcohol, advising pregnant women to drink no more than 1-2 units per week (the advice previous to 2007) seemed meaningless so the Department of Health, in order to give the most prudent

Alcohol content of popular drinks

| Drink | % of Alcohol/Volume | Number of units |
|---|---|---|
| Wine | 13%/175ml (standard glass) | 2.3 |
| Beer | 6%/568ml (1 pt) | 3.4 |
| Bacardi Breezer | 4%/750ml | 3 |
| Vodka and Coke | 37.5%/25ml + mixer | 0.9 |
| Champagne | 12%/125ml | 1.5 |
| Margarita cocktail | 40%/75ml | 3 |

advice, changed its recommendation to a total ban. This decision was not based on any new research about the dangers of alcohol during pregnancy and there remains much indecision about whether small (or even moderate) amounts of alcohol are safe.

To calculate the alcoholic units of a bottled or canned drink, multiply the percentage of alcohol by the volume (both displayed on the label) and divide by 1000 (see also the chart, page 17).

## WHAT THE RESEARCH SAYS

### Risk of fetal alcohol syndrome

It is well documented that drinking large amounts of alcohol is most definitely dangerous for a baby. The varied physical, mental and neurological consequences of drinking alcohol in large amounts while pregnant are known as fetal alcohol spectrum disorders (FASD), the most well known of which is fetal alcohol syndrome (FAS), which is the number one cause of non-genetic disability in infants. It results in brain damage and facial and physical malformations and is entirely a result of a mother drinking alcohol whilst pregnant.

The chances of producing babies with FASD are highest in those who drink too much whilst pregnant, but what constitutes too much? Nobody knows. Metabolic activity amongst pregnant women varies considerably so the ability of one pregnant woman to remove alcohol from her blood stream is likely to be very different to another pregnant woman so it's impossible to give a blanket recommendation on safe consumption. Worse still, a developing baby will take in any alcohol at least twice. Alcohol in a mother's blood stream crosses the placenta and enters her baby's blood stream. A fetus has a very limited ability to break down alcohol so it exits in his urine and enters the amniotic fluid, which he will then ingest thus restarting the cycle. A fetus depends on his mother's metabolic activity to eventually remove it.

Complicating the ability to recommend 'safe' drinking levels during pregnancy is that the concepts of 'light' and 'moderate' drinking are not universally agreed. Many researchers have their own definitions, which makes consensus difficult to find. As a very general guide, light drinking is anywhere from one to three drinks per week and moderate drinking is roughly three to eight drinks per week, spread throughout the week. These amounts are not to be confused with binge drinking, which, generally, is defined as drinking four or more drinks in one sitting. Binge drinking that lies within the light-to-moderate range [i.e. no more than twice a month] does not appear to have any adverse effects on IQ, 'executive' abilities (planning and organising skills) or attention span (Kesmodel US, et al and Underbjerg M, et al, 2012). However, there is a link with preterm delivery (before 37 weeks) even if bingeing is stopped before the second trimester. Binge drinking has, however, been associated with ADHD and learning problems (although the research is inconclusive). Moreover, both IQ and attention span have been shown in other studies, not specifically looking at binge drinking, to be affected by drinking low amounts of alcohol.

Another issue is that researchers can refer to a host of other quantitative alcohol measures including the number of units or drinks or grams or ounces of alcohol. Here, a drink means roughly one unit.

The research appears to claim that drinking eight drinks or less per week does not produce birth defects or heart malformations, or affect motor functions, speech development or executive abilities (Skogerbø Å, et al and Kesmodel, DS, et al, 2012 but that drinking nine or more drinks per week does.

These findings are due to the way the researchers divided pregnant women into groups, and should be interpreted as meaning that a pregnant woman should ensure she drinks far less than the nine drinks per week cut-off. What the researchers are claiming is that a couple of drinks a week will probably cause no problems; they're not claiming that if you drink eight drinks every week everything will be fine.

**Effect on conception**

The research would suggest that drinking while you're trying to get pregnant won't cause any problems for your baby (but you will be at least two weeks pregnant before you even discover it, so caution may be wise). However, it can affect the length of time it takes you to get pregnant. The evidence is conflicting as to whether drinking delays or actually speeds up the waiting time (Jensen, TK, et al, 1998 and Juhl, M, et al, 2003). Perhaps having the odd drink after a stressful day will indeed assist in getting pregnancy underway.

**Greater risk of miscarriage**

During the first trimester, women often 'go off' alcohol completely. This is definitely a sign that the body knows what it's doing as research shows that alcohol-related risks are greatest in the first trimester. In the first twelve weeks of pregnancy, drinking more than two drinks a week does increase the chances of miscarriage (Andersen, AM, et al, 2012) but drinking one to two may not. However, with increasing amounts comes increasing risk, and if three drinks a week has been associated with miscarriage, it may be prudent to go nowhere near that limit.

**Effects of preterm birth**

A baby born before 37 weeks is known as preterm and is at a greater risk of many health and developmental problems. He may have trouble breathing, feeding and fighting off infections and may be required to stay in hospital for an extended time. Depending on how early a baby is born, he may have more severe complications. Preterm birth may also affect his later growth and health

**Lower birth weight**

A number of studies have been conducted on birth weight and the effects of light-to-moderate drinking. Generally, if light drinking constitutes one to two drinks a week, there doesn't appear to be any effect on birth weight. However, if a pregnant woman drinks more than this (becomes a 'moderate' drinker), her baby's weight will lessen (Jaddoe VW, et al, 2007). Also, the detrimental effects will be more pronounced if the mum smokes or is carrying a boy.

**Reduced reflex behaviour and stillbirth**

There's good evidence to suggest that alcohol could inhibit the natural development of the startle or Moro reflex, which is apparent in all healthy newborns up to about four to six months of age. If a baby retains the Moro reflex for longer, stops exhibiting it before three months or is slow to develop it, it can mean that he is suffering from a birth injury or brain malformation (Hepper PG, et al, 2012). Light-to-moderate amounts of alcohol taken during pregnancy have been shown to delay the development of this reflex, and if mothers drink near to delivery, this can have a negative impact on their baby's sleep and breathing. With five or more drinks per week, the risk of stillbirth is 2–3 times more likely due to placental dysfunction (Andersen, AM, et al, 2012).

**Reduced brain development**

Drinking even low amounts of alcohol during pregnancy has been shown to affect brain development, particularly if alcohol is drunk at crucial developmental times, like the first trimester. Children of mothers who drink light-to-moderate amounts of alcohol have less grey matter in their brains compared to children whose mothers do not drink at all.

**Increased behaviour problems and learning difficulties**

Links have been made between light-to-moderate drinking and aggression, reduced attention span, memory problems and other learning difficulties and behaviour and mental health problems. B. Sood et al (2001) also found a child's temperament and his ability to sleep, eat well and handle stimuli, are also negatively affected. Other research, however, has found that women who drink low-to-moderate amounts of alcohol are not likely to have children with behavioural problems, and in 2010, M. Robinson et al one found that it may even have a mildly protective effect (see box below).

To date, there is no conclusive evidence that moderate drinking during pregnancy contributes to a risk of ADHD in a child (Burger, PH, et al, 2011) although P. Kim et al (2013) have found an association between moderate alcohol intake and hyperactivity in animals such as mice and primates.

**Decreased intelligence quotient (IQ)**

S.J. Lewis et al (2012) uncovered some evidence that light-to-moderate amounts (as low as 2 drinks per week) of alcohol drunk in pregnancy could have detrimental effects

**Understanding the research**

The study by Robinson et al that claimed that drinking low-to-moderate amounts of alcohol was mildly protective of 'bad behaviour' in children highlights the need for critical reviews of all research, and the fact that the media can sensationalise evidence. A critique of the investigation (Todorow, M, et al, 2010) explains that of the 1860 pregnant women initially brought into the study, one third dropped out. All these women were from a lower socioeconomic background and removing them from the data makes it impossible to view the whole picture and reach meaningful conclusions. Robinson et al's findings, therefore, are probably skewed and it's unlikely that alcohol offers protection from behavioural problems.

on a child's IQ although other experiments have found no detrimental link between IQ and drinking (Falgreen Eriksen HL, et al, 2012).

**Increased chance of osteoporosis**

M.E. Simpson, et al (2005)found that moderate drinking during pregnancy could increase the chances of a child developing osteoporosis in later life because alcohol has a negative effect on bone development.

**Future drinkers**

There's fairly conclusive evidence that drinking alcohol can produce babies with a liking for it. A.E. Faas et al (2000) found that babies exposed to low amounts of alcohol in the womb (roughly 2½ units wine at least once a week) were more 'interested' in the smell of ethanol than babies who were not exposed to alcohol. This interest was still present seven to ten days after delivery. Consuming large amounts of alcohol during pregnancy is generally accepted to create adolescents who are more predisposed to drink, even when other factors that could encourage teenage drinking are taken into account. Whether moderate or even light amounts of alcohol cause a similar effect is as yet not known, but the Faas et al study does show a link.

## Recommendations from the research

There is a growing bank of evidence that says even drinking light-to-moderate amounts of alcohol could have negative effects on a baby. Although many papers have found no adverse effects from drinking, it is difficult to rule out any risk because so much of the research contains errors in how it was conducted. Harm from drinking cannot be ruled out until these errors are addressed. Hence the government's blanket ban – better safe than sorry.

Although every woman must make her own decision, I feel it's fair to say that alcohol will never be good for your baby. However, if you can be confident about knowing your units and stick to seldom and light drinking, your baby will probably be fine. One key thing to remember is that stress is also not good for your baby or for you (see page 00) so if you drank too much before reading this article or before you knew you were pregnant, try not to worry, as worrying can potentially cause more problems than the alcohol itself. I drank a very small amount during my first pregnancy. After reading the research and writing this chapter, I intend to abstain for my next pregnancy....well, after I know I'm pregnant, anyway.

# Coffee and caffeinated drinks

Investigations focusing on the effects of caffeine have been criticised because they detract attention away from the far more harmful effects of alcohol and smoking on a baby; the repercussions of low-to-moderate amounts of caffeine are relatively mild in comparison. Criticisms aside, there is a good case for being careful about caffeine intake, particularly as the amount of caffeine can rack up quite quickly. Caffeine can be found in a surprising number of places. The obvious ones are tea, coffee and colas, but it can also be found in cocoa and therefore chocolate bars, and even in over-the-counter cold and flu remedies. A pregnant mum's body is slower at eliminating the caffeine from her blood than usual, and the caffeine passes freely to her unborn baby who lacks the ability to break it down properly. The effects that this caffeine, particularly in small amounts, can have on a baby have been researched voraciously.

## CURRENT RECOMMENDATIONS

According to the UK Food Standards Agency (FSA), caffeine intake during pregnancy should be less than 200mg per day – that's roughly 2 cups of tea and 1 chocolate bar. A breastfeeding mum should avoid drinking a lot of caffeinated drinks.

### Caffeine units

A recent Glasgow study (Crozier, TW et al, 2012) conducted on the amount of caffeine in popular coffee drinks found some worrying results. According to the NHS website, a mug of 'filter coffee' contains 140mg of caffeine. However, this study found one coffee shop where an espresso had 322mg of caffeine and another four shops had espressos over the 200mg limit for pregnant mums. These differences are due to the type of bean or mixture of beans used and how the beans are prepared. So, by having just 1 cup of coffee made from a shot of espresso, a pregnant mother could inadvertently go beyond her threshold.

It's also worth noting that a cup of decaffeinated coffee can still have up to 20mg of caffeine. Therefore, the great stratagem of switching to decaf coffee, so you can continue guiltlessly with your 8 cups a day, could put you over your limit if you throw in a measly chocolate bar.

## WHAT THE RESEARCH SAYS

### Effect on conception

The research is mixed as to whether caffeine affects how long you have to wait. Some research has found that drinking more than three cups of coffee a day can delay conception by up to a year. Other research concludes caffeine has no effect on getting pregnant nor extends waiting time.

Drinking caffeine before pregnancy may actually protect against gestational diabetes (when a mother develops diabetes during pregnancy) (Adeney KL, et al, 2007).

### Risk of miscarriage

Although findings are conflicting, many reviews considering all relevant research have found little evidence to link moderate caffeine intake (up to 300mg per day) with miscarriage. However, some research suggests that the more you drink, the greater the risk (Weng, X et al, 2008). Unfortunately, much of the research in this area fails to properly take into account other elements that may cause miscarriage (see below for more details).

I. Grant et al (2012), however, suggests a link between caffeine and how well an embryo implants itself. If an embryo fails to implant correctly then the pregnancy will miscarry. Failure to implant is thought to be caused by the effect that caffeine has on the protein that contributes to implantation success.

### Excessive caffeine intake

Drinking more than 8 cups of coffee a day has been associated with restricted blood flow to the baby and an irregular baby heart beat, which results in less oxygen getting to the fetus. Drinking this much caffeine has been shown to result in miscarriage (even after 20 weeks (Bech, BH et al, 2005)), stillbirth, low birth weight and a greater risk of SIDS (from animal studies). Babies exposed to high levels of caffeine in the womb also experience classic withdrawal symptoms like irritability, jitteriness and vomiting.

### Restricted growth and lower birth weight

There is good reason to suppose that a baby's growth can be restricted by caffeine and, as a result, her birth weight can also be affected. Generally, the more caffeine a woman takes, the greater the effect on her baby's weight. A baby born with a lower-than-expected birth weight (predicted from genetic and environmental factors) is more likely to suffer from obesity, diabetes, heart disease and high blood pressure as an adult.

In 2008, the Food Standards Agency commissioned two linked projects that looked specifically at low birth weights. They found that even at doses between 200–300mg there was evidence for a negative effect on birth weight (hence the 200mg of caffeine per day recommendation for pregnant women. J. C. Konje et al (2008) also found that even with intakes of less than 200mg, there remained a 2% risk of low birth weight. While this is considered extremely low, it does show that, even in small amounts, caffeine has an effect.

**Risk of birth defects**

Caffeine, it would appear, can also play a part in the development of neural tube defects (resulting in brain or spine problems, commonly causing death) if the baby is genetically susceptible to this kind of defect. Whether defects develop is dependent on how fast the mother breaks down the caffeine (specific enzymes perform this task in the body). It's actually the substances that are produced as a result of the caffeine being broken down that cause the problem, not the caffeine itself. As with alcohol (see page 18), the ability of one pregnant mum to break down caffeine can be quite different to another. Unfortunately, a mother cannot know how fast she breaks down (metabolises) caffeine without being tested. Testing requires saliva analysis, which is not routinely carried out. Oddly, the faster the mother breaks down the caffeine, the higher the risk of neural tube defects.

**Breastfeeding**

Drinking more than 300mg per day of caffeine whilst breastfeeding could have an effect on how well your baby sleeps at night. Whilst the findings were not significant enough to be conclusive, there was some evidence that women who drank more than 300mg a day were more likely to have babies who woke 3 times a night at 3 months old than women who drank less caffeine. (Santos, IS et al, 2012)

The amount of iron found in breast milk can also be affected by a mother's caffeine intake if she drinks more than 450ml of coffee a day. This could result in her baby suffering from mild anaemia.

**Comparison of caffeine-containing items**

| Item/quantity (approx) | Estimated amount of caffeine/source |
|---|---|
| **Instant coffee** | |
| 1 mug (260 ml) | 100* |
| 1 cup (190 ml) | 75* |
| **Black tea, infusion** | |
| 1 mug (260 ml) | 50-75* |
| 1 cup (190 ml) | 33-50* |
| **Drinking chocolate** | |
| (Made as per pack instructions [200 ml]) | 1.1-8.2* |
| **Filter coffee** | |
| Small (225 ml) | 160** |
| Medium (350 ml) | 240** |
| Large (450 ml) | 320** |
| **Cappuccino or latte** | |
| Small (225 ml) | 75** |
| Medium (350 ml) | 100** |
| Large (450 ml) | 150** |
| **Mocha** | |
| Small (225 ml) | 90** |
| Medium (350 ml) | 125** |
| Large (450 ml) | 175** |
| **Flat white** | |
| Small (225 ml) | 150** |
| **Americano coffee** | |
| Small (225 ml) | 75** |
| Medium (350 ml) | 150** |
| Large (450 ml) | 225** |
| **Energy drink (Red Bull)** | |
| 473 ml can | 151^ |
| **Cola drinks** | |
| 330 ml can | 11-70* |
| **70% dark chocolate** | |
| 40 g | 6*** |
| **Milk chocolate** | |
| 40 g | 3*** |
| **Cocoa powder** | |
| As made 190 ml cup | <1*** |

## Recommendations from the research

The effects of caffeine have been difficult to measure because of the number of other influences on the results. If, for example, one considers how caffeine affects the risk of miscarriage, how can other things that might cause miscarriage, such as genetic problems resulting in a non-viable embryo be ruled out? One major problem with many studies is that they don't take into account what's referred to as the 'pregnancy signal'. Many pregnant mums recall a rather sudden aversion or nauseous reaction to coffee – even its aroma – in the first trimester. Feeling nauseated is often a good sign of a healthy pregnancy and a viable embryo. If this rather common (about 65% of women) side effect, which causes women to naturally reduce their caffeine intake, isn't taken into account, then the results of a study looking into miscarriage risks, for example, will tend toward being invalid. A study would need to appropriately cater for women who don't have a viable embryo in the first place, in which case miscarriage is an unfortunate inevitability.

Another difficulty is how much to rely on the ability of a pregnant mum to remember how much caffeine she had last month (or even last year in some cases!) or whether to prescribe a certain amount of caffeine to her each week of her pregnancy. The latter evokes huge ethical issues because a pregnant woman would be prescribed a drug that could potentially be damaging to her unborn baby; she may, of her own accord, have ordinarily limited her caffeine intake. So, a sensible conclusion based on this research would be that drinking lots of caffeine could affect your fertility but you're probably okay to drink lower amounts – the recommended maximum of 200mg caffeine daily is based on good research and is a good goal to aim for. However, first cut out things that are definitely harmful to a baby then consider cutting down on caffeine.

# Peanuts

Allergies, in general, are on the increase particularly in the West, but peanut allergies can result in anaphylaxis (a potentially fatal allergic reaction, which occurs very quickly). It's natural, therefore, to try to abate the rise in peanut allergies but nobody actually knows what's causing the increase. Many theories exist and the best seem rooted in the notions that we don't let our children play in the mud enough (the hygiene hyposthesis, see page 74) and that we eat too many processed foods.

## CURRENT RECOMMENDATIONS

In 2008 the Committee on Toxicology (COT) reviewed its initial research and stated that pregnant and breastfeeding mums could eat peanuts as part of a healthy, balanced diet, if they so desired. If, however, they were allergic to nuts, it would be prudent to avoid them .Previously, in 1998, COT had looked at the link between allergies and peanut consumption during pregnancy and breastfeeding and based on their findings, pregnant and breastfeeding women were advised not eat peanuts if there was a history of allergy in the immediate family.

## WHAT THE RESEARCH SAYS

The initial (1998) advice resulted in two common misconceptions. The first was that pregnant women should avoid all nuts even though only peanuts were specified. The second was that all pregnant women should avoid peanuts. The recommendation, however, clearly stated that a pregnant woman should avoid eating peanuts only if there was a history of allergy in the 'immediate family' – the pregnant women herself or the father or any brothers or sisters of the baby. Further confusion arose when eating peanuts during pregnancy was also linked, not only with peanut allergies, but with asthma and other allergies.

COT's recommendation to avoid peanuts came from an understanding of how an allergy begins in a person. The development of an allergy has 2 stages. First, a person must become 'sensitised' to a particular substance, in this case peanut protein, by coming into contact with a small amount of it. Sensitisation has no symptoms that we would normally associate with an allergy; it is simply the first step in developing an allergy. It is sometimes referred to as 'priming'; the body waits, prepared and ready to meet the substance again. When it does so, an allergic reaction is triggered. This can be anything from hives to full-blown anaphylactic shock. As it had been commonly observed that some children who encountered peanuts for the first time had an immediate allergic reaction (or in other words, were already 'primed' having encountered the allergen somewhere

## Recommendations from the research

Making a recommendation based on cuurent research is largely pointless as it is still unclear whether exposure to peanuts in the womb or whilst breastfeeding will cause an allergy, protect against it or have no impact whatsoever. And to add to the confusion, it looks like the amount of peanuts could have a major impact.

There is a lot more research being conducted on this topic at the moment but the results will not be ready for another couple of years so the best course of action would be to adhere to the Government's advice to eat peanuts as normal unless you are allergic but to keep an eye out for the research that's due in the next few years.

before, the most obvious assumption was that these children had become sensitised in the womb, or whilst being breastfed.

Not a great deal had been written on the subject when COT made its initial recommendation, but at that time, it knew that food allergens were found in breast milk and could be passed to an infant through breastfeeding thus potentially causing a baby to become sensitised. COT also knew that most blood from umbilical cords, taken after babies had been born, showed that the babies had been sensitised to allergens already. So these key ideas were the basis for the recommendations.

In 2009 COT conducted a review of all the research carried out in the intervening years. Of the papers found to be satisfactory (chosen by relevant topic and good research methods), none supported the concept that peanut sensitisation and allergy was caused by a mother eating peanuts (or peanut products) during pregnancy.

Moreover, more information had come to light about how children could become sensitised to allergens via other routes, such as contact through a graze or broken skin. This shed doubt on the idea that allergens passed through the umbilical cord and in breast milk were the culprits. Also, it became clear that allergens transported through the cord have a far more complex effect on sensitisation than was initially thought and that allergens encountered by a mother do not necessarily affect her fetus in the way that was first assumed.

Some animal studies (Strid, J, et al, 2004) found that larger doses of an allergen (in this case, egg) were found to actually protect against allergies. E. Maslova et al's 2012 study of Danish mothers found that eating nuts during pregnancy was protective against their children having asthma. The more peanuts they ate during pregnancy, the less likely their toddler would become asthmatic. However, S.H. Sicherer et al's 2010 study, unfortunately found the opposite but only in women likely to have atopic (allergy-prone) children.

# Stress and its effects

The suggestion to 'just relax' from well-meaning family and friends often has the unfortunate effect of achieving the very opposite. Some stress can be a good thing but we're far more familiar with too much stress, which is generally known to be bad.

Stress is triggered by a variety of factors, any of which places some kind of demand on our bodies or lives, and with which we are (or think we are) incapable of dealing. It may be a result of personal circumstances; our work or vocation; lifestyle choices such as smoking, drinking alcohol or caffeine, being overweight or under exercised; or personality factors such as whether we suffer from depression or anxiety.

Stress has well-known mental and physical effects and can affect how long it takes a woman to get pregnant and once she's pregnant, her unborn baby. A expectant mum's stressful experiences are shared, in part, with her baby, who will be born into her current lifestyle. These result in a kind of 'programming', which dictates a baby's physical and mental health for the rest of his life. Scientists are now sharing with us the life-long implications that the nine months in utero entail.

## CURRENT RECOMMENDATIONS

The National Institute for Health and Care Excellence (NICE) suggests that couples be informed that stress can affect their relationship and that this can decrease libido; those undergoing fertility treatment should be offered counselling. It also advises against timing intercourse with ovulation because that may be stressful; instead, they recommend engaging in sexual intercourse every 2–3 days. NICE also recommends that in preparation for conception, women drink no more than two units of alcohol (see page 18) per week and that men be informed that excessive drinking can affect their sperm. It is also known that smoking or passive smoking can affect fertility, as can a BMI of over 29 (in both women and men).

## WHAT THE RESEARCH SAYS

### Effect on conception

#### *Lifestyle and health stressors*

There is good evidence that smoking can affect how quickly a couple conceive (Augood, C, et al, 1998); it can delay the chances of getting pregnant by up to one year. A delay in conception is also possible from passive smoking.

R. Kunzle et al found (2003) that a man who smokes can produce sperm with an increased chance of a genetic disorder (see also oxidative stress and sperm health, below) while a woman who smokes can have lower hormone levels (hormones are needed for an egg to be released and nourished prior to fertilisation).

Being overweight or underweight can also affect the chances of getting pregnant; either condition can upset the fine hormonal balance that is responsible for the normal functioning of the menstrual cycle and the release of eggs from the ovaries and the more extreme the difference from the norm (either under or over), the greater the upset. D.C. Gesink Law et al found (2007) that the average time normal-weight women took to get pregnant was 3 months, while many overweight and underweight women took up to 8 months and some even longer. F. Bolumar et al, however, found that being overweight only affected fertility in women if they were also smokers (2000). Men who are overweight or underweight (BMI less than 20 or above 25) have been shown to have fewer 'quality' sperm, which could also affect fertility.

Too much exercise can delay the time it takes to get pregnant according to a recent study. Five or more hours of vigorous exercise, such as running or swimming, per week, can reduce the chances of conceiving by about 50% every month. However, the researchers found that this was true only in normal-weight women. The fertility of women who were obese or overweight was not affected by vigorous exercise. The research is somewhat limited in this area, but it is fairly well observed, that menstrual cycles are commonly disrupted in very athletic women. This is due to fat levels (women need a certain amount of fat to have regular periods) but it could also be due to a disruption in the hormones needed for ovulation and the monthly cycle. Doing some moderate exercise, such as brisk walking or gentle bicycling, for any number of hours each week, however, increases the chances of getting pregnant but only by a small amount (Wise, LA, et al, 2012).

There are many chemicals, known as endocrine disruptors, which can affect the normal balance of hormones in the body. Over the last 50 years, declining sperm quality has been fairly well documented in most industrialised countries and blamed on the increased usage of endocrine disruptors in our day-to-day live, which are found in industrial chemicals such as lubricants and solvents, plastics (BPA and phthalates), pesticides, diesel exhausts and heavy metals. Their effect on female fertility is less clear cut, but many have been shown in animal studies to cause an array of problems including polycystic ovarian syndrome, endometriosis and ectopic pregnancies.

Some chemicals in sunscreens, cosmetics and food preservatives have also been shown to disrupt the normal hormonal functioning of adults (De Coster, S and van Larebeke, N, 2012).

Oxidative stress caused by free radicals is particularly destructive to cells. The body uses vitamins C and E and the mineral selenium to mop up these particles but if some remain, damage can occur. Oxidative stress is increased if there is a lack of vitamins C and E in the diet; by smoking; having a high BMI (above 25); contact with pesticides, fertilisers or petrochemicals; not getting enough sleep or exercise or having too much alcohol.

Oxidative stress can affect sperm in a number of ways. It can damage their tail, making them less efficient at swimming; it can damage the membrane, through which a sperm enters the egg, and it can damage the sperm's DNA, which not only decreases the chances of conception but if fertilisation occurs, it can affect how a baby develops.

Ultimately oxidative stress can result in infertility.

*Psychological stressors*

Anxiety, depression and mental distress may also contribute to the time it takes to become pregnant, although the research is not conclusive. A. D. Domar et al (2000) inferred that psychological stress plays a part in fertility from their evidence of the positive effects seeing a psychologist and/or being part of a support care group has on shortening the time needed to conceive for couples undergoing fertility treatment). J. Boivin et al (2011), however, showed that emotional distress does not affect the chances of getting pregnant, at least in cases of fertility treatment. Clearly the role that psychological stress plays on fertility is not well understood; it's possible that it affects some aspects of fertility but not all or only some people.

*Stressful events*

A death in the family, a new job or even a deadline can be detrimental to the time it takes to conceive. Such events can produce longer menstrual cycles and delay ovulation.

Even trying to conceive – particularly for those women undergoing fertility treatment – produces stress with knock-on effects such as delayed ovulation and reduced fertility. Men can also suffer from stress when sex is timed for certain crucial times of the month and this can result in ejaculation problems (Byun, JS, et al, 2012). Fertility treatments themselves are also stressful and counselling and support groups have been shown to increase the chances of a conception.

*Combined Stressors*

Finally, a combination of stressors could culminate in reduced fertility. For instance, if you and your partner exhibit only 4 stressors (that's 4 in total, not 4 for you and 4 for your partner), statistically there is only a 38% chance that you will conceive in 1 year. M. A. Hassan and S. R. Killick have shown (2004) if you or your partner smoke more than 15 cigarettes per day, drink more than 7 cups of tea/coffee daily or are over 35 for women or 45 for men, then the combination of these could cause decreased fertility.

**Effect on pregnancy**

*Lifestyle and health stressors*

Smoking during pregnancy can result in a baby suffering from a variety of conditions including low-birth weight, preterm birth, poor neuro-development, which increases the risk of later ADHD, and asthma and wheezing in childhood. Passive smoking during pregnancy also increases the chances of childhood asthma.

The effects of alcohol and caffeine are discussed in more detail on pages 17 and 22, but if a pregnant woman drinks more than 2 units of alcohol each week or more than 200mg caffeine per day, she could miscarry or her baby's birth weight could be lower than expected or he could suffer from reduced brain development.

Women who gain a lot of weight during pregnancy have babies who are more likely to become obese themselves as children and develop diabetes later on. However, light-to-moderate exercise during pregnancy can reduce the chances of mum developing gestational diabetes (diabetes brought on by the pregnancy), which in turn decreases the risk of baby developing diabetes in later life (Capra, L, et al, 2013).

Exposure to pollutants, commonly found in cities, can double the risk of a mother having a small baby and can increase the chances of having a preterm birth by a factor of 5 (Choi, H, et al, 2008). Some pollutants can affect a child's cognitive development and consequently his IQ, and increase his chances of developing asthma and cancer in later life.

The fetus and placenta are particularly sensitive to oxidative stress (see above), particularly in their early stages of development. Complications associated with a poorly functioning placenta include early miscarriage and pre-eclampsia.

***Psychological stressors***

The possibility of a baby developing an allergy is thought to have something to do with the number of antibodies the baby receives from his mother via the placenta before birth. The more antibodies passed on, the greater the chance an allergy will develop. Psychological stress during pregnancy has been shown to increase the number of antibodies, so stress could increase the chances of a baby developing an allergy – particularly if the mother already has allergies of her own (Peters, JL, et al, 2012). The effects of stress on a developing baby's immune system could also affect how well his immune system functions throughout his life; D. L. Bellinger et al (2008) found the likelihood of an autoimmune disease such as coeliac or diabetes type I may be increased.

Anxiety about being pregnant (especially in the first and third trimesters) is common – just over half of all pregnant woman experience anxiety at some stage. If this anxiety goes unchecked, it can cause a mum to increase smoking and/or alcohol intake or eat a less healthy diet.

An emotionally stressed mother may have abnormal hormone levels and inflammatory responses, which in turn can affect how long she carries her baby (Ruiz, RJ et al, 2003). It looks as though the more stress experienced by mum, the more it affects her child. Following hurricane Katrina, there was a large increase in the number of preterm births in the areas most affected. Both preterm and low-birth-weight babies can suffer from physical, emotional or cognitive delays or impairments.

Depression during pregnancy can also affect the development of the part of a baby's brain that controls memory and attention span, and can make a baby more at risk of developing behavioural and emotional problems. L. Capra et al (2012) found if medication is taken to treat depression, it can lead to a child developing heart malformations, high blood pressure and withdrawal symptoms.

Stress at key developmental times during pregnancy, particularly in the third trimester, can have a negative impact on brain development, particularly areas that control language, comprehension and reasoning.

***Stressful events***

Although the reasons are not clearly understood, if a pregnant woman experiences stressful events, it can have long-lasting effects on her child; A. C. Huizinket al (2004) found that prenatal stress can result in children who develop psychiatric disorders such as schizophrenia, anxiety or depression, and that neurodevelopment of teenagers and adults can be affected. Military women in front-line and other stressful positions often deliver preterm babies.

## Recommendations from the research

Stress undoubtedly affects how long it takes to get pregnant and the health and well-being of a baby. Mild stressors, such as an unhealthy lifestyle, a high BMI, smoking and a high alcohol and caffeine intake can delay conception (particularly if combined), or result in miscarriage, or lead to a preterm birth, a low birth weight or an increased risk of obesity. Moderate stress, such as moving home, workplace difficulties or personal circumstances can also play havoc with a woman's menstrual cycle and make tracking ovulation very tricky; it could also result in a preterm birth and a baby's low birth weight and increased chances of him developing an allergy. Severe stress (such as that experienced through mental health problems or a dangerous occupation) could lead to complete infertility or a baby being born with impaired brain development.

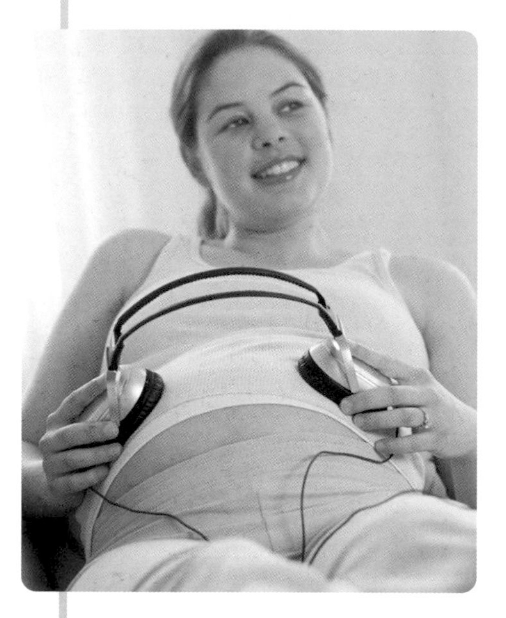

Living a healthy lifestyle, such as giving up smoking, decreasing alcohol intake and taking up exercise are all stress reducers. Whilst there's little you can do to prevent most stressful events, counselling, being part of a support group or taking measures to reduce stress (mastering relaxation techniques, for example) can be beneficial in combatting their effects.

Many mothers-to-be experience stress as anxiety about being pregnant, or become concerned about labour and delivery or their baby's health but there's lots they can do to alleviate their concerns. As a first port of call. discuss any anxieties with your midwife or GP and keep them informed about how you're managing. As mentioned above, receiving support can be beneficial.

Mind-body techniques, particularly those that prepare a mum for birth, have been shown to reduce anxiety (see page 43). For those undergoing fertility treatments, being part of a support group is also a very good idea.

Above all, don't try to 'not be stressed' nor to 'stress about being stressed'. If necessary, get so you are able to relax and try to do lots of things that you enjoy doing.

# Planned Caesarean

Every woman in the UK has the right to choose how her baby is born. A vaginal birth is heavily recommended, but for up to a quarter of women, a C-section is the best and healthiest option. A planned Caesarean is carried out before there are any signs of labour, usually between weeks 37 and 40. An emergency Caesarean is done as a result of complications during the labour, but specifically, after labour has begun, and the baby has started her descent (or is engaged) in the birth canal.

Controversy surrounds women who opt for a planned Caesarean having no medical grounds – the so-called 'too posh to push' type – as against those choosing one because they previously had a C-section and don't want to risk possible complications from a vaginal birth.

There are also a number of credible non-medical reasons why a woman may opt for a planned Caesarean, ranging from concerns over the physical state of her body after a vaginal birth (such as urinary incontinence or worse) to a deep-rooted fear for the safety of her baby or herself. Fear of childbirth appears to be a driving force behind most non-medical reasons. If a woman had a negative experience with a previous birth, she is more likely to ask for a Caesarean second time around. Medical professionals recognise that, for some women, fear should be taken very seriously and counselling offered.

Statistics are difficult to find on the actual number of women opting for Caesareans without a medical reason (somewhere between 3 and 7%). A common misconception is that being 'too posh to push' has caused the increase in Caesareans in the last 20 years. Research shows that this is highly unlikely, and that the vast majority of women opt for a Caesarean due to medical grounds (Bragg, F, et al, 2010).

## CURRENT RECOMMENDATIONS

The National Institute for Health and Care Excellence (NICE) currently recommends that a woman desiring a Caesarean (for no known medical reason), who cannot be dissuaded, should be allowed to have one. Any doctor, however, who feels uncomfortable performing an unnecessary procedure, has the right to decline and to pass the case on for a second opinion.

## WHAT THE RESEARCH SAYS

### Protection from incontinence and better sexual health

Having a Caesarean has been shown to protect against pelvic floor disorders such as bladder and bowel incontinence following birth (Koc, O, et al, 2012). About a third of woman suffer from urinary incontinence and 10% experience bowel incontinence following a vaginal delivery. Doing pelvic floor exercises during and after birth (and beyond) has been shown to reduce the effects of these disorders (Hay-Smith, J, et al, 2008). However, pelvic floor muscles are only protected if a Caesarean is done before labour starts.

Women who have a planned Caesarean appear to experience less discomfort during sex in the first three months following the birth, compared to those who have a vaginal delivery. At six months there is little difference between the two modes of delivery and sexual health (Barrett, G, et al, 2005).

**Risks of abdominal surgery**

Despite huge improvements in the last 20 years, there are still significant risks connected with the procedure including the effects of anaesthetics (see page 46), blood loss and blood clots (causing a pulmonary embolism). Adhesions (scar tissue causing organs to be held together) are also a common side effect of most abdominal surgeries. They can be painful and, if they occur near the Fallopian tubes, could block the release of eggs and consequently affect future fertility.

As with any surgery, there's a greater potential for infections after having a Caesarean that won't necessarily present until some weeks after the operation. This means it's more likely that a mum will have to go back into hospital at some time during her baby's first month for further treatment. Infection can occur at the incision site, the uterine lining or in the urinary tract (usually as a result of the catheter needed for the surgery). Moreover, a woman can experience discomfort from the incision for several days or even weeks following the surgery. As a result, it is not always safe to drive for up to six weeks after a Caesarean.

Occasionally a baby is accidentally cut by a scalpel during the procedure. This usually is a very minor injury.

On average, women who have a Caesarean stay in hospital for longer – 4 days against under 3 for a vaginal birth.

**Risks to future babies**

Having a Caesarean for a first pregnancy can affect following ones. There is an increased risk of stillbirth or serious blood loss (haemorrhage), for a second pregnancy when the first delivery is by Caesarean (Smith, GC, et al, 2003). Following a previous procedure, there is a greater chance of the placenta

### Medical reasons for a planned Caesarean

1. If a baby is in a difficult position for delivery – in a breech position (feet or buttocks first) or transverse (lying sideways). Many breech babies and all transverse babies must be born by Caesarean.
2. If a pregnant woman has a high-risk condition, such as bleeding, genital herpes, diabetes, pregnancy-induced hypertension or eclampsia.
3. If there's more than one baby. Some twins are successfully delivered vaginally if they're in favourable positions.
4. Small-for-dates baby or mother or baby has a high-risk condition.
5. The baby is very large. Some babies are just too big to be born from a mother's birth canal.
6. Fetal distress as a result of the stresses and strains of labour. This can be detected by fetal monitoring and special tests.
7. Two or more previous Caesarean sections. After an initial Caesarean, a VBAC (vaginal birth after Caesarean) may be attempted.

attaching next to or covering the cervix (known as placenta praevia). Or it can attach too deeply and won't detach properly after delivery (known as placenta accreta). If attempting a vaginal birth following a Caesarean, there is an increased chance of the uterus rupturing along the scar of the previous operation. The more Caesareans, the greater the risk of any of these complications, although all are relatively rare.

Having a Caesarean has also been shown to affect how easily a woman can become pregnant again and of having an ectopic pregnancy or miscarriage. Whilst very rare, there is a chance that a new embryo could implant itself in the Caesarean scar. This rare type of pregnancy (referred to as ectopic even though the more common type is when the embryo implants in the Fallopian tube) means the embryo cannot survive. It is also very dangerous for the mum as it can result in severe blood loss.

**Effect on breastfeeding**

Once established, there's no difference in success with breastfeeding between those mothers who delivered by Caesarean and those who delivered vaginally. However, babies born by Caesarean tend to be less efficient at sucking. The sucking reflex is often strongest 45 minutes after birth and starts to decline after two hours. Generally following a Caesarean, there is a delay before a mum can begin breastfeeding (the length of the delay depends on the hospital's guidelines). Such delay can lead to breastfeeding being reduced during the first days, so a baby born by Caesarean is more likely to lose weight and not re-gain it in the first week. His mother is also likely to experience pain from baby lying on or near her scar, which can further curtail feeding. (Pain can be avoided by holding a baby in a different position or using pillows to cover the scar.)

Failure to gain weight is also a result of mum's breast milk being slightly altered in content and volume as a result of her Caesarean. Hormones that signal a mother's body to produce milk after delivery are at a lower level after a Caesarean than a vaginal birth (Nissen, E, et al, 1996).

Other factors that can affect milk supply are blood loss during surgery (particularly if it was substantial), the timing of the first feed post surgery, any further delay due to mum feeling groggy or being in pain and pain medications given to mum after the surgery.

Planned Caesareans also result in lower amounts of pain-relieving endorphins being released by a baby during birth and into her mum's breast milk (or colostrum, at this stage). The baby is therefore denied natural pain relief after what is a mechanically stressful experience. This can further delay the establishment of breastfeeding.

All that being said, if a mum is determined to breastfeed and is given the right support in the early days, she should be able to do so successfully.

**Preterm birth**

A baby born after 37 weeks is said to be fully mature. However, babies born by Caesarean between 37 and 39 weeks (but before any signs of natural labour), are 120 times more likely to need help with their breathing.

There is a risk with a planned Caesarean that a baby can be born before she is 'ready' and this can have very serious consequences (Madar, J, et al, 1999). An otherwise healthy baby should only be born by planned Caesarean after 39 weeks.

If a baby does have problems with her breathing, there is a good chance she will need to be separated from her mum in a special care baby unit (SCBU) in order to be properly cared for. This separation can be distressing for both mum and baby, and also can affect how quickly breastfeeding is established.

Early planned Caesareans also have been shown to increase the chances of a child having special educational needs (SEN). The earlier a baby is born, the greater the risk; effects can be seen, even if small, at a 39 weeks delivery (MacKay, DF, et al, 2010).

**Vaginal delivery prepares one for life**

There are a number of theories connecting a normal vaginal delivery with priming a baby for life outside of the womb. Babies born by Caesarean are more likely to have problems with breathing, specifically with the amount of air they can take in. This is thought to be as a result of the babies not being able to properly clear their lungs of fluid. Studies on animals have found that chemicals are released during a vaginal delivery that stop the production of this fluid and cause the lungs to begin to absorb it. So a vaginal birth results in a smaller amount of fluid needing to be cleared from the lungs, resulting in a decreased risk of breathing problems.

Babies born via Caesarean are also less able to control their body temperature in the hours after birth. After 90 minutes, a lower skin temperature is found in babies born by a planned Caesarean than those born vaginally or by an emergency Caesarean (Christensson, K, et al, 1993). Although findings do differ, some evidence shows that babies born by Caesarean have lower blood glucose levels after birth.

When babies are born they are suddenly plunged into a world of bacteria and other microorganisms from which they have previously been largely protected. In order to defend themselves from an onslaught of bugs, they must have a well-functioning immune system. Studies comparing babies born by Caesarean with those born vaginally have found that the former have less effective immune systems. They also have significantly lower blood pressure for up to three weeks afterwards. Interestingly, there doesn't appear to be any real differences until after 24 hours post-delivery when the blood pressure of babies born vaginally increases more than that of babies born by Caesarean.

There have been some suggestions that babies born by Caesarean are at greater risk of developing asthma. Unlike vaginally born babies who encounter their mother's bacteria in the birth canal and whose early exposure to bacteria helps them have a healthy immune system, those born by Caesarean don't experience the same levels or types of bacteria. An immune system not used to encountering bacteria is thought to be connected with a greater incidence of allergy. One study found that babies born by Caesarean are 20% more likely to develop asthma than those born vaginally, though the researchers felt this to be a result of a mother's breast milk (which is where babies encounter most bacteria) being altered by a Caesarean. There is still more research needed for the reasons to be conclusive (Thavagnanam S, et al, 2008). It's also not entirely clear whether the increased risk of

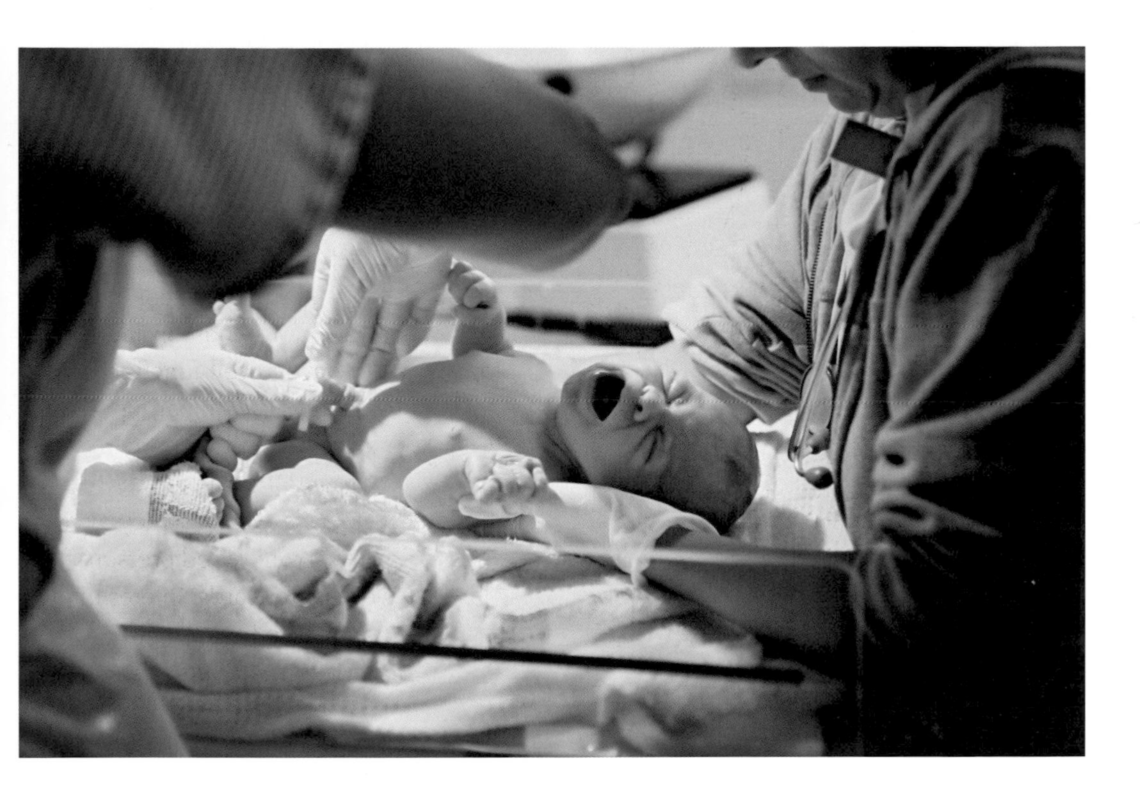

***Shock of the new***
*Having just been delivered, this infant expresses her feelings about being suddenly born into a completely different environment.*

asthma is due to problems with the immune system or with breathing problems at birth. There is also a growing bank of evidence, which suggests that if there is a family history of allergy, a baby born by Caesarean is more likely to develop a food allergy such as cow's milk intolerance. She is also more likely to develop type-I diabetes, which is also related to the immune response.

Adults who were born by Caesarean have an increased risk of obesity although researchers aren't exactly sure why the delivery method would affect weight. The situation is complicated by the fact that these differences only appear in early adulthood; there is little weight difference amongst children.

Babies born by Caesarean are less reactive a few days after birth and show fewer optimal responses to neurological tests (such as those for normal reflexes). At their eight-week immunisations, Caesarean babies cry for less time than vaginally born ones. This may sound like a good thing but it does show that responses to stress are still altered eight weeks after birth.

**Postnatal depression**

The risk of postnatal depression is slightly higher in mothers who undergo a planned Caesarean (rather than emergency procedure) compared to a vaginal birth, although the research is not conclusive. The risk diminishes after six months post delivery (Yang, SN, et al, 2011).

## Recommendations from the research

Vaginal births are safer and better for your baby's development, health and well-being. They are also better for you, despite the risk of incontinence.

Should you wish to proceed with a planned Caesarean, try to make sure it's not too early; the optimal time is at 40 weeks. If you are afraid of a vaginal birth, your healthcare provider should treat your feelings seriously. Make sure you talk everything through with him or her. You are still entitled to a Caesarean if your anxieties are not allayed.

**Greater risk of death**

The chance of either mother or baby dying as a result of a planned Caesarean is difficult to measure. This is because the statistics are usually combined with deaths from emergency procedures, which are more dangerous and occur as a result of complications. Despite this, M.H. Hall and S. Bewley (1999) found that the chances of either mum or baby dying is almost 3 times more likely for planned Caesareans than for vaginal births. That being said, the death rate for any type of delivery is very low in the UK; for planned Caesareans, it is approximately 58 per million.

**Repeat Caesareans**

Women who have had 2 or more previous Caesareans will be scheduled for another. However, those whose first birth was by Caesarean will be advised to try for a VBAC (vaginal birth after Caesarean) the second time round but also encouraged to make an informed decision. Labour following a Caesarean, particularly if there is less than a year between deliveries, carries a greater risk of the uterus rupturing during the birth, which can cause serious blood loss for mum and could result in a hysterectomy and problems for baby. I. Al-Zirqi et al (2012) found the chances are increased if the labour has to be induced. It is very slightly safer that a second birth should be by Caesarean if the first was also (12 deaths per 10,000 births compared to 35 deaths per 10,000 for a vaginal birth).

# Natural or medicated pain relief

A very small number of women (about 1 per cent) report experiencing no pain in labour; the other 99 per cent unfortunately do, albeit in varying degrees.

Societal expectations about pain also vary considerably, but at one end of the spectrum, 'working with pain' sees pain as a necessary part of a normal labour and that a woman, if given enough support, can manage it (Leap, N, et al, 2010) and at the other, 'pain relief', promotes the idea that in a developed society, it is barbaric to expect a woman to go through labour in pain.

Whether a mum decides to manage the pain or to kill it completely will have consequences for herself and often for her baby. They are important decisions and she should have all the facts.

### Why labour hurts

The uterine muscles have to work hard in labour and require adequate oxygen and nutrition to do so. if these are lacking, the muscles produce pain signals. Fear and anxiety can cause muscles to tense, which also results in pain, so pain in the first stage of labour may indicate that the body needs extra oxygen, nourishment or to be more relaxed. Just as you would change an activity if you suddenly developed pain while exercising, labour pain may be a signal to change breathing patterns, increase nourishment or relax muscles, in order to help the uterine muscles to do their job. In the later stages of labour, pain will be due to the pressure of the baby's head and body directly on the cervix but as this stage is shorter, a woman may find that natural endorphins and the knowledge her baby will be born soon will help to spur her on and enable her to bear the discomfort without medications.

## CURRENT RECOMMENDATIONS

The National Institute for Health and Care Excellence (NICE) recommends that a mother be involved in planning her care during labour and delivery and that the plan should be tailored to her needs and those of her baby. Women should also be given the opportunity to make informed decisions about their care. NICE recommends the use of the birthing pool as a method of pain relief and that relaxation techniques – breathing and massage – should be encouraged. Women should also be able to listen to music and take advantage of acupuncture and hypnosis if provided by external professionals (but not in-house healthcare workers) if they so desire. Gas and air, opioid drugs and epidurals should be available but women should be informed of all the risks associated with these pain relief methods. TENS should not be offered once labour is established.

## WHAT THE RESEARCH SAYS

### The importance of support and setting

A women's satisfaction with her birthing experience depends more on the support of her caregivers and the relationship she has with them than the amount of pain she experiences. Satisfaction is also related to her feeling of being involved in decision making and her personal expectations prior to birth. E.D. Hodnett's 2002 study found that the influence of caregivers is far more important than pain relief.

There's good evidence to suggest that having continuous support during labour leads to a shorter than normal labour and a vaginal birth free from interventions (like the use of forceps) and painkillers. Continuous support by a midwife or doula can be a form of pain relief but unfortunately, this level of support is not always available for labouring mums-to-be. Well-meaning partners, friends and close relatives are far less effective in supporting a labouring woman than a trained professional, present only for that purpose but partners can be more effective at giving support if a trained professional is on hand to guide them.

Birth supporters who are not employees of the hospital or institution where a birth takes place are more effective. Although midwife-led units and the support they provide are beneficial to labouring women, midwives are not always continuously available. It's worth noting, however, that support from a partner or friend, with or without a professional on hand, is still effective at improving a woman's birth experience (Hodnett, ED, 2011).

In addition to support, the setting also plays a key role in coping with labour. Privacy and protection from disturbances can help a woman's body to produce endorphins (chemicals released by the body in response to pain or stress), which provide pain relief in a way similar to that of opioid drugs. This is backed up by observational studies that find women report having less painful home deliveries when compared to hospital births (Leap,N, et al, 2010). Birth centres, either within a hospital or stand-alone, also aim to provide privacy and protection from disturbances as well as some home comforts, such as comfortable furnishings and music players.

Stages of labour

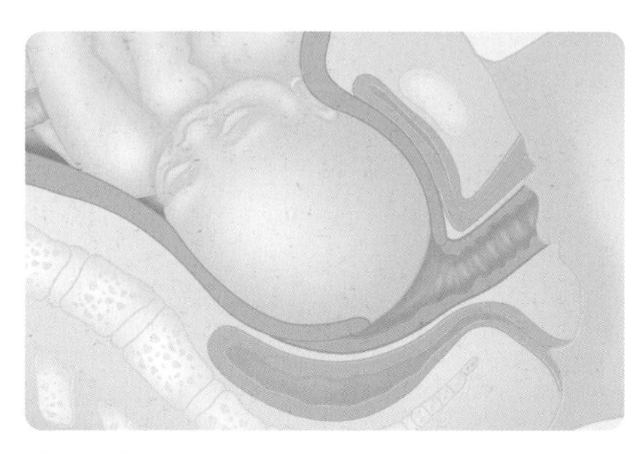

**The cervix starts to open**

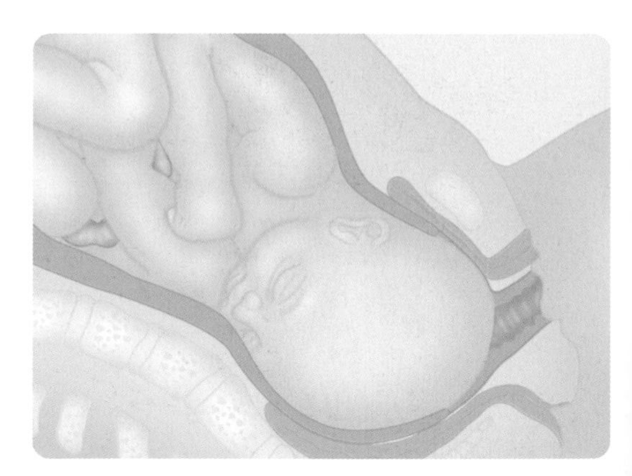

**At full dilation**

## The importance of preparation

Educating oneself about labour, knowing what is happening and what pain management is available, will increase the satisfaction of childbirth (Henry, A, et al, 2004). K. Crowe (1989) demonstrated that women who attend birth preparation classes tend to be less anxious and experience less pain in labour, and A. Mehdizadeh et al (2005) found they are less likely to have a Caesarean delivery. Women who describe themselves as anxious particularly benefit from childbirth preparation classes.

Whilst there is a lot of evidence to suggest that being prepared for labour and delivery can decrease anxiety and pain levels, women who attend antenatal training are more likely to have an epidural because they become better informed about the pain relief available (Fabian, HM, et al, 2005).

## Beneficial positions

There is evidence that adopting an upright position during labour – sitting, standing, squatting, kneeling or using birthing stools or balls – can shorten both the first and second stages, although the research is less clear cut for the second stage. Position (particularly the use of birthing stools), however, doesn't seem to affect the need for

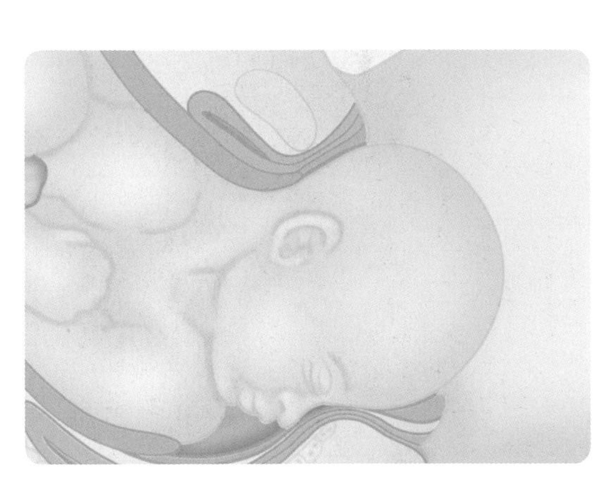

**Your baby passes down the vagina**

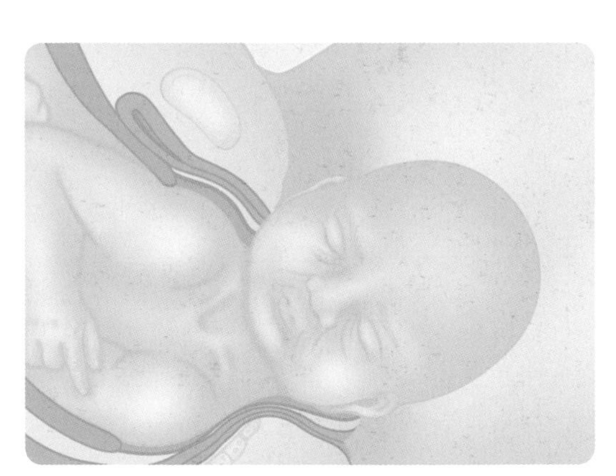

**The head is born**

interventions like forceps or a Caesarean, which can arise if a woman has been pushing for a while and becomes exhausted (baby being in a difficult position, for example) or if a baby isn't coping well with labour. There is still some controversy surrounding the best position for the second stage because being upright can result in greater blood loss. However, evidence suggests that a woman should choose any position in which she is most comfortable (Lawrence, A, et al, 2009 and Gupta, JK, et al, 2012).

The perineum is most likely to tear if a woman is in the lithotomy position – her legs in the air on stirrups. A tear is also more common if a woman is lying on her back. The best position for avoiding a tear is lying on the side, although fewer tears are also experienced by women who squat. It doesn't seem to matter to a baby in which position a woman chooses to give birth; babies tend to be fine whichever.

**Hypnosis**

A prepared mother can perform self-hypnosis if it has been taught to her as part of her preparation for labour. She will be fully aware of what is going on but will be very relaxed. Unfortunately, there hasn't really been enough research to reliably say whether hypnosis is helpful. Of the research that has been conducted, there's no evidence to suggest hypnosis reduces pain or the need for an assisted birth or satisfaction with the childbirth experience.

**Biofeedback**

This is another method that must be learned from a practitioner. It works on the principle that by controlling your thoughts and emotions, you can affect how your body functions – your heart rate, blood pressure, muscular tension and body temperature, etc. A practitioner would use various pieces of electrical equipment such as an electromyograph (to measure muscle tension) and a skin temperature gauge, and offer relaxation techniques in an attempt to alleviate pain associated with muscle tension. Similar to hypnosis, there's not enough research to say whether it is useful but of the research that has been done would lean towards the process having little or no effect.

**Water birth**

This is when a mum uses a larger than normal tub or small pool to immerse her lower half in water during labour. It is thought to be beneficial for a number of reasons. Water makes it easier for a woman to adopt the best position for birth, and as the water is warm, it can ease the pain of contractions, reduce blood pressure and encourage relaxation, which leads to the release of endorphins. It can also result in a woman feeling more in control, which limits the fear factor. Less fear means more oxytocin is produced by the body and contractions become more effective.

Labouring women in a tub have been shown to experience less pain than those not in water and when birth is all over, are found to be more satisfied with the whole experience, but this is only true of the second stage of labour. The evidence is less clear on the effects for the first or third stages, although women are less likely to use other methods of pain relief (such as anaesthesia) during the first stage. There is also evidence that the first stage of labour is quicker when in water. Women who use a bath have lower blood pressure.

**Aromatherapy**

Essential oils from plants can be massaged into the skin or inhaled with steam or from a burner. They are thought to improve mood and anxiety levels, and have been shown to increase the production of the body's neurotransmitters, chemicals that are sedative, stimulating or relaxing.

There isn't a lot of evidence (nor is the evidence sufficiently conclusive), to say that aromatherapy is effective at relieving pain, increasing birthing satisfaction, or decreasing the need for interventions. It might make the room smell nice though.

**Relaxation techniques**

A variety of techniques – breathing, imagery, yoga, progressive muscle relaxation, audio analgesia — exist for women to use in labour; it doesn't appear to matter which is employed as simply doing something relaxing appears to bring benefits including a reduced chance of needing an assisted birth (i.e. with forceps) but the research isn't conclusive, as yet. While there isn't yet enough research to say how relaxation techniques compare to other methods of pain relief, they can certainly be part of a woman's arsenal of options. Of the following techniques, controlled breathing and yoga have been shown to reduce pain and increase the satisfaction of the birth experience.

Controlled breathing techniques are designed to encourage relaxation in response to labour pain. More efficient breathing can also improve oxygen levels in the blood (better for mum and baby) and can interfere with how the body transmits pain signals. To work effectively, these techniques need to be practised before delivery.

Guided imagery is when a woman is encouraged to remember an enjoyable and relaxing experience. It is thought to affect pain felt in labour by reducing stress. Woman can also be encouraged to imagine pain felt in the body being replaced with either hot or cold sensations.

Yoga incorporates both breathing and visualisation as well as physical postures and meditation. The different postures encourage strength, flexibility and balance and coordinated imagery and breathing.

Progressive muscle relaxation is when a woman tenses and then relaxes groups of muscles in her body in order to release tension within the muscles.

Audio analgesia is the use of white noise or music to reduce pain. It most likely works by distraction but has been found to be quite effective during dental procedures on children (Baghdadi, ZD, 2000). Unfortunately, there's no evidence to suggest that it's effective at reducing labour pain but it may be effective if used in conjunction with other techniques.

**Acupuncture**

An ancient Chinese practice, this consists of inserting fine needles into particular points on the body (ears and lower back, for example) to alleviate pain. How this works is not well understood. One theory is that it blocks pain through the spinal cord; another is that it encourages the release of pain-killing endorphins.

Although findings are conflicting, there is some evidence to suggest that acupuncture and the related therapy of acupressure or shiatsu (a pain-relief technique whereby hand and finger pressure is applied to meridian points) are effective at alleviating labour pain. They have also been shown to reduce the need for interventions such as an assisted delivery. Unfortunately, the research isn't of a sufficiently high quality to produce firm conclusions and to add to the difficulties, there are different styles of acupuncture.

### Massage

Massage has been shown to help women experiencing backache during the first stage of labour, which can occur if a baby isn't positioned properly or is lying with his back against his mother's back. It's thought that massage relieves pain by encouraging relaxation and inhibiting the transmission of pain signals. It has little effect in the second and third stages of labour nor on whether a Caesarean or forceps delivery will be needed or on the well-being of the baby, when he arrives.

### Reflexology

According to practitioners, reflex points on the feet correspond to the organs and other body parts so that applying pressure to a specific part of the foot will dull pain experienced in another part of the body.

There is some evidence to suggest that reflexology can decrease the length of labour and the pain experienced during it. There is, however, some doubt about the calibre of these studies and further high-quality research is needed to prove whether reflexology is effective.

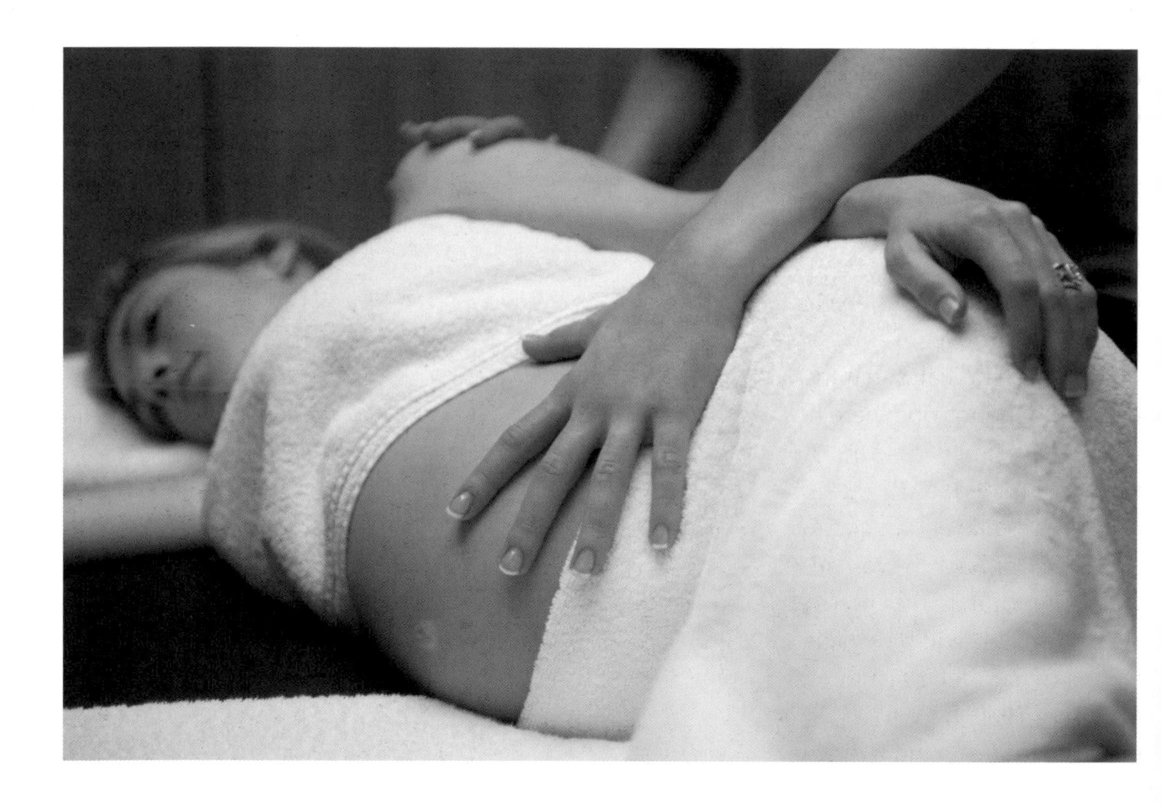

**TENS**

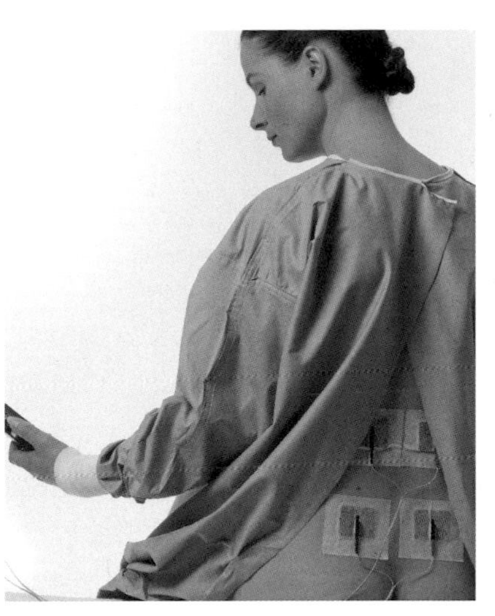

Transcutaneous electrical nerve stimulation depends on a machine, which has wires with pads that are stuck to a woman's back, acupuncture points or head (but only by trained professionals) during labour. The TENS machine delivers low-voltage electrical impulses to the skin and while nobody really knows exactly how it works, two theories are that it interferes with how pain is transmitted along the central nervous system or that it encourages the body to release endorphins. Another theory is that because the timing and intensity of the impulses can be altered by the mother, it gives her a sense of control because she sets the duration and intensity of the impulses. This feeling of being in control and the general distraction of the device reduce anxiety and therefore, pain. It's also thought that TENS may shorten the length of labour because it suppresses the release of chemicals that inhibit the womb contracting. Or, it could be a combination of all of the above.

Research indicates that the effectiveness of TENS depends on where the pads are attached. If simply attached anywhere on the back there doesn't appear to be any pain relief but if attached to acupuncture points on the lower back, a woman may experience a marked improvement in symptoms and will be satisfied with it as a means of pain relief. Unfortunately, attaching TENS pads to acupuncture points usually requires a professional being on hand to do so. Moreover, when compared to an epidural (see below), TENS was rated largely ineffective. It is a pain management technique not a pain reliever.

An interesting aspect of the TENS research was that even when attached to the back and reported as not being very effective at pain relief, roughly 66% of women reported that they'd use it again in a future labour. What was even more interesting was that 40% of women who had been given an inactive device (a control) reported they, too, would use it during a future labour. Pain in labour, therefore, has a psychological as well as a physical component. It perhaps shows that mums-to-be, whilst in labour, need distraction and to feel in control.

Beyond pain relief, TENS doesn't appear to have any impact on whether an intervention will be needed and research is yet to be conducted on whether using TENS can delay the time at which a woman goes to hospital. Delaying admission to hospital means that interventions will be less likely and stronger pain relief not needed. TENS has been shown to be quite safe for mum and baby (as long as it's used correctly.

**Nitrous oxide**

Nitrous oxide or 'gas and air' is the most common and safest form of inhaled analgesic. A woman in labour can breathe in the mixture when she feels a contraction coming on. Scientists are a little unsure as to why gas and air works, but it's thought to

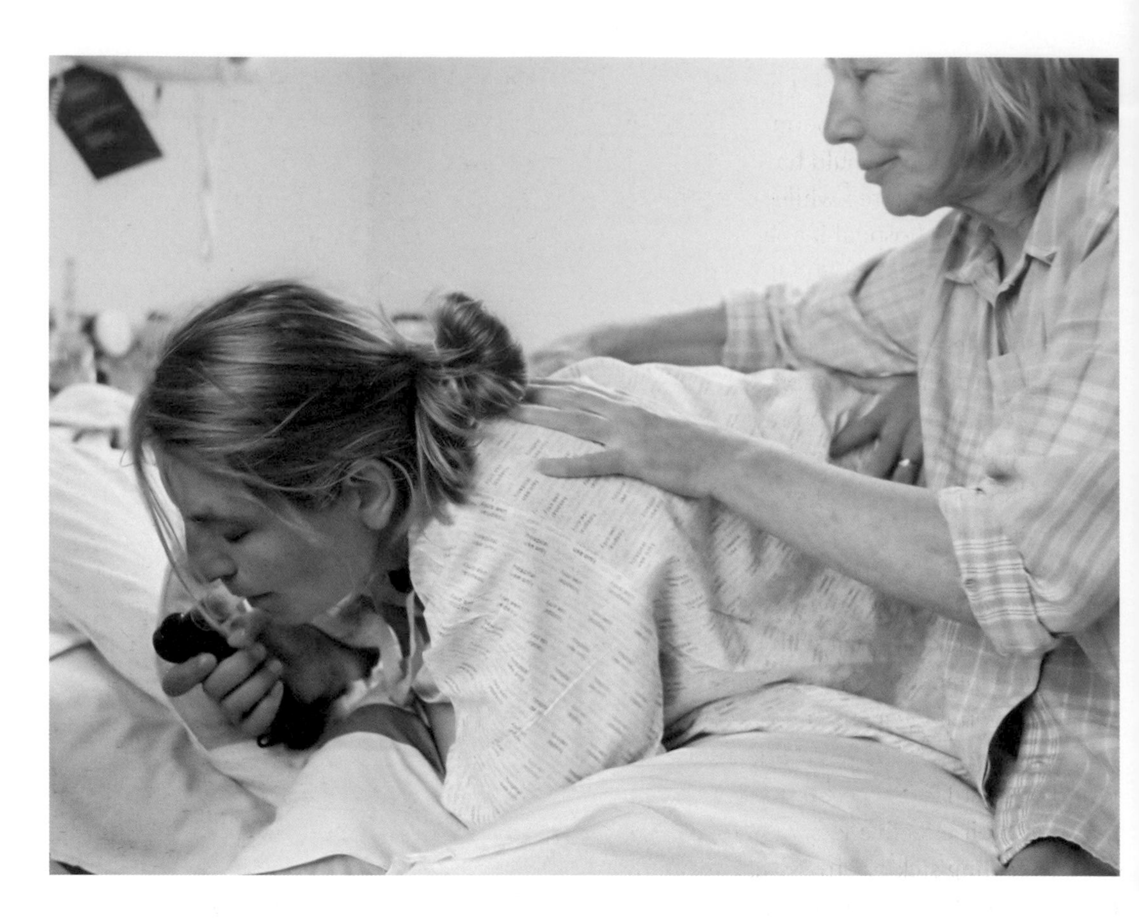

affect how the brain perceives pain. Although effective at relieving labour pain, common side effects include nausea, dizziness and drowsiness. The drug does reach the baby but because it is rapidly exhaled by a newborn, it is not considered to be dangerous and hasn't been shown to affect, in any way, a newborn's health (as demonstrated by Apgar scores).

However, gas and air should not be inhaled continuously (G. Ahlborg, Jr, et al [1996] found that health professionals who have a lot of exposure to nitrous oxide suffer from higher rates of infertility and miscarriage. Continuous use can also cause hallucinations, vomiting and hyperventilation and in extreme cases, a baby who has been deprived of oxygen.

**Pethidine, diamorphine and meptazonal (Meptid)**

These are opiate drugs derived from the poppy and synthetic versions of these drugs are referred to as opioids. They are generally injected by a midwife during labour to effectively dull the pain of labour and can be used in conjunction with gas and air. Concerns about these drugs include the inability of a mother to make decisions during her labour, vomiting or nausea and problems with breathing. A. Lawrence et al (2009) showed it can also make a woman feel very drowsy and make her want to stay lying down, which can make her labour longer and her contractions more painful. Opiates are also easily absorbed by a baby, and they can cause problems with his heart rate whilst in the womb, and once born, with breathing and hypothermia. This could mean that a woman would have to be continuously monitored while in labour or be transferred to hospital if at home. Also, M.I. Hogg et al (1977) found these drugs stay in a baby's system for up to six days making him less efficient at suckling (which delays the establishment of breastfeeding and reduces its success) and less alert.

Two-thirds of women find that even after an injection of an opiate drug, they still experience moderate-to-severe pain. However, when compared to a placebo, they are effective at relieving pain. More women feel sick or vomit with pethidine than with a placebo (and even more with meptid), even if given an anti-sickness drug.

Interestingly when opioid drugs were tested against TENS, no real difference was found in their effectiveness, other than more women feeling sick or drowsy using the opiate drugs. Researchers compared the two methods on things like satisfaction with pain relief, pain scores and the need for interventions. However, it is unclear from the research whether other pain relief methods were employed, such as gas and air. This finding may make sense if both groups (opioid drugs and TENS users), also inhaled gas and air.

**Paracetamol and other non-opioid drugs**

Non-opiate tablets can relieve mild-to-moderate pain in the early stages of labour and, as long as they're not taken for long periods of time, are very safe. They are, however, limited in their effectiveness and won't do much for severe pain. While they don't have the same effect as opioids like pethidine, they will do something.

**Epidurals**

An epidural involves injecting the lower back with a local anaesthetic to numb the area then inserting a tube into the spine through which anaesthesia is continuously fed. The anaesthetic is often combined with an opioid drug. Traditional epidurals consist of high dose anaesthetic, which blocks all sensations, pain included. This means a woman won't feel contractions, be able to move around or know when it's time to push her baby out.

An epidural with a lower-dose anaesthetic, known as a mobile epidural, will partially block pain but should allow a woman to move. She can be comfortable but still feel contractions, the urge to push and encourage labour to progress by changing position. (Lower-dose epidurals may not be available in all hospitals).

Another type is the combined spinal-epidural (CSE), which combines a spinal anaesthetic and a low-dose epidural.

The differences between spinal anaesthetics and epidurals are to do with the type of drugs used and where, precisely, they are injected. Spinals are a one-shot only injection of low-dose anaesthetic whereas epidurals are topped up when needed. Spinal analgesia is fast working and effective but because it is short lived, it can require a low-dose epidural to be used in conjunction with it. An epidural by itself can take up to 10 minutes to start working. The advantage of a CSE is that it is fast acting and effective but still allows for movement, sensations like contractions and the urge to push, and less of the anaesthetic is transferred to baby. In studies looking at maternal satisfaction with epidurals, the CSE was a favourite.

Traditional epidurals have been shown to prolong the second stage of labour and are much more likely to result in an assisted delivery because mum can't push as effectively. Other side effects include low blood pressure, fever, problems with breathing and heart rate, shivering, retention of urine and tinnitus (a ringing in the ears). Generally, however, research has shown that epidurals are not likely to increase the chances of a Caesarean being necessary.

In approximately 1% of epidurals, the needle accidentally enters the intrathecal space in the spine, which can cause a post-dural puncture headache or PDPH. This severe headache typically gets worse when the mother tries to sit or stand upright. Many women (and their babies) will need to remain in hospital for an extended time until the headache eases or is treated.

Low-dose epidurals that include an opioid drug will have some of the same side effects as those listed for pethidine (itching, problems breathing, poorer establishment of breastfeeding) as well as some of those associated with traditional epidurals (low blood pressure, PDPH), but they enable a woman to push better, because she is still able to feel contractions, and is less likely to need an assisted delivery. The CSE, too, also has some of pethidine's disadvantages (itching, problems breathing, poorer establishment of breastfeeding) although itchiness is more common amongst women who have had the CSE compared to the low-dose epidural. Except for the CSE being faster acting, women haven't found much difference between it and a low-dose epidural in effectiveness for pain relief. But compared to the traditional epidural, where sensation is totally blocked, the CSE is less likely to result in an assisted delivery because a mum still retains the ability to know when to push.

Rare, but still worth a mention, are the extreme effects of any epidural. These include infection from the injection, such as meningitis; damage to the nerve roots causing pins and needles or weakness; or accidental injection into a vein, which can cause convulsions and the need for resuscitation and urgent delivery. The risk of these is higher for the CSE because there are two needles entering the spine. There's also a higher risk of a baby needing to be resuscitated and of having a slower than normal heart beat.

Despite the adverse effects there is considerable evidence to show that epidurals are highly effective in managing labour pain.

**General anaesthetics**

Sometimes, when a woman undergoes a Caesarean, she may need to be put under completely. This is usually due to pre-existing health problems or to complications in pregnancy or labour. No differences exist between general or regional (epidural) anaesthesia in the effects on a baby's health when born or the need for resuscitation (Afolabi, BB, and Lesi, FE, 2012).

## Recommendations from the research

Consider wisely the setting where you choose to have your baby and who you have to support you through it. These can make a big difference to how you manage pain during labour. It may also be prudent to encourage your birthing partner to get involved beforehand in order to be prepared. A doula, if you can afford one, is a good idea as is taking an antenatal class that focuses on labour and pain relief. If nothing else, the added confidence you may feel can alleviate anxiety and lessen pain in labour.

Aim to be mobile and upright for at least the first stage of labour and if you have to lie down, make sure it's on your side. Do some research into relaxation techniques and alternative methods of pain relief, perhaps taking a yoga class and practising deep breathing. Put some thought into what makes you relax, and plan to have any necessary items nearby during your labour. At the very least, they will serve as useful distractions.

If you like the idea of being in control, try using a TENS machine (borrowed or rented from where you intend to give birth or bought relatively inexpensively).

If your midwife suggests using paracetamol, changing your position, going for a walk or taking a bath in the early stages, follow her advice even if you think it won't do any good! If you can, ask for a water birth. Gas and air, if it doesn't make you sick, seems to be the best painkiller because it has the fewest side effects for you and baby. Carefully consider the other pain-relief options and make sure you're aware of the possible consequences. Only you and your partner can make the decisions for what's best for you and baby. Above all, have a well-thought-out birth plan, and get it written down and circulated well before you start labour.

# Spacing pregnancies

Many things – age, fertility, economics, etc. – influence when parents start trying for another child but what is best for you and your existing child(ren) should be top of the list. Research can lend a helping hand in coming to the right decision.

## CURRENT RECOMMENDATIONS

The World Health Organisation recommends that for all mothers, the minimum time between a birth and attempting to conceive another child should be 24 months. This means that there would be an age difference of at least 33 months (2 years, 9 months) between children. The reason for this recommendation is largely due to the recovery time needed to restore a woman's body to a healthy enough state to support another baby.

## WHAT THE RESEARCH SAYS

### Health issues

It was once believed that a fetus acts like a parasite taking whatever nutrients he needs whether his mum can spare them or not. This is not strictly true as studies have now shown that when mum is mildly restricted in nutrients, baby will be too. Studies based on famine conditions show that if a mum loses 3% of her body weight during a pregnancy, her baby will be, on average, 10% smaller than he should be, and that when a mum is low in nutrients, for example protein, it will not automatically be given to her baby but instead the little that is available will be shared between them. It is only when there is a severe lack of nutrients that a baby will be favoured over his mum but, even then, only to a degree.

Being pregnant requires large maternal resources so if a woman chooses to have births very close together, there is a chance that she can deplete her resources, especially if she breastfeeds as well. Maternal health directly affects that of the unborn baby and when a mum's resources are low, this can result in complications, which can be severe and occasionally fatal. Following a pregnancy, women tend to have low levels of many nutrients but most return to normal quickly afterwards. Iron and folate, however, require more time to recover. If a woman falls pregnant again quickly, she can become deficient in these important nutrients, and her baby could have a lower-than-expected or low birth weight (less than 2500g), be preterm or suffer from neural tube defects and/or schizophrenia (King, JC, 2003 and Gunawardana, L, et al, 2011). Low birth weights and prematurity lead to an array of physical and mental problems for a child, including restricted growth and insufficient neurodevelopment.

E. Fuentes-Afflick and N.A. Hessol (2000) found that women who allowed only 8 months between pregnancies were up to 47% more likely to have premature babies than women who left a gap of 18 months. Other studies have shown that gaps of about

6 months increase the chances of having not just a preterm birth but also lower-than-expected birth weights and low birth weights by up to 50% compared to 18 months or more between pregnancies.

It would appear from a number of studies on this subject that a 'safe' time for both a mum's and baby's health, is not to start trying until 18 months after a birth (Nabukera, SK, et al, 2008 and Conde-Agudelo, et al, 2006) and there is the lowest risk of complications if one waits to conceive between 18 and 23 months after a previous birth (Zhu, BP, 2005).

However, leaving too long – more than 6 years between births – also puts a pregnancy and birth at an increased risk. The reason for this is not entirely understood but the main theory is that a woman's body returns to its pre-pregnancy state after six years and her risks in having another child become similar to that with her first. First births are usually riskier than subsequent births, provided the later births happen within a reasonable time frame.

Longer intervals between pregnancies also mean that a woman may approach an age when either attempting a pregnancy can be more difficult or the birth becomes complicated (more likely to produce a baby who is preterm or of a low birth weight. S.K. Nabukera et al found that for women older than 35, the lowest risk was when they tried for another baby between 12 and 17 months after a previous birth (2008). This is sooner than the 18 months recommended for younger women, and is based on the increased risks of conceiving for the over 35s outweighing those associated with insufficient maternal resources.

Taking supplements between pregnancies can replenish a woman's store of vitamins and minerals, making her better prepared for another healthy pregnancy. B. Caan et al (1987) compared women who took supplements for up to 2 months with those who took them for up to 7 months. The latter had healthier (larger weight) babies. The study did not specifically look at shorter times between pregnancies.

With Caesarean deliveries, D.M. Stamilio et al (2007) concluded that a short inter-pregnancy interval can increase a woman's risk of uterine rupture if she attempts a vaginal birth afterwards (VBAC). If the previous delivery was by Caesarean delivery, the minimum recommended gap is 1 year, 9 months (i.e. waiting a full year before trying for another baby), but the longer the scar is allowed to heal, the less likely there are to be complications with a future vaginal delivery. According to this study, waiting less than 6 months before trying again will triple your chances of uterine rupture during VBAC.

**Sibling relationships**

There's quite a lot of evidence to show that older children regress – have more frequent mishaps if toilet trained and/or return to bottle feeding – when a new baby is born. The effects are greatest if the gap between siblings is between 1 and 4 years.

Separation anxiety (a normal developmental stage during which a toddler experiences anxiety if separated from his primary caregiver) is greatest when a toddler is between 15 and 30 months old, so the introduction of a new baby at this time (i.e. becoming pregnant between 6 and 21 months after the birth of the first) makes the regression more pronounced. Before the age of 1 or beyond the age of 4, a child is better able to cope with a new sibling. This is because a baby has a limited capacity to remember the increased contact time with either parent and a 4 year old is more secure in his identity separate to his mum; he is likely to have his own routine, friends and space. To avoid regression, therefore, you should plan to be pregnant roughly 3 months or 3¾ years after a previous birth.

Children who are close in age can be very good at playing with each other as they share similar interests and abilities but they are also more likely to become jealous and fight. Jealousy is more pronounced when children are less than 3 years apart and sisters, apparently, can be worse (Koch, H, 1960). Common interests lead to competition and a child can become insecure if he constantly does less well than his sibling. While the most competition occurs when children are spaced between 15 and 22 months apart, things don't necessarily improve as siblings age; 5 to 16 year olds who are only 2 years apart, tend to have more negative feelings towards each other.

Even though a lack of shared interests means siblings with a larger age gap play less often together, they do appear to enjoy each other more. Those who are 4 or more years apart tend to accept help from each other more readily (usually the younger from the elder,) and be more willing to teach and praise each other. M.E. Wagner et al (1985) found that children who are more widely spaced also tend to have better communication.

## IQ

There is some evidence that first-born children tend to have higher IQs than their siblings. One theory is that this is due to the extra time parents spend with this child before the arrival of any siblings. Whether siblings can affect IQ is a matter of great debate (Black, SE, et al, 2007 and Kanazawa, S, 2012). J.L. Rodgers et al believe it is simply a trick of statistics (2000). In the research that found a difference in IQs amongst siblings, gaps of less than 3 or 4 years between children result in lower IQ scores for both, and the second child usually has a lower IQ than the first. Vocabulary also appears to be affected; the larger the gap between siblings, the better the vocabulary of the younger child. K.S. Buckles et al (2011) showed that reading ability, too, is better for the first child if the second is not born until at least 2 years later; it gets better the greater the gap.

Most research shows that only verbal abilities are affected; mathematical ability is largely not. Moreover, the effects of spacing on IQ and vocabulary are more noticeable if the siblings are boys. Gaps of at least 24 months between children appear to have advantages over shorter gaps, but when children are 4 years apart, the older sibling has no effect on the younger's IQ or vocabulary (Wagner, ME, et al, 1985). So, in terms of IQ, it's better to leave at least 2

## Recommendations from the research

In terms of health, an issue that many would see as the most important, it is best to wait at least 18 months before trying for another baby (giving a gap of at least 27 months) but to also not wait longer than 6 years. If you're over 35, however, it might be best to start trying after only 12 months. Every woman should consider post-natal vitamins after a birth particularly if she wants to conceive again sooner than 18 months and is breastfeeding.

In terms of how well siblings will get along and become secure, well-adjusted and confident children, there is an argument to say it's best to have them at least 4 years apart. Siblings who are 4 or more years apart tend to have better relationships with each other, although they probably won't play well together. When older, they are likely to appreciate each other more.

If IQ is affected by having siblings, then a bigger gap is also better for both children's IQ as well as vocabulary and reading ability.

In terms of ease for parents, the best is a 4-year gap or more, followed by a small gap of less than 18 months; the most challenging gap is one of 18 months to 4 years. While gaps of less than 18 months or over 4 years are easiest for parents, an 18-month gap between children is not good for health reasons, so it is better to opt for a 4-year gap.

years between children and the longer the gap, the better.

**Effect on parents**

Based on observational studies of parents and how well they coped with young children, M. E. Wagner et al (1985) reported that it was easier to care for children if there was either a gap of less than 18 months or one of more than 4 years. If the gap is small, a parent can treat and speak to both babies similarly (though this can stunt the older child's development), but if the older child is between 18 months and 4 years of age, he is too old to be treated similarly to his younger sibling but is also not capable of caring for himself. Beyond 4 years, a child is more able to look after himself.

Not surprisingly, parents who have children close in age are more tired at the end of the day and have less time for their spouses and housework! However, parents who are able to adjust their desires to their reality, in that they are able to match the number and spacing of their children to their desires, tend to have greater marital success regardless of how close the children are in age (Christensen, HT, 1968).

# PART 2
# PARENTING ISSUES

# Breastfeeding

Every mum is faced with the decision of whether to breastfeed or not, and her decision will be heavily influenced by her background, knowledge base and experiences. If you choose to breastfeed, further influences will determine whether you will be successful. The following section aims to improve your knowledge base – only one of the many influences affecting your decision.

## CURRENT RECOMMENDATIONS

Both the World Health Organisation (WHO) and the U.K. Department of Health state that breast milk is the best form of nutrition for infants. Babies should be exclusively breastfed for the first six months of life and breastfeeding should continue beyond six months while other foods are introduced.

## WHAT THE RESEARCH SAYS

**The optimal food for a developing baby**

Breast milk contains the appropriate amounts of nutrients, such as easily digested proteins, fats, carbohydrates, vitamins and minerals, a baby needs to develop. Moreover, as a baby grows, breast milk's composition changes to meet his needs. For example, two key proteins – whey and casein – change over time. Whey protein is digested and absorbed quicker than casein, so is better suited to an infant but his tummy will feel empty faster, whereas casein is digested more slowly, so it allows a baby to feel fuller for longer. At first, whey proteins are far more prevalent in breast milk, but then casein levels increase until a 50:50 ratio is reached much later on.

Formula milk, however, can't manage this change as smoothly. Formula milks for young babies have more whey than casein and designated 'hungry milk', has a lot more casein (up to 80:20 ratio for casein:whey).

Both breast and formula milk contain minerals such as calcium, iron and zinc but the proteins in breast milk are thought to make them more easily absorbable than is possible with formula milk. There is also some evidence that these proteins facilitate the absorption of vitamins such as $B_{12}$.

Some breast milk proteins have been found to protect against a range of 'disease-causing bacteria', viruses and fungi; others have been found to stimulate the growth of 'friendly' bacteria in the intestines of babies and to help develop effective immune systems, which are associated with fewer childhood illnesses and allergies.

Long chain polyunsaturated fatty acids, also known as omega-3 fatty acids, have been shown to boost brain development. Breast milk is a good source of these fatty acids but they are also added to formula. Other constituents of breast milk may potentially help in the proper growth and development of the lining of a baby's intestine (Lönnerdal B, 2003).

There is a difference in the pattern of weight gain between breast- and formula-fed babies. Breastfed babies put weight on quickly during the first 2 months but then more slowly during the rest of their first year. Toddlers who have been breastfed, therefore, tend to be leaner. Slower weight gain later in infancy is considered to be healthier because R. W. Leunissen, et al, (2009) have shown that fast weight gain during this period can increase the chances of a child becoming obese in later life or of developing type 2 diabetes or heart disease.

**Protection against various maladies**

***Upset stomachs:*** Evidence suggests that babies who are breastfed exclusively for at least three months are much less likely to develop a gastrointestinal tract infection – even in countries where there is ready access to clean water supplies and sanitation (Kramer, MS, et al, 2001). Babies who are breastfed for at least six months are even less likely to develop one and those breastfed for a shorter time (13 weeks) are still protected from stomach upsets beyond the period of breastfeeding.

***Constipation:*** The chances of a young baby becoming constipated are also decreased if he is breastfed as opposed to being formula fed (Savino, F & Lupica, MM, 2006).

***Allergies:*** Some studies show that breastfeeding can protect against the development of eczema, and children who are exclusively breastfed for at least their first three months are much less likely to develop eczema in their first year of life (Kramer, MS, et al, 2001).

There are a number of theories as to why breastfeeding might protect against allergies (although protection against asthma and hayfever is yet unproven). Firstly, breast milk contains traces of proteins from the foods a mum eats so her baby becomes tolerant of these proteins and limits his chances of becoming allergic to them. Other theories are that breast milk encourages the normal development of the gut, promotes the right balance of 'friendly bacteria' and helps the immune system to develop appropriately.

However, some studies have found that breastfeeding actually increases the risk of certain allergies (Wegienka G, et al, 2006). The chances of a young child developing atopic dermatitis are increased by breastfeeding and the longer he is breastfed, the greater the risk (Purvis, DJ, et al, 2005). .

Unfortunately, due to the nature of breastfeeding, and the many influences on a woman's choice to breastfeed or not, the research cannot be of top quality (i.e. a randomised controlled trial or RCT). One particular investigation (Kramer, MS, et al, 2007), which made a good attempt at being high quality (a strict RCT is more or less impossible for breastfeeding trials mainly because of the ethics involved), found no real difference in breastfeeding and either increased or decreased risks, but this trial was carried out in Eastern Europe where allergy rates are lower.

There is some evidence that breastfeeding provides protection against the development of coeliac disease (intolerance to gluten). In one study, groups of breastfed and formula-fed children were introduced to gluten at the same time during weaning but fewer of the former group developed coeliac disease (Kelly, DA, et al, 1989).

***Infections:*** Breastfeeding has been shown to protect against infections of mucosal membranes, the tissues that line certain areas of the body including the lungs. Exclusive breastfeeding for at least 15 weeks protects against respiratory illnesses (partial breastfeeding with supplementary formula was nowhere near as protective [Wilson, AC, et al, 1998]). However, due the complexities of the immune system, there is also good evidence that breastfeeding protects against other infections; Mårild, S, et al (2004) found evidence that breastfeeding also protects against urinary tract infections in children aged 1–6.

***High blood pressure, obesity, diabetes, cancer and inflammatory bowel disease:*** There is some evidence to suggest that breastfeeding plays a part in preventing high blood pressure in adults (Fewtrell, MS, 2003) but there is really only very weak evidence that breastfeeding reduces BMI in later life (Owen, CG et al, 2005). In terms of any long-term protection against high blood pressure and obesity, there does not appear to be any difference when a baby is exclusively breastfed for three or six months – the protection and benefits are much the same (Kramer, MS, et al, 2009). Children who have been breastfed, however, are less likely to develop type I or type II diabetes (Gouveri, E, et al, 2011).

Breastfeeding may offer a very small amount of protection from certain types of cancer (such as leukaemia) in young children and it looks as though the longer breastfeeding continues, the greater the protection (Martin, RM, et al, 2005).

There also seems to be a decreased risk of developing an inflammatory bowel disease such as Crohn's disease or ulcerative colitis in later life, if a baby is breastfed for more than six months; but this association is still debated (Hansen, TS, et al, 2011).

***SIDS:*** Babies who are breastfed have a lower risk of sudden infant feath syndrome (Moon RY and Fu, L, 2012). This clear association could be caused by a number of factors, such as more protection from illnesses so less chance of breathing difficulties, and better development of the muscles involved with breathing and swallowing. It is also possible that breastfeeding mums are more likely to keep their babies in the same room as them for longer, which has also been proved to reduce the risk of SIDS.

**Cardiovascular disease in later life**

A study conducted in 2001 (Leeson, CPM, et al) reported a connection with breastfeeding and artery stiffness in adults who had been breastfed as babies. The longer they had been breastfed, the stiffer

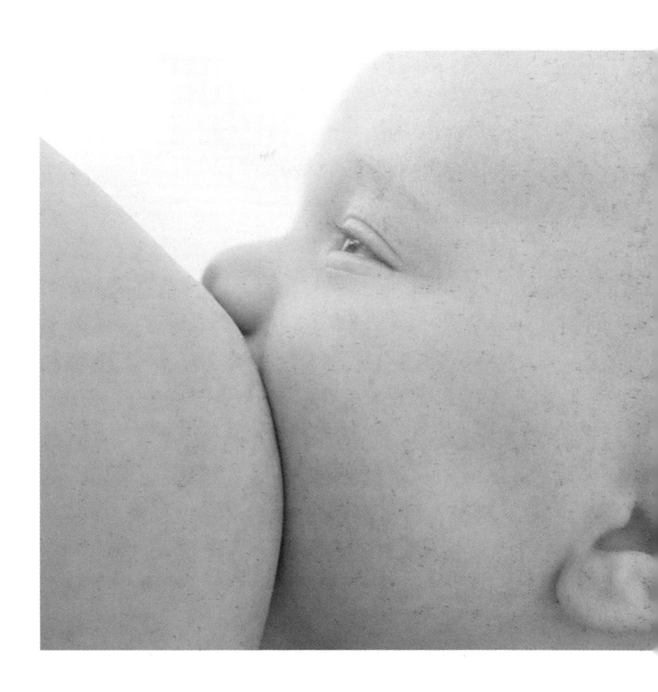

the arteries. Arterial stiffness, which is likely to be a marker of vascular disease, is further affected by high blood pressure, cholesterol levels and BMI. The initial study wss backed up by another one that showed men who were breastfed for one year were more likely to develop heart disease 60 years later (Fall, CH et al, 1992).

A breastfeeding baby has a higher concentration of cholesterol in his blood than a formula-fed baby but adults who have been breastfed have very slightly less cholesterol than those who were formula fed. The question is whether these levels of cholesterol in early life during a period of rapid development programme the body to be at higher risk of cardiovascular problems later in life. The higher levels of cholesterol found in babies may affect the development of their artery walls, and if breastfeeding continues for over a year, a higher proportion of cholesterol may remain in the artery walls. This could be further exacerbated by a western diet with its higher amounts of saturated fats. It would appear that if a baby is breastfed for longer, it could impact upon his cardiovascular health later in life.

This area of research is complicated and unfortunately no firm conclusions can be made based on the evidence so far. This is because of what is referred to as a dose-response – the longer a baby breastfeeds has an impact on the results, as does whether he is exclusively breastfed or at what age solids are introduced. The studies also need to take into account what exactly was in the formula milk (Leon, DA and Ronalds, G, 2009).

**Increased IQ**

There's reasonably good evidence to suggest that breastfeeding can influence a child's cognitive development. One particular study (Kramer, MS, et al, 2008) looked at IQ levels for six and a half year olds and their teacher's evaluations of their reading, writing and mathematical abilities. Children who were breastfed exclusively, even if only for a relatively short time (e.g. 12 weeks), had higher IQ scores and were found to be more competent in reading, writing and mathematics. The longer children were breastfed, the higher the IQ (Jedrychowski W, et al 2012).

Despite these findings, the researchers are still unclear as to why breastfed children have greater cognitive development. It could be due to the long-chain polyunsaturated fatty acids in mum's milk, which are thought to boost brain development. However, studies that looked at formula-fed children who were supplemented with the same fatty acids weren't able to confirm that they had any effect on IQ. There are many other constituents of breast milk which may also have an effect on IQ, such as vitamins and peptides.

Another theory is that it is the actual contact between mum and baby that boosts cognitive development. Studies on rats (Weaver, IC, et al, 2004) found that this interaction accelerated neurocognitive development. Breastfeeding is also supposed to decrease feelings of stress or depression by eliciting beneficial hormone responses; this could mean a better mother-infant bond or better care for baby (Daniels, MC and Adair, LS, 2005), which could positively impact on how well he does academically.

In terms of an increased IQ, It doesn't appear to matter whether a baby is exclusively breastfed for three or six months (Kramer, MS, et al, 2009).

**Better speech and teeth**

There is a lot of evidence suggesting that breastfeeding – in which a baby sucking on a breast uses his tongue, jaws, lips, cheeks and pharynx in complex and repetitive motions in order to extract the milk, encourages the proper development of the muscles and organs used in speech (and breathing, chewing and swallowing ([Barbosa, C, et al, 2009]). Not using a bottle until nine months of age is protective against speech disorders. A study comparing breastfed babies to formula-fed ones, found that by six months, the brains of the former responded differently to speech sounds. This study also suggests that the differing brain responses were unrelated to the naturally occurring polyunsaturated fatty acids found in breast milk (Ferguson, M, and Molfese, PJ, 2007). The researchers concluded that the act of breastfeeding rather than the contents of breast milk was an advantage for a baby's later linguistic skills. A previous study had found that boys who had been breastfed had much better clarity in their speech and better reading ability compared to boys who hadn't been breastfed. However, the effect of breastfeeding for girls was less pronounced (Broad, FE, 1975).

Breastfeeding has also been proven to prevent the misalignment of teeth (Labbok, MH, and Hendershot, DE, 1987). Breastfeeding that continues for 12 months affords the best protection, but benefits can be seen for any length of breastfeeding (Romero, CC, et al, 2011). The muscles of the tongue, lips and cheeks affect bone growth and dentition. If a child is predominantly bottle fed, this could lead to either an imbalance in or the incorrect development of these muscles. Bottle feeding promotes less mouth muscle development than breastfeeding, which also could result in the misalignment of teeth and could interrupt the proper functions of swallowing, chewing and breathing. It could also affect the resting positions of the tongue or lips. However, sucking on a dummy or finger are much more likely to cause misaligned teeth than bottle feeding. Breastfeeding decreases the likelihood of a baby developing a sucking habit in the first place (Ngom PI, et al, 2008).

Mouth breathing is more common in bottle-fed babies; they are more likely to

**Breastfeeding, sleep and depression**

There is a potential relationship between postnatal depression and lack of sleep. Some studies have shown that symptoms of depression are improved when a mum is able to get more sleep (Armstrong, KL, et al, 1998). Although not definitive, the research suggests that mums who breastfeed exclusively, or who formula feed or do a mixture of both, all get the same amount of sleep per day or night. Breastfed babies, however, do tend to wake more frequently during the night, but possibly because of the low levels of disturbance associated with breastfeeding, mums still manage to achieve the same levels of sleep because they fall asleep quicker afterwards. Some mum's even manage to breastfeed whilst sleeping (Montgomery-Downs, HE, et al, 2010)!

half-open their lips while feeding, whereas breastfed babies tend to breathe more through their noses. Mouth breathing, if not corrected early, can cause an array of problems beginning in childhood and extending into adulthood. These include abnormal teeth, changes to the shape of the face and poor sleeping habits and levels of oxygen in the blood. If air travels through the nose before reaching the lungs, microbes are filtered more effectively and the air is slightly warmed.

As for the chances of tooth decay, the research is mixed. Some investigations show that breastfeeding has a protective effect, others show that it increases the risk of tooth decay and still others show no effect whatsoever. However, a commonly reported risk factor for tooth decay is allowing a child to take a bottle to bed (nursing bottle caries, see also page 38).

**Increased brain activity**

An EEG measures brain activity. When the EEGs of breastfed and bottle-fed babies were compared, those of the former showed significantly more brain activity during a feed than the latter. The same areas of the brain were stimulated by both types of feeding but there was more activity in breastfed babies (Lehtonen, J, et al, 1998). This brain activity is thought to demonstrate a baby's emotional response, which is evidence that breastfeeding increases the mother-infant bond. A breastfeeding mother's brain activity is also increased in response to her baby's cries during the first month as compared to a bottle-feeding mother. Brain 'activations' are thought to encourage a greater sensitivity to a baby (Kim, P, et al, 2011).

**Effect on temperament**

Research shows that mothers who breastfeed are more likely to rate their babies as being difficult than formula-feeding mothers during the first six months (differences even out after this). Breastfed preterm babies cry for up to an hour longer than formula-fed babies during the early months (Thomas, KA, 2000). Breastfed babies are reported to be more distressed and difficult to soothe, and to smile, laugh and vocalise less than bottle-fed babies. However, the researchers explain that this is due to formula-fed babies usually being over-nourished, which is what causes them to appear content. The so-called irritability of breastfed babies is an important communication technique that enables an infant to signal his needs. It also could be due to babies finding the initiation of breastfeeding more stressful: it is more difficult to learn to suck at a breast than at a bottle (de Lauzon-Guillain, B, et al, 2012).

**Maternal benefits**

Although the research is inconclusive, about two-thirds of the investigations carried out found that breastfeeding can protect against breast cancer. From the research that did find a protective effect, it's apparent that the longer a woman breastfeeds, the better protection she has. The risk of cancer decreases by 4.3% for every 12 months of breastfeeding (Rea, MF, 2004). Breastfeeding also reduces the risk of cancers of the ovarian epithelium.

During breastfeeding, women produce up to 1 litre of milk per day. This would naturally lead to the conclusion that the calcium taken from mum to make this milk would more likely result in her becoming deficient in this mineral, which is needed for healthy bones. However, the opposite appears to be true. A mum's calcium levels are replenished when her baby is weaned and, as a result of breastfeeding (the longer the better), she will be have more mineral bone density than if she hadn't breastfed (Wiklund PK, et al, 2012). This means that she is less likely to have bone fractures as a result of osteoporosis in later life.

Although the evidence is quite weak and often contradictory, breastfeeding may offer some protection against rheumatoid arthritis (Lahiri, M, et al, 2012). Women who exclusively breastfeed are also less likely to suffer from anaemia, as menstrual bleeding doesn't tend to occur during this time.

Breastfeeding (which requires over 600 calories per day to make milk) can result in quicker pregnancy weight loss. Women who breastfeed exclusively during the first six months lose weight faster and those who do so for 12 months, tend to have the lowest BMIs. However, by about 1 year post birth, there is little difference in weight between those who breastfed and those who didn't, or even between those who had children and those who hadn't (Motil, KJ, et al, 1998).

**Bottlefeeding is expensive**

One estimate is that a formula-fed baby will drink up to £550 of milk in one year, and that doesn't include the cost of feeding and sterilising equipment and the energy costs required for upkeep). Breastfeeding is therefore cheaper, even if mum requires more calories!

**'Improper' bottlefeeding can cause health problems**

The risks associated with formula milk – adding too much or too little formula powder, which can result in vomiting, diarrhoea, stressed kidneys, low nutrient intake or seizures; or infections due to feeding equipment not being sufficiently clean – largely stem from a parent not following instructions properly or making mistakes, which is easily done at 3 a.m.

## Recommendations from the research

Breastfeeding is far the best for baby, both nutritionally and emotionally and it's also best for mum's health, in the short and long term. There is really only minor evidence against it. Successful breastfeeding, however, is far more complex but, if given proper help when needed, most women will achieve it.

# Circumcision

Circumcision dates back thousands of years as both a religious practise (Jewish and Muslim) and a possible health precaution (Australian Aboriginals). Until the 1950s, it was common in the UK but today, only about 5% of baby boys are circumcised for medical reasons and it's unclear how many for religious considerations. Elsewhere, about a third of all men are circumcised making it the most common surgical procedure in the world, although countries differ markedly in uptake. In the US, circumcision uptake is relatively large although rates are falling, largely due to the lack of free medical aid to cover costs.

There are passionate advocates both for and against. Common arguments cite the antiquity of religion and culture and a desire for a son to look the 'same' as his father or brothers as against the notion that baby boys are being unnecessarily mutilated or forced into a religion they have not themselves chosen. When deciding on whether or not to have your baby boy circumcised, there will be a host of medical, ethical and religious ideas to consider but science can only deal with the medical aspects.

The NHS does not provide routine circumcision but some primary care trusts may carry them out on request. Otherwise, parents will need to pay for it privately.

Operations are usually conducted as a treatment for an illness but newborn circumcision is preventative and may not be necessary for later life. Although it might be beneficial later on in life (as are routine immunisations), the risk of acquiring measles,for example, is far greater than the protection offered by circumcision.

## CURRENT RECOMMENDATIONS

The NHS does not currently recommend circumcision for preventing infection because of the risks of bleeding and infection from the procedure. However, the American Academy of Paediatrics (AAP) recently stated that the benefits of circumcision warrant the medical acceptance of a religious choice. Although the benefits outweigh any risks from the procedure, they were not significant enough to justify all babies being circumcised. The decision should lie with the parents.

## WHAT THE RESEARCH SAYS

### Protection against irritation

It is commonly understood that the foreskin is designed to protect the glans or head of the penis from excretions, rubbing or other trauma and, therefore, serves a useful purpose, which, with proper care, can continue throughout life.

Historically, circumcision was most often performed in sandy, arid conditions (Middle East and Australia). If sand becomes trapped between the glans penis and the foreskin, the latter can become inflamed and painful. Whilst there may still be an argument for those living in sandy conditions today to have the procedure as a preventative, it is not relevant to most western countries.

**Reduction in urinary tract infections (UTIs) and inflammation**

More commonly, bacteria can accumulate under the foreskin, which can act as a reservoir for infection. UTIs are less common in circumcised boys (Simforoosh, N, et al, 2012) and they have fewer microbes around the penis. However, the evidence is limited, particularly in terms of the number of infections that would be avoided. There is only about a 1% chance of a healthy boy getting a UTI in the first place whereas infections from the procedure are on average about 2%, so as a method of avoiding infections the numbers don't add up. That being said, boys who are at a greater risk of UTIs (they've had the condition before), benefit more from circumcision than other healthy boys.

No top quality research has been carried out investigating a link between circumcision and UTIs (Jagannath, VA, et al, 2012). Given the nature of circumcision, randomised controlled trials are unlikely as this would require one child to be randomly assigned to circumcision and another not, a situation few parents would agree to.

There are a variety of conditions, which can cause inflammation of the penis. Some, such as psoriasis, are skin diseases, others result from an infection. The risk of a boy needing treatment for such a skin conditions is greater in those, aged between 1 and 8, who have not been circumcised (6.5% for circumcised compared with 17.2% for uncircumcised according to an AAP Technical report, 2012). B. J. Morris et al (2012) estimates that a third of uncircumcised men will suffer at least once from a foreskin-related medical problem during their lifetime.

**Protection against sexually transmitted disease (STD)**

There is significant evidence that a circumcised man is better protected against HIV during heterosexual relations. N. Siegfried et al (2009) demonstrated that in African countries, circumcision has been shown to reduce the chances of acquiring HIV by between 38 and 66% over a 2-year period, but in developed countries, the protection is significantly less, closer to 10%, although there is a lack of recent high quality back-up evidence.

A. A. Tobian et al (2010) found proof that circumcision also offers women partners protection against the human papillomavirus, herpes simplex virus type 2 and bacterial vaginosis but there is less clear evidence that it protects against syphilis, chlamydia or gonorrhoea. The mechanism of protection is not entirely understood but some believe that the foreskin is at risk of small tears, which expose the body to microbes, or that, similar to UTIs, the foreskin traps the microbes and gives them time to grow. Removing the foreskin makes it less likely that a man will become infected.

**Decreased risk of cancer**

Cancer of the penis is a very rare condition, and its incidence has been decreasing during the last forty years. There is some evidence that circumcision is mildly protective against invasive cancers but not against a precursor, which usually gives rise to cancerous tumours (Schoen, EJ, et al, 2000). Penile cancer is related to a condition known as 'phimosis' whereby the foreskin cannot be retracted fully but the rarity of this cancer makes it highly illogical to carry out routine circumcisions in order to prevent it. The human papillomavirus

## Recommendations from the research

There is some evidence that circumcision can decrease the risk of UTIs and some strong evidence that it can decrease the chances of acquiring STDs, including HIV. It may also help to decrease the chances of very rare penile cancer. Pain during the procedure can be managed effectively but probably not eliminated completely, and complications are usually minor and rare. Generally, circumcision protects against skin diseases; however, with proper cleaning these can also be avoided.

may also increase the risk of cancer in the uncircumcised man.

**Pain during and complications after**

Creams that numb the area can be used to relieve any pain associated with circumcision but in low birth weight babies, there's a greater chance that the cream will irritate the skin and other methods such as a nerve block will be needed. However, skin reactions are rare and usually minor. Analgesia is effective particularly if given alongside a dummy dipped in sucrose. No pain relief method can completely remove all discomfort caused by the procedure but methods have been shown to be safe for healthy newborns (Brady-Fryer, B, et al, 2009).

Complications are rare and relatively minor in healthy newborns. Numbers differ widely between studies and range from less than a 1% chance to just under a 5% chance (Weiss, HA, et al, 2010 and Pieretti, RV, et al, 2010). Bleeding and infection are the most common short term, but longer-term problems, such as the removal of too little or too much foreskin can occur. Longer-term complications during childhood, are slightly more common (one third are short term, two thirds are longer-term). More serious complications are extremely rare and there are only a few reported cases but they can be devastating. The risks and complications increase with age; the older a child is when circumcised, the greater the risk of complications.

### Effect on sexuality

Although it's a long way off, whether their son's penis will function as 'well' if circumcised is an important, if not slightly uncomfortable, parental consideration. It is an area of great debate. In general, the research states that there is no significant evidence of any difference in function, sensitivity or sexual satisfaction between circumcised or uncircumcised men (Kigozi, G, et al, 2008) and the majority of men, circumcised in adulthood, report being satisfied with their circumcision (Fink, KS, et al, 2002).

# Disposable or reusable nappies

Making a decision as to whether to go with disposable or reusable nappies can be frustrating. It's not easy to find reliable unbiased sources of information. There are many passionate viewpoints but ultimately the decision must rest with the parents as there is little research on the subject.

## CURRENT RECOMMENDATIONS

There are no official UK recommendations on which type nappy is best. Many county councils currently offer samples or vouchers in order to encourage the use of reusable nappies and cut down the number of disposables on land-fill sites.

## WHAT THE RESEARCH SAYS

### Skin conditions

Today's disposable nappies use an absorbent gelling material to hold urine away from the skin. They also have a limited ability to control the pH of the urine. Prolonged wetness and an increasingly alkaline environment can cause skin conditions such as nappy rash or dermatitis. R. L. Campbell et al (1987) compared disposables with reusables and found that disposables were far better at reducing skin wetness and maintaining normal skin pH. Fewer babies wearing disposables had nappy dermatitis. In addition, G. N. Erasala et al (2011) believe the fit and comfort of disposables compared to reusables helps limit the chances of dermatitis. In fact, F. Akin et al (2001) have shown that actual yeast infections on baby's nappy area, can be reduced by up to 50% by using modern disposable nappies. It is unclear, however, whether more frequent changing of reusable nappies can be just as effective at controlling nappy rash as it's the prolonged contact of the urine on baby's skin that causes the problem.

### Infection

The passage of faecal matter to the mouth can cause illnesses and M. Kamat and R. Malkani et al (2003) found it's more common for poo and urine to leak from reusable nappies compared to disposables.

### Environmental considerations

Due to the increasing popularity of reusable nappies, in 2005 the Environment Agency conducted an independent review to determine which type of nappy was the most environmentally friendly. They considered the entire life-cycle of each nappy type, from source of materials and manufacture through to disposal and decomposition as well as an average use of 2½ years per child. It concluded that there was no real difference in environmental impact between the two. However, the review was heavily criticised for excluding key pieces of information and including some other, rather odd ideas (that parents iron reusable nappies, for instance). So, in 2008 the Environment Agency wrote an addendum to its 2005 review, which included more up-to-date information, such as the energy efficiency of modern washing machines and the weight reduction of disposable nappies (the lighter the nappies the less energy needed to transport them). In the addendum the researchers report that using disposable nappies for an average of 2½ years will result in a total of 550kg of carbon dioxide – a decrease of 13.5% since the 2005 report – which indicated that disposable nappy manufacturers are making attempts at reducing carbon. Disposable nappies also require a considerable amount of landfill and take a very long time to decompose. Landfill represents a serious environmental concern, not only are they deemed ugly but they also facilitate the concentrated release of gases and chemicals into the soil and air.

Levels of carbon dioxide, produced over the 2½ years, for reusable nappies was found to be 570kg, which is relatively similar to disposables, although this depends on how the nappies are cleaned. If resusable nappies are washed at 90°C as opposed to 60°C (as recommended) and are dried in a tumble dryer, carbon emissions are greatly increased, well beyond those of disposables. However, washing nappies in full loads in a washing machine, line drying them and re-using them for subsequent siblings was found to produce significantly less carbon dioxide emissions than disposables. If you're prepared to follow these guidelines for washing and drying reusables then they are the more environmentally friendly option.

There is also the consideration of resource. Disposable nappies are generally made from wood cellulose fibre and a mixture of polymers. The wood cellulose fibre is taken from trees, which causes deforestation and the polymers are made from crude oil. The sheer number of disposable nappies required means that these resources are needed in significantly large amounts. Reusable nappies also require resources, in particular cotton which requires a lot of water and some pesticides (unless organic) but because fewer are needed, the impact is far smaller.

**Costs**

Reusable nappies require a costly initial outlay; however, there will be few subsequent costs. Compared with disposable nappies, they have significant cost benefits. These are further increased if the same reusable nappies are used for subsequent siblings (Wong, DL, et al, 1992).

There are not just cost saving incentives for you but also for your local council. One council estimates that it costs up to half a million pounds to dispose of a county's worth of nappy waste per year.

**Exposure to chemicals**

Nappy manufacturers are not obligated to reveal the ingredients in their nappies but it is likely that most disposable nappies contain a superabsorbent substance known as polyacrylate. While not hazardous, exposure to the dust, for instance if the nappy layers are separated, can cause the eyes to burn or itch, irritate the lungs or aggravate pre-existing skin conditions.

This substance and similar super-absorbers have in the past been removed from tampons as they were associated with toxic shock syndrome (an illness caused by bacteria). To date, however, there is no reason to suspect it can cause a similar reaction in babies from nappy use. In fact, K. Lintner and V. Genet (1998) reveal that related substances by depriving bacteria of water. have acted as disinfectants in cosmetic products.

Many nappies have been bleached to make them a more presentable colour. This could expose a baby to cancer-causing substances known as dioxins, although little research has been conducted in this area. In response to public pressure, many nappy companies have opted for unbleached or chlorine-free bleaching methods, which do not produce dioxins. Passionate advocates for reusable nappies fail to point out that this bleaching usually occurs in reusables as well. However, M. J. DeVito and A. Schecter looked at dioxins (2002) in nappies and tampons and and found that neither contained the most harmful type of dioxin and that levels were far below those likely to cause any harm.

Using animals, R. C. and J. H. Anderson (1999) looked at the potential impact of some chemicals routinely used in disposables. Mice were exposed to the emissions of 3 different types of disposable and one reusable nappy. The researchers found that some disposable nappies emitted a mixture of chemicals that caused the mice breathing problems and irritation. The reusable nappy caused significantly fewer problems. The researchers suggested that the use of disposable nappies may cause or exacerbate pre-existing asthma in some children.

Many nappies also contain dyes used to create the pictures often found on the surface of the nappy and glues and rubbers used in fasteners. In rare cases, these substances, when coming into contact with

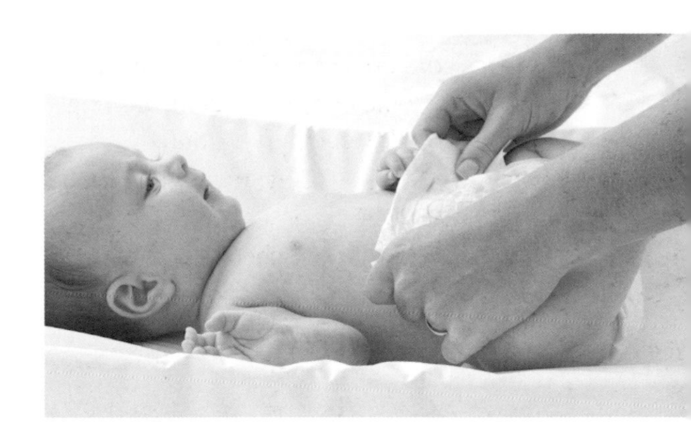

a baby's skin, can trigger an allergic reaction (Alberta, L, et al, 2005). If your baby has an allergic reaction to a disposable nappy, it may be worth trying one without dyes.

**Eco-disposables**

For the environmentally conscious, 'eco' nappies promise a guilt-free alternative. (Eco-disposable nappies tend to be friendlier to the environment because they are made from renewable resources, e.g. corn, will decompose more quickly than normal disposables and have usually been treated with fewer chemicals. They may also be more hypoallergenic for baby as they contain fewer dyes or perfume.)

However, they may not be as great as many hope. The use of chlorine-free bleaching or no bleaching at all may be an advantage in terms of a baby's exposure to dioxin (see above) and for the environment as these chemicals 'leak' out during decomposition. They are also likely to be freer of dyes and other chemicals than normal disposables, making them more hypoallergenic. In terms of compostability some eco disposables are supposed to decompose more quickly. However, landfill is quite often tightly packed and lacks air, a necessary ingredient for successful decomposing, which means that it is unlikely that these eco-disposables will decompose as effectively as they could. Eco nappies do, however, tend to be made from renewable resources, which makes them a greener option.

**Sperm production**

There's a possibility that disposable nappies have contributed to the decline in male fertility over the last 70 years. E. Carlsen et al (1992) found that both volume (from 3.4 to 2.75ml) and concentration (from $113 \times 10^6$ per ml to $66 \times 10^6$ per ml) to the power of 6 decreased from 1938 to 1991. In order to produce sperm, the testicles must remain at a relatively cool temperature. P. J. Partsch et al (2000) found that baby boys wearing disposable nappies had significantly higher scrotal skin temperatures compared to those in cotton nappies. Whilst sperm production doesn't begin until puberty, there is the potential that a raised temperature at a young age could affect the testicles' normal development. Nobody knows exactly the effects of increased temperature, but studies of undescended testicles, which develop under higher body temperatures similar to those caused by disposable nappies, have a negative effect on sperm production later in life. In their study, Partsch et al also compared traditional cotton nappies with disposables; however, today's modern reusable nappies may not facilitate low temperatures as sufficiently as traditional cotton ones.

## Recommendations from the research

The limited good-quality research seems to support the notion that reusable nappies are better for the planet, from source to finish, but only if you aim for environmental friendliness when you wash and dry them; they're also cheaper in the long run. Disposables, however, lead to less nappy rash and prevent the spread of infection. Less conclusive are the notions that disposables may affect male reproductive health in later years and contribute to or exacerbate asthmatic symptoms. Bleaching of either type of nappy is potentially not great for baby or the environment but is becoming more easily avoided through chlorine-free bleaching mechanisms. Eco disposable are a nice alternative but are probably not as 'green' as may be hoped.

The choice has to lie with the individual parent as no researcher can weigh up the advantages of time and convenience for new parents. On the whole it looks as though, if you can manage it, reusable nappies are the best option. However, don't rule out a compromise, using a mixture can be beneficial for both parents and the planet.

# Infant supplements

Worrying is part of parenting. When you are solely in charge of how your baby is nourished, the responsibility can sometimes feel a little overwhelming. It's comforting to know that you can occasionally reach for a few vitamins or minerals to ensure against any deficiency in your little one. But do you really need to? Is it likely your baby will be deficient in anything?

## CURRENT RECOMMENDATIONS

The Department of Health currently recommends that all newborns receive a vitamin K supplement and that children are given supplements of vitamins A, C and D – but only from age 6 months until they are 5 years old.

## WHAT THE RESEARCH SAYS

There are, however, many reasons a baby may need supplements earlier than 6 months, these include if she has darker skin, was born preterm, doesn't eat well or if mum's nutrient stores were low during pregnancy, for instance if she didn't take a vitamin D supplement.

### Vitamin K

Newborn babies all receive this vitamin either as an injection or orally during their first few days. This is because vitamin K is not transferred from mum to baby well before the birth nor does breast milk contain enough to meet baby's needs. Although rare, a vitamin K deficiency can result in spontaneous bleeding and possibly death. B. Lauer and N. Spector (2012) have found a lot of strong evidence for giving vitamin K. There was a scare that linked the vitamin K injection with the development of childhood cancer but J. Golding et al (1992) and P. Ansell et al (1996) found no link.

### Vitamin D

Vitamin D is needed for the body to absorb calcium and deficiencies can give rise to rickets where the bones become soft. An array of other conditions can result from a vitamin D deficiency, including delayed development, multiple sclerosis, diabetes and cancer. The body is able to make its own vitamin D but it must have sunlight on

the skin to do so. Unfortunately, in the UK, many people do not have access to enough sunlight to make sufficient amounts, particularly in winter, and according to the Department of Health (2012), pregnant and breastfeeding mothers, as well as children under 5 are in the high-risk category for deficiency. Adding to the problem is the sometimes over precautious use of suntan lotion and staying out of the sun.

Pregnant women are advised to take a vitamin D supplement to ensure their baby starts off with the right amount and to take a supplement whilst breastfeeding. Formula milk is fortified with vitamin D. Currently, parents are advised to supplement their baby from age 6 months old (and a formula-fed baby once she is drinking less than 500 ml per day) until the age of 5.

Recently, whether supplementation taken by mums whilst pregnant and breastfeeding is sufficient to cover their baby's needs for the first 6 months has been questioned. According to B. Lauer's and N. Spector's research (2012), many countries supplement from birth, including the USA, Ireland, France and Germany. Ireland, a country with a similar climate and latitude to the UK, has recently changed its recommendation because insufficient amounts can be made by the body due to lack of sunshine (Health Service Executive, 2010). However, if a mum received sufficient amounts of vitamin D during pregnancy and is taking sufficient amounts during breastfeeding, it's likely that her baby will have enough vitamin D until she is 6 months old (Marshall, I, et al, 2012). The Department of Health's Scientific Advisory Committee on Nutrition is currently reviewing guidance on this matter.

A. A. Litonjua (2012) has highlighted that within the last couple of years, evidence has come to light that too little or too much vitamin D can increase the risk of asthma or eczema. We understand the minimum most children need to avoid vitamin D deficiency but unfortunately we are not sure of the quantities needed to trigger allergies.

K. Kneller (2012) has also found evidence that vitamin D supplementation should continue throughout a person's entire life.

**Vitamin A**

Needed for a healthy immune system and for us to see in poorly lit areas, a severe deficiency (usually In developing countries) can result in death from infectious disease. Vitamin A is found in dairy products, carrots and dark green vegetables. It is recommended for babies from 6 months until 5 years because young children require a lot of it (More, J, 2007) and many children don't eat a balanced diet. However, if a mum was deficient during her pregnancy, supplementation can start at 1 month. Formula milk has added vitamin A so a formula-fed baby shouldn't need a supplement until she drinks less than 500ml of formula milk per day.

Despite the Department of Health's recommendation that children be supplemented, A. A. Leaf (2007) found there is little evidence that it is needed, as long as children have a balanced and healthy diet In fact, according to J. R. Gregory et al (1995), levels of vitamin A are actually slightly too high amongst young children.

**Vitamin C**

Vitamin C is also needed for a healthy immune system but is especially for the absorption of iron. It can be found in oranges, tomatoes and strawberries. Despite the Department of Health recommending that children be supplemented from 6 months, A. A. Leaf (2007) found, as with vitamin A, there is little evidence that it is actually needed if a healthy balanced diet is eaten. Due to their higher consumption of fruit juices, it is likely that young children actually have too much vitamin C.

**Iron**

Iron is needed for oxygen to travel around the body and is also important for energy levels and brain development. A deficiency is rare in children. Provided her mum had adequate iron levels during pregnancy, a healthy full-term newborn baby is likely to have enough stored iron, which combined with either iron in breast milk or formula milk, is plenty to keep her going for at least 6 months. At 6 months, she can be introduced to iron-rich foods such as meat, leafy green vegetables, apricots, beans and fortified cereals. If iron-rich foods are not introduced to the exclusively breastfed baby at 6 months, her chances of a deficiency begin to rise. Iron deficiency can also occur if a child is moved onto cow's milk too early (before 1 year old). As long as a child maintains a healthy diet, there is no need for supplementation, but if you have picky eater it may be worth being aware of symptoms of iron deficiency – repeat infections, loss of appetite, tiredness and dropping percentiles in weight and height – as levels are crucial are levels in the first few years.

Supplementation of iron from about 1 month has been investigated because there is some evidence that it may boost a child's development. J. K. Friel et al (2003) discovered higher visual acuity in toddlers who had received supplements from 1 month of age. However, H. Szajewska et al who reviewed the subject in 2010 found no effect on development. It's also worth noting that too much iron can cause constipation in babies.

**Prebiotics and probiotics**

Probiotics are 'friendly' bacteria that can drive down the number of harmful bacteria, whilst prebiotics are components of the food we eat, but cannot digest (oligosaccharides), that stimulate the growth of beneficial bacteria, such as probiotics, in the colon. Both are commonly added to formula milk in order to better mimic the naturally occurring substances found in breast milk. They have a range of theoretical benefits for baby including prevention of infections, especially in the stomach and intestines, and of allergies. They are also very safe in otherwise healthy children.

Prebiotics in formula milk may help prevent against the development of eczema in young children but D. A. Osborn and J. K. Sinn (2007) found this is really only true for babies who were high risk of developing it anyway; they have a parent or sibling with the condition. As for the role of prebiotics in preventing other allergies, there is little convincing evidence.

The jury is still out as to whether probiotics help to prevent against allergies. There are a number of studies that show one strain, L. rhamnosus, protects against eczema in 'high risk' children but the above research found that not many studies are conclusive of its benefits for other allergies nor are all studies conclusive of its benefits for eczema. S. Rautava et al (2012), however, demonstrates a beneficial link between probiotic supplements given to mums whilst pregnant and a lack of eczema in high-risk children So if either mum, dad or a sibling have an allergy, these probiotics, either during pregnancy or in formula (breast milk has probiotics naturally) may be beneficial to baby.

Certain probiotics may be helpful in preventing diarrhoea caused by an infection. The best strains have been found to be Lactobacillus GG and S. boulardii, although how much should be given is still debated. The research is not yet conclusive, but S. Guandalini (2011) believes if the best strains and doses can be determined, it may well prove to be effective at diarrhoea prevention. Probiotics may also help prevent diarrhoea caused by antibiotics, which is fairly common when given to young children (Johnston, BC, et al, 2011).

There has been some promising, but not conclusive, research looking at the benefits of probiotics (taken through the mouth or a nasal spray) on middle ear infections, a common infection in young children. There is more research needed on the best strains of probiotics needed but L. Niittynen et al (2012) listed a number of research papers (though not all), that found probiotics helpful in preventing middle ear infections.

Probiotics may also be useful in preventing urinary tract infections (UTIs). These are fairly common in young children (5-10%) with a high chance of suffering repeat infections. G. Williams and J. C. Craig (2009) found some evidence that probiotics may mildly protect against UTIs. E. K. Vouloumanou et al (2009) have found some good evidence to suggest that probiotics may help to lessen the symptoms of respiratory tract infections (such as the common cold) and to help them clear up sooner. Q. Hao et al (2011) uncovered some evidence, but far less conclusive, that they may help to prevent respiratory tract infections all together.

**Long chain polyunsaturated fatty acids (PUFAs)**

I. Romieu et al (2007) found some evidence that omega-3 supplementation (a type of long-chain polyunsaturated fatty acid) for mums whilst pregnant (see page 14) may decrease the chances of their children becoming allergic. However, N. D'Vaz et al ( 2012), who looked at supplementing young babies, found that whilst it increased the levels of fatty acids in their bodies, it did little to prevent the development of actual allergies Another study by the same researchers found that supplementation during the first 6 months caused a decrease in the immune response that is typically associated with allergies

Essential fatty acids may also be beneficial for language development. S. J. Meldrum et al (2012) looked at brain and language development in toddlers who had been supplemented with PUFAs from a young age, and found some improvements in communication (gestures etc...) but not in their brain development. However, P. Willatts et al (1998) found beneficial effects such as better visual acuity and better problem solving skills at 10 months. K. Simmer et al (2011) believe that the timing of supplementation (for instance a full year's supplementation may be needed) as well as the dose, and even the source of the PUFAs, may all be relevant when it comes to maximising the benefits.

Supplements of PUFAs, specifically omega-3, commonly come from fish oils, which could contain contaminants such as heavy metals. To prevent the accumulation of metals in the body, they are often purified. If you intend to give a supplement to a young child make sure it is of high purity.

**Multivitamins**

Some parents are comforted to think that with a single supplement all bases are covered. However, multivitamins with nutrients bring added complications; L. Waddell (2012) found that up to half of children taking such a supplement didn't actually need one. If you give your toddler a multivitamin, there is a chance she may actually take in too much, particularly if she already eats a healthy balanced diet. R. Briefel et al (2006) found that a significant percentage of American toddlers were getting too much vitamin A, zinc and folic acid. High levels of vitamin A can impact on bone health if they continue for a long time. Too much zinc can cause nausea and vomiting and in the long-term can affect copper levels, which can lead to anaemia. Too much folic acid can disguise a vitamin $B^{12}$ deficiency, which could eventually lead to a damaged nervous system.

## Recommendations from the research

All newborns need to have vitamin K, no matter how healthy mum was during her pregnancy. Vitamin D must also be supplemented from 6 months until at least 5 years. Many vitamin D drops include other vitamins, such as the recommended vitamins A and C, and although they may not be needed for a child who eats a balanced healthy diet, they are unlikely to cause harm, and could give peace of mind to parents of fussy eaters (see also page 136). Iron is unlikely to be needed but if mum's levels were low during pregnancy or you have a fussy eater, keep an eye out for symptoms of anaemia. The research on prebiotics and probiotics is inconclusive but promising, particularly if there are allergies in the immediate family. When breastfeeding and formula milk feeding is finished, it may be worth trying to supplement your child with these, with GP guidance, of course. Long chain polyunsaturated fatty acids have also shown some promising results. Found naturally in breast milk and added to formula milk, they may also be worth considering when your baby is older (remember to choose high purity examples) because of the many potential benefits. As with anything you give a child, always ensure that it is safe for her age group and that you give the correct amounts . And, of course, always keep any supplements out of reach of any children.

# Hygiene and cleanliness

Something has caused an increase in allergies in the last 40 years (the rates of eczema and hayfever have almost doubled). The reason(s) why still evades us although there are a number of promising theories and observations. Researchers are now leaning towards the idea that there are a few causes rather than just one.

## CURRENT RECOMMENDATIONS

In the UK, almost a million people a year get food poisoning  of which up to 500 may die as a result, so the NHS promotes measures designed to eliminate all disease-causing germs. These include food storage and preparation tips and the regular cleaning of household items and toys.

## WHAT THE RESEARCH SAYS

### The Hygiene Hypothesis

This theory, first proposed by Dr. David P. Strachan of St. George's Hospital Medical School, London, in 1989, claims that the rise in allergies is due to children not being exposed to as many microbes (bugs) as they used to be. Strachan discovered that larger families had fewer allergies (hay fever and eczema) than smaller ones and he believed that was because children in smaller families had a less chance of coming into contact with microbes (fewer siblings to pass them on). Children who don't encounter microbes have under-developed immune systems and are more likely to overreact to things that are not dangerous.

Dr. Strachan's work has been backed up by other studies that found that allergies are more common in smaller families (Bodner, C, et al, 1998). But it's not a case of the more often a child comes into contact with a 'bug', the less likely she is to have an allergy to it; protection from allergies is not related to how often a young child is ill (a child who is often ill is much more likely to develop an allergy). So how exactly does having lots of siblings protect against the development of allergy?

One theory is that the more children a woman has, the fewer antibodies she passes to her unborn babies via the placenta so decreasing the chances of a later babies developing an allergy later in life (Karmaus, W, et al, 2004). Studies of umbilical cord blood have proven that after each subsequent pregnancy, fewer antibodies are passed on to the babies. Researchers don't really understand why this is so, although they have noticed that where there's a family history of allergies, more antibodies are passed from mum to baby, which increases the chances of a baby developing an allergy. So it stands to reason that if fewer antibodies are passed between them, the chances of allergy decrease.

Another (unproven) theory is that children who attend large nurseries come into contact with a wider variety of bugs so are more protected from developing allergies. This is backed up to a certain extent by the observation that young children who attend nursery tend to have more colds and wheezing when they are very young but by the time they reach 6–13 years, they have fewer colds and wheezing than children who have been looked after at home (Ball, TM, et al, 2000). A related study (Svanes, C, et al, 1999) looked at the possible protective effect of children sharing a bed (with the understanding that these children would pass illnesses between themselves more frequently); it found that these children were less likely to develop an allergy, thus backing up the hygiene hypothesis.

An observation that would also appear to lend credence to the hygiene hypothesis is that asthma rates in developing countries are about 2-3% compared to as much as 40% in developed countries.

M. Ben-Shoshan, et al (2012) found that allergies tend to be more common in children whose parents are well educated and a recent report by the American College of Allergy, Asthma and Immunology remarked on a link between an increased chance of allergy in children and affluence. A major conclusion of these studies was that a cleaner home gives rise to more allergies.

Other theories, however, propose other reasons for this link. Affluent and better educated parents may be more likely to seek medical help for symptoms and be given a diagnosis of allergy or they may avoid known allergens during pregnancy or when their children are young (once thought to minimise allergies but now believed to exacerbate their risk).

More evidence for the hygiene hypothesis was provided by a large investigation of families with 15 month olds, rated as having either a high or a low hygiene level, that looked at the children's risk of developing asthma. The children in families with a very high hygiene level were more likely to develop asthma symptoms and eczema by two years of age (Sherriff, A, et al, 2002). However, there may be other factors at play. M. Nickmilder et al (2007) found no link between the amount of household cleaning products used and an increase in allergies.

Interestingly, several studies that investigated household hygiene found that we're not quite as effective at cleaning as we may think we are. Bacteria were still present despite washing-up with soap and wiping surfaces clean. Hours after bathrooms had been 'cleaned', the levels of bacteria quickly returned (Scott E, et al, 1984). This is mainly due to the use of cloths (which spread the germs), air-borne bacteria and leaving surfaces damp, which facilitates the growth of bacteria. On the other hand, washing hands is effective at limiting the spread of bacteria.

Our homes are still full of bacteria and, since we spend more time in them than we used to, we are even more exposed to bacteria so it can't be that allergies are on the increase because our homes are too clean. However, keeping our children 'clean' may still have a part to play in the rise of allergies but this has yet to be proven.

**Antibiotics and vaccinations**

It is thought that the increase in successful vaccinations and the widespread use of antibiotics has left our immune systems with less to do. However, greater antibiotic use and vaccinations predate the increase in allergies by a number of decades, and there's still plenty of work for our immune systems: some illnesses, such as tuberculosis, are prevalent once again, there has been a

steady rise in campylobacter, a type of bacteria that causes food poisoning, and no real changes have been seen in the number of colds and flu to account for the rise in allergies. Nor is there enough information to claim an association between the decrease in parasitic worms and the rise in allergies.

And yet, there still may be a link between antibiotic use and allergies. Several studies have shown that children who require antibiotics during the first two years of life are much more likely to develop allergies. Nobody is sure whether the antibiotics are causing the allergies or if antibiotics are needed by children more likely to develop an allergy (Farooqi, IS, and Hopkin JM, 1998 and Celedón, JC, et al, 2004).

**Farming**

Children of farmers have fewer allergies than children who grow up in the countryside or in towns; they are less likely to be allergic to cats or to suffer hayfever but are no more protected against asthma than anyone else. Farming's protective effects only exist if a child actually grows up on a farm; adult farmers who grew up elsewhere are not more allergy protected.

Endotoxins are present in the cell walls of certain bacteria and are released when the bacteria is destroyed. The level of endotoxins may represent the number of bacteria that a child has come into contact with or the endotoxins themselves may affect a child's immune system. A greater number of endotoxins are found in the dust of farm houses and in the mattresses of children living on farms. There is also a difference in the types and numbers of 'friendly' bacteria in the intestines of children who suffer from allergies and those that don't. It's thought that the endotoxins that a farming child encounters from an early age encourage the development of specific friendly bacteria and that this protects them against allergies (Bloomfield, SF, et al, 2006).

**Allergy and specific illnesses**

Studies have shown that people who suffer hepatitis A and salmonella as children are less likely to develop allergies. The causes for this are unknown. There are two main theories. The first is that early exposure to microbes protects against allergies (hygiene hyposthesis), and these illnesses are possibly just a representation of the many bugs to which these children were exposed. The second theory is that certain people have a genetic code, which protects them from allergy when they have caught specific diseases. A study that compared hospitalised children suffering from salmonella and those with a non-bacterial stomach inflammation showed that the former were much less likely to develop allergies (Pelosi, U, et al, 2005). It would appear, therefore, that only specific illnesses have the power to correctly 'tune' an immune system; as mentioned above, a child that is often ill is more likely to develop an allergy.

Other studies show that an infestation of intestinal parasitic worms caused by eating poorly washed food, undercooked meat or mosquitos. can be protective against allergies. How the worms protect against allergies is not fully understood but it is thought to be something to do with how they decrease a healthy immune system's normal activity (Yazdanbakhsh, M, and Matricardi, DM, 2004).

Mycobacteria are a group of microbes often found in soil and water (modern sanitation techniques have ensured they are less prevalent today). It has been suggested that children who contract mild forms of bacteria from this group but are not made very ill by them, may have fewer allergies. However, no pattern can be seen between the rise of allergies (since 1970) and water sanitation (started in 1910). But what has changed in the last forty years is the food we eat and it's possible that different bugs living within our modern diet have changed our immune systems. But with little information on the bugs that previously lived on our food, this theory can't be proven.

**Probiotics**

Products that increase the number of friendly bacteria in the intestines have been shown to have an effect on our immune systems. If probiotics are taken when the immune system is developing (during pregnancy) and when still young (during the early years of life), they could affect the immune system in a positive way and thus prevent the development of an allergy (Pelucchi C, et al, 2012).

**'Old friends'**

Probably one of the best backing for the hygiene hypothesis, is a paper written by GA Rook et al (2004). Rook suggests that the rise in allergies is due to a lack of 'background exposure' to not very dangerous bacteria such as mycobacteria, endotoxins and parasitic worms, which he calls 'old friends'. These old friends enable our immune system to function correctly thus decreasing the chance of allergies.

Some promising results are apparent in tests in which an inactive form of a mycobacteria was inhaled by asthma sufferers. The results show this type of bacteria may eventually work as a treatment or even a prevention for allergies like asthma (Camporota, L et al, 2003).

**Possible other causes**

Not all theories fit in with the hygiene hypothesis. Our diets have changed a lot in the last forty years; we tend to eat far more fatty, salty and sugary food and fewer vegetables, milk, fibre and foods rich in vitamin E. As mentioned above, this change in diet could alter the types and number of friendly bacteria in our guts.

## Clean but not too clean

*Here are some things you can do that may help to maintain 'good' bacteria while removing the 'bad':*

* Wash hands with soap and running water or use an anti-bacterial hand-gel especially if: handling uncooked meat or eggs; changing nappies and after sorting the dirty laundry.
* Wash hands more often if someone in the house with whom you have contact has a cold, diarrhoea or is vomiting.
* Thoroughly clean surfaces where food is prepared with soap and hot water. A disinfectant spray can also be used.
* Sterilise with heat (boiling in 90° water for 2 minutes) or disinfectant (or both) the following: cleaning cloths, shower heads, humidifiers, and any utensils used for cleaning e.g. a scrubbing brush. 
* All items should be dried following sterilising and anything used for cleaning should not be stored when damp.
* Thoroughly clean bathroom surfaces, ensuring they are left dry as dampness can encourage the growth of bacteria.
* Although the risk of infection is low in the bathroom, toilets should be disinfected regularly, particularly under the rim and following a bout of diarrhoea/vomiting to prevent the spread of disease.
* U-bends in drains can also contain bacteria and these could be sterilised sporadically.
* Laundry that comes into contact with poo or skin pathogens should be washed at 60° and separately to cleaning cloths.
* Laundry should be dried immediately.
* Once a week the washing machine should have a cycle at a high temperature.
* The risk of catching a bug outside the kitchen and bathroom is low; clean floors with a detergent and vacuum regularly and disinfect tiles or areas that are often damp.
* Maintain good ventilation in the home to prevent the build-up of dust mites and prevent damp areas caused from humidity

*Taken from the consensus document by the Scientific Advisory Board of the International Scientific Forum on Home Hygiene entitled: "Guidelines for prevention of infection and cross infection in the domestic environment"*

## Recommendations from the research

There is no good evidence that we should change recommended hygiene procedures. In developed countries, infections cause widespread severe illness and about 4% of deaths every year. Good hygiene practices are the most effective defence we have against infectious diseases, particularly with some strains of bacteria becoming resistant to antibiotics. A study of UK hygiene practices when preparing food, however, found that only 5% followed the recommended practices and 4% actually prepared food in a way that put them at serious risk of an infection (Griffith, C, et al, 1998).

While it would appear that exposure to certain types of bacteria is necessary to prevent children developing an allergy, there are plenty of nasty bugs that can make children and adults very poorly and in some cases kill them. If we're talking about increasing the risks of death from disease to offset a possible increase in say, hay fever, then it's probably not worth it! Unfortunately, researchers cannot identify 'good bacteria' from 'bad ' with certainty, as yet, nor can they determine ways to kill off one type and keep the other. What is important is to appreciate the difference between 'looking clean' and 'being clean'. A surface can look clean but be full of germs. Certain parts of the home (kitchen and bathroom) need to be cleaned well; we can be more relaxed about other areas.

# Sleep interventions

Reaching the period in a baby's life when he is 'sleeping through the night' is a highly longed for and often coveted goal. If a baby can fall asleep without parental help when put to bed, then he can manage any subsequent awakenings in the night without mum and dad having to help.

Being able to self-soothe or get oneself back to sleep usually develops somewhere between 6 and 12 months although many parents will intervene to ensure it happens earlier. The ability to sleep for a 5-hour stretch during the night (the strict definition of 'sleeping through the night') can begin from about 3 months of age and it is advisable that prior to this age parents

### Common techniques

**Prevention** Educating parents before baby comes along on 'normal' sleeping patterns. Carrying out any or all of the following activities: allowing baby to settle to sleep by herself; giving a twilight feed between 10pm–12am; stretching time between night feeds after 3 months, by carrying out other activities such as rocking, cuddling, etc; emphasising the differences between day and night; limiting interactions – changing nappies only if necessary at night; responding to baby only if properly crying; allowing a time for fussing – and decreasing activity when baby starts to show sleepiness.

**Unmodified extinction** Allowing baby to 'cry it out' and only going in to make sure she is physically well. Little or no contact at night time.

**'Controlled crying'** Allowing intervals of time before comforting baby. For example, first waiting 5 minutes and then 10, and so on.

**'Camping out'** A parent stays in the room with baby but has limited or no contact with her until she is asleep. Over 2–3 weeks, the parent gradually moves further and further away, until he or she is out of the room or gradually reduces the time spent in the room waiting for baby to fall asleep.

**Scheduled awakenings** Waking a baby before her usual waking time at night to either feed or comfort her. Once a child reaches a point where she no longer wakes up by herself during the night, the parental awakenings can gradually stop.

**Positive routines or faded bedtime** A set of child-pleasing activities conducted before bedtime such as changing a child's bedtime to when she is naturally sleepy, giving her a bath or some milk, reading her a story or singing, cuddling or tickling her, etc. Once she successfully falls asleep by herself, bedtime can be gradually brought earlier.

respond immediately to their baby's cries in order to reduce overall crying time. After this time, according to F. Cook et al (2012), parents could choose to follow an intervention except for 'prevention', which should be followed from birth.

## CURRENT RECOMMENDATIONS

The NHS choices website suggests using controlled crying after 6 months of age for putting baby to sleep and for night awakenings, as long as things like hunger, unusual distress possibly caused by fears/bad dreams or temperature have been ruled out as the causes for waking.

## WHAT THE RESEARCH SAYS

### The most effective treatments

During the first couple of months, the most effective treatment to reduce how often and how long a baby cries is the speed of a parent's response. S. M. Bell and Salter Ainsworth, MD (1972) found that responding promptly to crying is more effective than any care given.

J. A. Mindell et al reviewing effective sleep interventions (2006) found, generally, that any intervention was enough to have an impact on most children. 94% of studies report that interventions are effective. The most effective treatments for reducing sleep problems are prevention followed by unmodified extinction; controlled crying is effective but to a lesser degree. Positive routines, faded bedtime and scheduled awakenings are also usually effective.

V. I. Rickert and C.M. Johnson (1998) compared scheduled awakenings with unmodified extinction to determine which was most successful at reducing night-time awakenings in babies and toddlers. Within 1 week of treatment, the latter was found to be the most successful at reducing night-time awakenings, however, at 8 weeks both were equally successful.

T. I. Morgenthaler et al (2006), however, showed that prevention was shown just as effective as camping out and unmodified extinction.

The above is simply a synopsis of what researchers have found to be most effective at 'teaching' a baby to self-soothe, however, the consequences of, in some cases, a lesson taught too quickly or harshly, are discussed later on and they need to be considered alongside the facts listed here.

### Effects of disrupted sleep

Teaching babies to self-soothe and get back to sleep by themselves is a key ingredient to unbroken, restful sleep for the whole family. In order to properly weigh up the pros and cons of controlled crying, one must consider the effects of never having a full night's sleep on both children and parents.

Insufficient sleep can affect a child's cognitive development, health, attention span and overall quality of life. B. Zuckerman et al (1987) showed that children with persistent sleep problems have more tantrums and behaviour management problems. 10 years later, another study of theirs revealed that about 40% of babies with a sleep problem that persists beyond 8 months may continue to have a sleep problem until at least 3 years of age. Some sleep problems may even extend into adulthood. There's also about a 20% chance that a baby who slept well at 8 months may develop a sleep problem at age 3. However,

H. Hiscock et al (2007) provided evidence that controlled crying and other interventions can halve the number of children experiencing persisting sleep problems into their second year.

Persistent infant crying has been shown to be particularly distressing to mothers, many of whom develop depression, feelings of low self-confidence and marital problems, as a result. What's more, the relationship between a baby and his mother is negatively affected (Papousek, M and von Hofacker, N, 1998).

H. Hiscock and M. Wake looked at the effects of lack of sleep on mothers who were suffering from postnatal depression (2001). They found that if these mums incorporated controlled crying (or similar interventions) into their night-time schedules, this greatly reduced sleep problems amongst their babies. As a result, the mums reported fewer depressive symptoms. The decrease in such symptoms was significant but small indicating that a variety of factors contribute to a mother's depression; her baby sleeping well is just a small part.

In a related longer-term study (2007), the same researchers found that controlled crying had a beneficial effect over a year later on maternal depression. This makes sense as persistent sleep problems that extend into the second year have been more closely associated with maternal depression.

Mothers who are consistently awoken by a young toddler experience anger about their child's requirements during the night and doubts about their competence as a parent. There is a chance that if these experiences are extreme, they could affect the mother-baby bond (Morrell, JM, 1999).

**Baby's short- and long-term mental health**

Controlled crying and its equivalents are effective at solving sleep problems, if the parents can carry it out but what of their effects on a baby? Controlled crying antagonists are considerably passionate about this topic.

Attachment theory, developed from the work of psychologist John Bowlby, states that a young child will form an attachment to her primary care giver. This attachment is not necessarily formed with the person who spends the most amount of time with a baby but rather with the person who responds most accurately to the baby's needs, referred to as 'sensitive responsiveness'. A baby will look to this person for security, comfort and protection. This person will act as a secure base from which the baby can confidently explore his world from about age 6 months to 2 years (E. Waters and E. M. Cummings [2001] have argued that this extends into early adolescence, although the secure base figure can extend to other people in a child's life). These bonds, formed early between parents and baby, are known to be critical for baby's healthy development

An important aspect of attachment is the responsiveness of a parent to hischild's distress. If a child is allowed to become distressed for long periods of time, it could result in him having long-term insecure attachment relationships, particularly if attachment wasn't formed well initially (Murray, L & Ramchandani, P, 2007). It is not known whether delayed parental responses to a child's cries can affect this attachment, but if it is affected by something like controlled crying, then there could be long-term consequences.

If a child does not form an attachment with an adult within her first five years it will have severe developmental consequences and can result in an inability to form emotionally meaningful relationships, cognitive impairment – particularly in language – and more aggressive behaviour. These are extreme consequences of non-attachment and really only occur in very specific situations (e.g. children raised in institutions or who are abused or have a mentally unwell parent), but they are mentioned often by controlled crying antagonists.

A. M. H. Price et al (2012) in a well-designed study, has shown that there is no negative effect on a child's later mental health (in 6 year olds) when controlled crying was carried out at about 7 months of age. They looked at emotional and behavioural aspects, as well as stress levels and parent-child closeness. H. Hiscock et al (2008) found no adverse effects on mental health from controlled crying at 2 years of age. No studies have looked at the effects of controlled crying on very young babies.

W. Middlemiss et al looked at the important link between a mother and her baby (2012). Following a normal day of interacting with each other, their cortisol (a hormone released in response to stress) levels were found to be synchronised (measured through saliva). Samples were taken before controlled crying was initiated and after the baby had fallen asleep. Cortisol levels in mum and baby followed the same pattern, initially. By day 3 of controlled crying, the baby no longer vocalised distress whilst trying to fall asleep, although her cortisol levels were still elevated. The lack of distress shown by baby (i.e. by not crying) had the effect of lowering mum's cortisol levels, even though baby was still distressed. This meant that mum and baby were no longer synchronised to the same extent. This is a relatively new study and the effects of this decrease in synchronicity are yet to be established. The fact that a baby's cortisol levels remained high even though he wasn't crying has been used as an argument against controlled crying. However, there is no evidence as to whether cortisol levels returned to normal or indeed remained high after a period of time. If a baby's cortisol levels return to normal in a relatively short period of time then the consequences of controlled crying would be minimal, but if they remain high for long periods, the consequences could be more severe.

M. C. Larson et al (1998) have found that the ability of a baby to produce cortisol in response to stress is greatest during the first few months of life. After this, cortisol levels begin to follow a natural rhythm and are not so easily elevated by stressful experiences, such as physical examinations. Cortisol levels continue to decrease until about 12 months of age, when it is thought that children become secure in the knowledge that their parents will protect them. Toddlers anticipate this protection so their cortisol levels are not as easily affected by minor stressors. However, there is evidence that if children experience less responsive and sensitive care from their parents, minor stressors will increase their cortisol levels. This tends to be true for children who are fearful or anxious or are easily frustrated. It is unclear whether controlled crying increases cortisol levels in the face of minor stressors as it is usually noted in children who experience long-term abuse. Salivary cortisol levels are also not a very reliable rating of stress according to M. R. Gunnar and B. Donzella (2002).

If there is a risk, most psychologists don't recognise it as severe and many advocate interventions (like controlled crying) to solve sleep problems (Blunden, SL, et al, 2011). That's probably because they recognise that the effects of sleep problems are detrimental to both parents and child, as well as other siblings, so it may, therefore, be wisest to pursue a lesser of two evils.

During the first 2 years of life, A.N. Schore (1996) found that a baby's brain development can be affected – positively or negatively – by her primary care giver. Whilst key parts of the brain are developing, prolonged episodes of stress impact negatively on maturation. The stress hormone affects the 'wiring' of the brain and can lead to greater vulnerability to developing psychiatric disorders. However, this is usually observed in extreme cases, such as abused children. It is unknown whether this theory can extend to an intervention like controlled crying.

R. Beijers et al (2012) looked at how the cortisol levels of 12 month olds who breastfed or co-slept with their parents to 6 months of age (the researchers allowed same room as well as same bed for this definition) were affected by a stressor. While a baby is very young, breastfeeding and co-sleeping are associated with lower cortisol levels. At 12 months of age, babies who co-slept with their parents had a lower cortisol response to the stressor and that of babies who had been breastfed returned more quickly to normal. This study lends credence to the idea that cortisol production before 6 months regulates cortisol production in later life as shown by less cortisol produced at 12 months of age. If a child experiences high cortisol levels early in life, for whatever reason, he may be more inclined to experience higher cortisol levels later on. Unfortunately, there are still not many studies dealing explicitly with this important

## Recommendations from the research

There are many effective strategies for solving sleep problems – though not every technique will work for every baby. The most effective and least invasive is prevention, but controlled crying, if properly carried out, has also been found to work. Whilst some good research has attempted to rule out any longer-term effects when children were treated with controlled crying at 7 months, this needs far more investigating in order to be conclusive, particularly the effects of this technique on babies younger than 7 months. All researchers agree that controlled crying should not be carried out on babies younger than 3 months as such young babies simply do not have the capacity to learn self-soothing or to even to go without milk for prolonged periods of time. After the first 3 months a method to encourage self-soothing can begin to be employed although it may take a baby some time to grasp the skill.

However, if a sleep problem is already causing stress within the family, this is also not good for baby. In this situation, many psychologists would recommend controlled crying (after 3 months) in order to improve well-being for the whole family.

It is important to stress that no intervention should be attempted if a baby is unwell.

issue and more research is needed to evaluate any effects of causing distress in babies. As a whole, the research is inconclusive.

**Incorporating interventions**

Sleep-deprived parents can often find it difficult to stick to the sometimes stringent guidelines of a specific sleep problem intervention. To disturb a sleeping baby for a twilight feed (about the time mum goes to bed) can feel detrimental to the goal and the thought of rocking, cuddling or singing in order to delay a feed at 3 a.m. is the last thing a parent wants to do, especially if he or she has to be up in a few hours for work. Many parents also have intense struggles with adhering to any intervention that involves letting their baby become distressed. According to I. St. James-Roberts et al (2001), difficulties with following a sleep problem intervention can result in parents not following them correctly leading to decreased effectiveness. It's also common for many parents to fail to adhere to a set of consistent rules for bedtime and night-time (Goodlin-Jones, BL, et al, 2001). The important thing to take from this is there's no real point in attempting an intervention, such as controlled crying, unless you're sure you can do it.

# MMR vaccination

Immunisation is the ability of our body to fight off microbes, even if it first meets them during the early years of life. This is because our immune system remembers microbes that it encounters and, even better, remembers how to fight them the next time it meets them.

Vaccination is the man-made attempt to allow our body to become immune to new microbes without actually having to become ill. A weakened or dead form of a microbe is introduced to the body, and the body identifies the microbe as foreign and begins producing proteins called antibodies to kill it. Like a key to a lock, each antibody is specific to one microbe. When the body encounters the microbe in the future, it is able to fight it off without illness. The vaccine has made the body immune to the microbe. No other treatment for disease, including antibiotics, has been as successful at saving lives.

Measles is a highly contagious and dangerous viral disease. It can cause high temperatures, coughing, vomiting and can make a child feel quite unwell for days. At its worst, it can cause blindness, brain damage or death.

Mumps is also caused by a virus and predominantly results in swelling of the salivary glands found on both sides of the face just below the ears. At its worst, it can cause complications such as viral meningitis and other serious infections or if caught later in life, can reduce male fertility.

Rubella, also known as German measles, is a relatively mild viral infection. It is characterised by a distinctive rash and cold-like symptoms and although not dangerous to children or adults, it can be devastating for unborn babies. If a pregnant woman catches the illness during her first 16 weeks of pregnancy, her baby could suffer from eye problems, deafness, heart abnormalities and brain damage.

Children in the UK are routinely vaccinated against measles, mumps and rubella, primarily to prevent an outbreak of measles. While WHO recommends a 95% uptake of the vaccination amongst the population, current rates are short of the goal. This puts your own child's health as well as the nation's at risk, which has recently been documented in certain areas in the UK where lower uptake of the vaccine has resulted in measles outbreaks.

## CURRENT RECOMMENDATIONS

As part of NHS vaccination schedule, all children are offered (free of charge) an initial MMR vaccination at 12–13 months and a second dose at 3 years and 4 months, or soon after.

### Wakefield controversy

The MMR vaccine was introduced in 1988 and has since been successful in decreasing the cases of measles, mumps and rubella. In 1998, Dr. Andrew Wakefield, a surgeon and senior lecturer at the Royal Free Medical School, along with 14 other researchers, published a paper in which they implied that there was a link between the MMR vaccine and autism and inflammatory bowel disease (enterocolitis). Although the researchers did not express a direct link, media coverage asserted that there was one (Wakefield, AJ et al, 1998, retracted 2010).

The same year Wakefield's paper was published, a Medical Research Council seminar consisting of 37 scientific researchers (including experts in the field of virology, gastroenterology, immunology and autism) reviewed all the published and unpublished evidence on a possible link between the vaccine and autism and inflammatory bowel disease. They found no link and no reason to discontinue the practise of immunising babies and children. Another large study, conducted that year, also found no link.

However, Wakefield's theory was again given prominence when an Irish researcher, in conjunction with Wakefield, discovered that a small number of autistic children were found to have the measles virus in their guts (O'Leary, JJ et al, 2000). Another paper, written a couple of years later, in which both Wakefield and O'Leary were involved, proposed a link between the measles virus and inflammatory bowel disease in children with developmental disorders (Uhlmann, V et al, 2002) although there was no specific mention of the MMR vaccine. A further U.S. study corroborates this finding but contains no evidence that the vaccine caused the condition. These inspired a new study to investigate if the immune responses were affected by the MMR jab; no difference was found in the immune responses of 10-12 year olds who had autism and those that didn't. Another study, which also included O'Leary, found no link between autism and the measles virus present in the intestines (Hornig, M et al, 2008).

Several researchers reported that the number of children with autism or bowel disorders had not significantly changed in the decades before or after the introduction of the vaccine. Furthermore, a Japanese study, found the opposite. They discovered a rise in autism after the MMR vaccine was withdrawn in 1993. Various, well designed studies, have now proven that there is no link between the MMR vaccination and autism (Hensley E and Briars L, 2010)

Closer investigation by B. Deer (2011) into Wakefield's original research found that this contained errors, was to be used as evidence in a lawsuit for a claim against a pharmaceutical company that made the vaccine and the solicitor in charge of the lawsuit had given Wakefield money years before the paper was actually produced.

Wakefield, after facing a General Medical Council hearing, was removed from the medical register and his paper was retracted in February 2010. However, media exposure of Wakefield and O'Leary's findings led to a sharp decrease in the number of parents opting for the MMR vaccination. In 2001, the rates of uptake in England for the MMR vaccine were as low as 84%, a drop of 8% from the 1995-96 rates of 92%, before the controversy.

## WHAT THE RESEARCH SAYS

### Recognised side effects

As with all vaccines, there are side effects after an MMR immunisation, although most are mild. In S. Pérez-Vilar et al large sample study (2012), just under half the babies suffered from fever and just over a third suffered swelling at the injection site although only 8% of reactions were classified as severe. Other more rare side effects include rashes, loss of appetite, a mild form of the mumps, seizures or severe allergic reaction. Despite all these possible side effects, within the study group, all children made a complete recovery. The researchers concluded that it is a safe vaccine.

**Individual jabs**

On the back of Wakefield's 1998 paper, he proposed that it might be safer for individual doses of measles, mumps and rubella vaccines to be given because of a possible overload on the immune system. His suggestion was not based on sound research

### Immunisation schedule

**2 months**
5-in-1 (DTaP/IPV/Hib) vaccine. This protects against 5 separate diseases: diphtheria, tetanus, pertussis (whooping cough), polio and Haemophilus influenzae type b (Hib is a bacterial infection that can cause severe pneumonia or meningitis in young children).
Pneumococcal (PCV) vaccine.
Rotavirus vaccine.

**3 months**
5-in-1 (DTaP/IPV/Hib) vaccine, second dose.
Meningitis C.
Rotavirus vaccine, second dose.

**4 months**
5-in-1 (DTaP/IPV/Hib) vaccine, third dose.
Pneumococcal (PCV) vaccine, second dose.

**Between 12 and 13 months**
Hib/Men C booster. This is given as a single jab containing meningitis C (second dose) and Hib (fourth dose).
Measles, mumps and rubella (MMR) vaccine, given as a single jab..
Pneumococcal (PCV) vaccine, third dose.

**3 years and 4 months, or soon after**
Measles, mumps and rubella (MMR) vaccine, second dose.
4-in-1 (DTaP/IPV) pre-school booster. This is given as a single jab containing vaccines against diphtheria, tetanus, whooping cough (pertussis) and polio.

**Around 12-13 years**
HPV vaccine, which protects against cervical cancer (girls only) – three jabs given within six months

**Around 13-18 years**
3-in-1 (Td/IPV) teenage booster, given as a single jab which contains vaccines against diphtheria, tetanus and polio

**Around 13-15 years**
Meningitis C booster.

and since the link between autism and the combined MMR has been proven to be wrong, the case for separate jabs is limited (Pearce, A, et al, 2008). When given separately, there must be a length of time between injections, during which a child is susceptible to catch any of the illnesses. Also, when the injections are given separately, up to 48% of children remain unimmunised as a result of parents not managing repeat appointments. The side effects of the MMR jab are no different to those experienced by toddlers for the individual jabs, they're just more spread out in the latter (Halsey, NA, et al, 2001). (A separate mumps vaccine is no longer available in the UK nor will the NHS give individual jabs.)

Another reason why some parents opt for single jabs is because they don't believe that all three illnesses are severe enough to warrant immunisation (Brown KF, et al, 2012). Whilst they may not always be severe for the child e.g. rubella, they can have dire consequences for developing fetuses; this is an important reason why children should be routinely immunised.

**Vaccine overload**

Another commonly held view differing slightly to Wakefield's theory is that a young child's immune system may become overloaded if given a combined vaccination. This 'overload' is thought to make the body more susceptible to disease in the months after the vaccine is administered. (Dr. Wakefield's connection was that the overload could trigger inflammatory bowel disease and autism in susceptible children.) Moreover, vaccines may be linked to autoimmune diseases, such as type 1 diabetes, and allergies. Animal studies, however, have shown no strong evidence that combined vaccines can cause the development of diabetes or allergies, such as asthma (Stratton K, et al, 2002).

J. Stowe et al (2009) looked at the numbers of severe infections in children in the months following the MMR vaccination in an attempt to see if a child's immune system was less able to fight off infection because of the combined vaccine – as a result of being 'overloaded'. Not only did they find no correlation between infections (viral or bacterial) and the vaccination, they concluded that there may even be a protective effect – children were less likely to be poorly from other illnesses in the months following the vaccine. This rules out the concept that the combined MMR vaccine would cause the immune system to become overloaded and less able to fight off other infections in the months following the jab.

**Thimerosal and autism**

Another common concern is not with the MMR vaccine but rather with its preservative which contains mercury or thimerosal. This was the topic of a second trial in an 8-year long US federal court case against a vaccination manufacturer on the grounds that vaccines( in this case the preservative, thimerosal) was linked to autism. The other trial was the MMR and autism link. Both trials were lost in 2010 and it is now widely accepted that neither the MMR jab nor the mercury preservative, thimerosal, found in vaccines causes autism (Kirkland, A, 2012).

## Recommendations from the research

The research against the combined MMR vaccine has been disproved and there is no chance that a child will develop autism as a result of the vaccine. There can be some side effects of having the combined MMR but in general these are considered far less severe than  the risk of an outbreak of measles, for example. The evidence for giving individual jabs for measles, mumps and rubella, is relatively weak and increases the risk of a child missing one of the injections and subsequently not being protected. Since mumps vaccine is no longer available in in the UK apart from the combined immunisation, it would mean a child undergoing individual jabs would not be protected against it. It is also unlikely that the combined jab will cause children to be more susceptible to illnesses because of an immune system overload. There may be some ethical dilemmas surrounding the use of vaccines but decisions need to be made wisely, in the best interests of the individual child and also for society as a whole. Hopefully, the public's trust in the 'vaccine' in general, which has saved millions of lives will continue to return particularly following the recent outbreaks in the UK and the clear disadvantages to missing vaccines.

**Objections to the source of the vaccine**
Viruses cannot reproduce alone; they require living cells to make new copies of themselves. While some, like the chickenpox vaccine, require human cells to carry out their reproduction ,others are able to successfully be made from animal cells, such as chicken embryos. In the 1960s, cells were taken from two naturally aborted fetuses and used to develop the rubella, chickenpox, hepatitis A and rabies vaccines. (The fetuses were not aborted for the purpose of making the vaccines.) These fetal cells are kept in laboratory conditions and they continually divide and replenish themselves so no additional cells are required. Today, many of our vaccines are manufactured using these cells (called cell lines), which have consistently proven to be a safe and effective in allowing viruses to reproduce for the purpose of making vaccines. Neither has technology developed to the extent whereby we can manufacture these vaccines in another way. There is also a cost issue involved as to whether it makes economical sense to pursue another method of manufacture when there exists a safe and effective method already. Ethicists and most religious groups advocate that vaccines should not be refused and no person opting to have one is complicit in the abortion of the original fetuses (Grabenstein, JD, 1999).

# Dummies

There are many topics on which parents (and grandparents) develop fairly vehement viewpoints but when it comes to dummies, everyone is ready to take a stand. Exhausted parents see them as vital to a good night's sleep but others, particularly when confronted with a dummy-sucking five year old, view them as a bad habit that should be instantly broken.

## CURRENT RECOMMENDATIONS

The World Health Organisation (WHO) and pro-breastfeeding organisations are not in favour, and The British Dental Health Foundation counsels avoidance. The Foundation for the Study of Infant Deaths (FSID), however, advocates dummies for sleeping and nap times.

## WHAT THE RESEARCH SAYS

### Breastfeeding may be curtailed

A mum who introduces a dummy to her baby is more likely to stop breastfeeding earlier than a mum who doesn't, which is why WHO have recommended refraining from offering a dummy. However, it's not quite that simple. WHO's researchers determined that it wasn't the introduction of the dummy that caused less breastfeeding but the mother's motivation (or lack of it) to continue. If a woman is struggling with breastfeeding and is on her way to stopping, she's more likely to introduce a dummy. So it's not the dummy, per se, that stops the breastfeeding even though the researchers (Jaafar, SH, et al, 2012) couldn't rule out the detrimental effect of introducing a dummy before 4 weeks of age, motivations aside. (This makes sense since the first 4 weeks of demand-led breastfeeding are important for starting the flow of milk.)

### Greater risk of middle ear infections

Babies who continue to suck on a dummy past the age of one develop more middle ear infections (acute otitis media) than those who don't use a dummy. One study suggested the risk was twice as high (Jackson, JM and Mourino, AP, 1999). The American Academy of Pediatrics recommends weaning a child from a dummy in the second six months of life in order to prevent her habitually suffering middle ear infections. However, if a child is limited in dummy use or only uses a dummy when falling asleep or sleeping, then the risk of developing a middle ear infection is all but extinguished (Niemelä, M et al, 2000).

### Oral thrush becomes more prevalent

Most of the population carries Candida albicans, the fungus that causes oral thrush or candidiasis, in their mouth or gut and sometimes in the vagina. It is easily treated and not usually considered to be harmful although if it affects the back of the throat, it can cause discomfort when swallowing, lead to an infection of a breastfeeding mum's nipple and result in a stubborn nappy rash.

The research suggests that children who suck a dummy are more likely to have fungi from the candida family in their mouths. That does not necessarily mean they will be ill – as most healthy children are able to fight off these microbes – but if a child is already not very well, she may be at greater risk of developing the symptoms of oral thrush. E. Comina et al 2006 found that the dummy acts as a reservoir of microbes, infecting the child over and over again. L. C. da Silveira et al 2009 research into dummy types – latex or silicone – showed that after 16 weeks' use, the surface of a silicone dummy remains relatively unchanged and smooth. However, a latex dummy will have significant fissures where microbes could escape cleaning and potentially re-infect a child. If your child uses a dummy, change it frequently to help reduce the chances of infection.

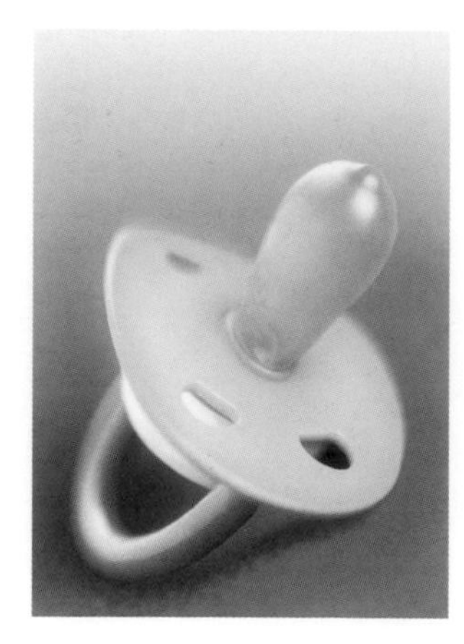

**General ill health**

There has been a connection made between dummy use and an increased incidence of colic, wheezing, vomiting, fever, diarrhoea and general poor health. But whether dummy use is to blame for this is not as straightforward. Younger mums and those who smoke, have lower levels of education or experience financial difficulties, are more likely to use dummies and researchers are unable to rule out whether other contributing factors, such as poor dummy hygiene or overcrowded housing, facilitate the spread of disease.

**Poor dentition**

Using a dummy can affect the arrangement of a young child's teeth and as a child needs baby teeth to chew and speak, misaligned teeth can cause problems with both. The extent to which baby teeth affect adult teeth is unclear but many researchers clearly state that poor arrangement of baby teeth can lead to poor arrangement of adult teeth. (Franco Varas V & Gorritxo Gil, 2012).

However, it looks as though a baby needs to suck on a dummy for at least 2 years before there are any adverse effects to her teeth. There is also evidence suggesting that, in some cases, poor alignment can spontaneously correct itself if dummy sucking ceases. Orthodontic dummies are very popular with parents but unfortunately, no research has shown that they are any less harmful to baby teeth.

**Early childhood caries**

This will be diagnosed if a child has one or more decayed baby teeth or baby teeth that have fallen out due to decay. Generally, dummy sucking is not to blame although prolonged dummy use can result in a higher incidence of salivary lactobacilli and candida in a child's mouth – conditions that are significant risk factors for childhood caries (Ollila, P, et al, 1997). Moreover, some parents coat their dummies in something sweet and this practice leads to tooth decay (Peressini S, 2003).

Dummy use can cause teeth to become misaligned and crooked teeth can be difficult to clean. If one tooth completely or partially blocks another, it may be almost impossible to brush the 'hidden' tooth properly, leading to decay and disease.

### Possible speech disorders

Misaligned teeth, which can be caused by dummy use, can lead to speech disorders. C. Barbosa et al (2009) found that speech disorders are 3 times more likely in children who use dummies for 3 or more years. There has also been a link made with poorer speech development and dummy use because a dummy limits the amount of babbling or talking practice a baby does. Also, as sucking on a dummy increases the risk of middle ear infections, recurring infections affect speech because a child can't hear very well.

### Accidental damage

Unfortunately, many untoward incidents are exacerbated when a child sucks on a dummy at the same time. Most commonly, if a child sucks whilst learning to walk, she invariably falls and the dummy in the mouth worsens what is usually a pain-free experience. Moreover, a dummy-sucking child is more likely to displace a tooth and bruise her mouth during a fall (although she is less likely to fracture a crown than a child who falls without a dummy, as the dummy broadens the force of the impact).

Cases have been reported of children swallowing the dummy's nipple, which resulted in a complete bowel obstruction (Rubi, SC, et al, 1990 and Neville, HL, et al 2008) and, much more grimly, children have been strangled by the cords used to attach their dummy or asphyxiated by inhaling both intact and shredded nipples.

### Latex dummies and allergies

Latex is a common substance that is difficult to avoid; it is found in both obvious and obscure items such as balloons and the buttons on your remote control. If your child is predisposed to allergy or considered atopic (there is a family history of asthma, hay fever, eczema, etc.), recurring contact with a latex dummy can cause her to become sensitised to latex or can trigger an allergy or allergic reaction.

### Help in preventing SIDS

R. Y. Moon and L. Fu (2012) found a strong link between using a dummy and a decreased risk of SIDS. So promising are the findings that the American Academy of Pediatrics has recommended that a baby use a dummy for the first year of her life when put to bed. But it advises not starting until after the first month if breastfeeding (to protect against possible interference). The risk of SIDS is highest between 2 and 3 months of age, so holding off until after the first month will still mean a baby is protected when she needs it most.

### Fewer sleep disruptions

Sleeping aids such as dummies or teddies seem to reduce the occurrence of sleep disorders but they don't appear to result in less crying or fussy behaviour. One study, albeit with a different primary focus, noted that the time it took to comfort a crying baby at 4 weeks and at 6 weeks with a dummy was much the same as the time taken to comfort her in other ways, e.g. rocking, stroking, tickling, etc....

The joy of using a dummy seems somewhat overlooked by the researchers; a dummy frees mum and dad from having to rock or tickle their baby and perhaps lets them have a well-earned sit-down! Whilst not conclusive research, C.M. Johnson's survey of parents (1991), found that offering a dummy was effective at quietening their children. Since crying has detrimental effects on a baby including an increased heart rate and blood pressure, less oxygen getting around the body and even brain injury, S.M. Ludington-Hoe et al (2002) recommended immediate interventions to stop it and dummy use is one of the few options that does not require a parent's presence for long periods.

**Shorter hospital stays for pre-term babies**

Using a dummy has also been found to reduce the length of time a pre-term baby stays in hospital. The length of hospital stay is dictated by the ability of a baby to gain weight (when no other complications exist), which is directly related to her sucking ability. Sucking on a dummy makes it quicker for babies to get their nourishment orally (instead of through feeding tubes), which also has a positive effect on their vital signs (heart rate and oxygen saturation) (Pinelli J and Symington A, 2000).

**Pain relief**

For very young babies, sucking a dummy has been found to be an effective pain reliever during procedures such as the heel-prick test. R. Carbajal et al (1999) found that pain (as best they could measure it in very young babies) was all but diminished if they coated a dummy in sugar for the procedure. Other research corroborated these findings but only for babies younger than 3 months. Babies older than this may need a stronger concentration of sugar.

## Recommendations from the research

Giving your baby a dummy may be a good idea for two main reasons:

**1** It's safer; there's less chance of SIDS and

**2** It's very effective at comforting her without you having to be around.

You can also avoid all of the negatives if you follow some simple rules:

* If breastfeeding, don't introduce a dummy before one month of age;
* Try to wean your baby off it, particularly during the day, between 6 and 12 months;
* If you have an early walker, don't let her suck a dummy and walk at the same time.
* If there's a history of allergy in your family, stick to silicone dummies and change them every month or so, even if you use a steriliser.

# Colic

Colic can be diagnosed by applying a rule of three: at least three hours of crying for no apparent reason, for at least three days a week, for at least three consecutive weeks. However, some parents find it more useful to refer to colic as 'PURPLE':

* Peak of crying (at 2 months)
* Unexpected
* Resists soothing
* Pain-like face
* Long lasting (up to five hours a day)
* Evening.

Colic affects between 10-30% of babies worldwide and there is no known cause. The leading theory is that there is not just one single aggravating factor but a few, so it is worth trying as many techniques as you can; something may work for your baby.

Breastfed and bottlefed babies suffer equally; there's no difference in the numbers. This would imply that if you are breastfeeding and are considering stopping as a way to treat colic, it's more than likely that won't do any good.

The comforting news is that there doesn't appear to be any detrimental effects on a baby suffering from colic. The vast majority tend to gain weight as normal and grow into healthy children. Studies comparing toddlers suffering from (or free) of colic when newborn, found no difference in either temperament or health. Nor are colicky babies more likely to develop allergies. Once the colic has abated, mum and dad also should make a full recovery from any depression or stress-related problems resulting from it. In severe cases, however, some lasting consequences such as difficulties in communication and unresolved conflict between family members may persist (Räihä, H, et al, 1996). Any differences between families whose children suffered from colic and those who didn't, all but disappear after three years.

Convincing parents that there isn't a serious disease or illness at play is understandably difficult. There are, of course, diseases and illnesses including gastrointestinal reflux, infantile migraine, meningitis and middle ear infections, as well as physical traumas such as something in a baby's eye or a bone fracture that could mimic colic and these need to be eliminated by a doctor's physical examination before colic is diagnosed.

It may not come as great comfort, but perhaps no surprise, to know that it may take over a year for parents to fully recover from their infant having colic. The greatest difficulties are suffered by parents who are unable to console their distressed baby. But all parents of colicky babies experience sleepless nights and stress, and some even feel rejected. When parents are tired and frustrated, babies are at greater risk of abuse and, in very severe cases, colic can interrupt the necessary bonding between parents and child. A failure to bond can lead to a child having behavioural problems later in life. Therefore, crying that persists or doesn't begin to subside by three months, should be investigated further. Most importantly, colic should not be dismissed. Whilst it is not serious for an affected baby, it is very serious for his parents.

## CURRENT RECOMMENDATIONS

The National Institute for Health Care Excellence (NICE) recommends, once a diagnosis of colic is confirmed and any other causes are ruled out, that parents should first be reassured that their baby is not rejecting them and that the condition will pass in time (by 3–4 months of age). Parents should be informed that holding their baby throughout the crying may be beneficial and they should get lots of support and individualised instructions on how to best combat it. Hypoallergenic formula may be considered but only under medical guidance.

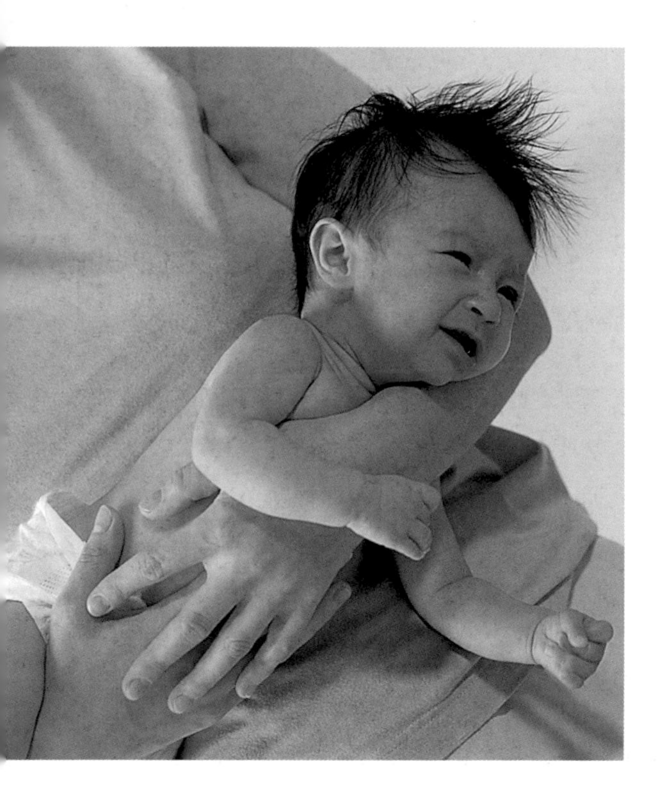

## WHAT THE RESEARCH SAYS

### Possible causes

From the available information, it would appear that infant colic is caused by a number of factors, each one affecting a suffering baby in varying degrees. All the causes fit with the natural ending of colic symptoms at around three months of age.

***Gastroinestinal abnormalities:*** Probably the most common theory is that colic is caused from a tummy ache. This is largely because of how a baby positions himself during an episode (brings knees to the abdomen and clenches the fists) and the fact that he usually passes wind towards the end. Radiographic images, however, have shown that this wind production is actually a result of the excessive crying rather than the cause of the colic itself.

A related theory is that colicky babies suffer from hyperperistalsis. Peristalsis is the natural movement of the intestines that helps food to move through the digestive system. Hyerperistalsis is when this movement is too rapid and causes cramping and pain. Motillin and ghrelin are gut hormones that increase peristalsis to clear food out ready for the next meal. Babies who suffer from colic have been found to have higher levels of these hormones in their blood than do non-colicky babies (Savino, F, et al, 2006). Furthermore, the levels of motilin have been found to be

***Gentle swinging***
*Many parents find gently moving their baby back and forth and slightly up and down helpful in easing crying*

slightly higher in formula-fed babies than in breastfed ones (Lothe, L, et al, 1990). Some babies may be intolerant to cow's milk or other allergens (if breastfed); if this is the case, a baby may produce lots of wind, followed by relief on expulsion.

***Parental care:*** If they have an inconsolable baby, parents often doubt their parenting skills. Studies show, however, that when a colicky baby is in the care of a trained occupational therapist, he will still cry for twice as long as a baby without colic. There is no evidence that colic is related, in any way, to a parent's anxiety levels or personality type although colic will dramatically increases stress and anxiety within a family.

***Feeding disorder:*** Babies who suffer from colic have been found to be less rhythmic with their sucking, show less interaction with their mothers during feeding and experience more discomfort after being fed than do babies who don't suffer from colic. C. Miller-Loncar et al (2004) also showed they are more likely to suffer from reflux (regurgitation of the stomach contents tino the mouth, nose or oesphagus.

***Lack of 'friendly' bacteria:*** The composition of bacteria that line the intestines may not be optimal during the first few months of life. F. Savino et al (2011) showed that colicky infants have fewer types of bacteria that promote the digestive process and larger amounts of gas-producing bacteria.

***A difficult temperament:*** There is no evidence to suggest that a colicky baby is simply a difficult child in the making. However, if colic symptoms persist well beyond three months, they could cause behavioural problems in toddlerhood and beyond. It's unclear what causes 'difficult' behaviour but unfortunately it may be due to an interruption in the natural bonding process between parent and child as a result of a long-lasting bout of colic.

***Exposure to cigarette smoke:*** I. Matheson (1995) found that mothers who breastfed and smoked more than five cigarettes a day were more likely to have babies who suffered from colic .

I. Milidou et al (2012) found that babies who were exposed to nicotine (as in nicotine patches) during pregnancy were also more likely to suffer from colic .

***Learning to sleep:*** There is a suggestion that colicky babies are less good at sleeping than non-colicky ones. In a study (White, BP, et al, 2000) that compared colicky babies with a control group, the latter slept, on average, for two more hours during the day and had less disrupted nighttime sleep than those who had colic. Colic, therefore, might be due to a delay in the natural establishment of the circadian rhythm (a 24-hour body cycle that controls when we sleep and can be affected by light and dark, as well as temperature).

***Incomplete breastfeeding:*** Breastfeeding mothers are no doubt familiar with the consistency of their milk (the first milk is more liquid and thirst quenching and the later milk is creamier and satisfies the appetite better). One theory for the cause of colic has to do with the possibility that some breastfed babies may get too much of the former and not enough of the latter milk. This can occur if a mother switches breasts before a breast is completely drained of milk so that her baby doesn't get enough of the fattier, more appetite-satisfying, milk and too much lactose from the thinner milk, which travels rapidly through the intestines and ferments in the large intestine resulting in a constantly hungry baby, frothy or explosive poo and frequent crying.

A study conducted to compare the effects of completely emptying one breast (i.e. reaching the creamier milk) with draining both breasts equally (a feed of mainly thinner milk), showed that the former was less likely to produce babies who suffered from colic (Evans, K, et al, 1995).

Feeding on demand rather than feeding on a 3-to-4-hourly schedule has also been shown to decrease the chances of colic occurring.

***Normal neurodevelopment:*** It's possible that the period between two weeks and four months of age is simply a stage of neurodevelopment in which crying is an aspect of behaviour. Some babies cry more than others within this time frame. On the 'normal crying' scale, colicky babies lie towards the extreme end of crying all the time while babies who rarely cry at all, are at the other end. This explains why the crying subsides after about three months.

**Treatments**

There is no proven cure-all for colic because there is not just one proven cause. Below you will find some methods that are used to reduce symptoms – although not all 'work'. You should talk through the strategies with a health professional but probably by the time you've tried all of them, your little one will have moved out of the crying phase!

***Simethicone medications:*** Very popular with parents as a weapon in the battle against colic and very safe for young babies, simeticone enables a baby to more easily pass wind. However, research does not tend to support the idea that this helps with the actual colic (Metcalf, TJ, et al, 1994), lending credence to the concept that colic is not caused solely by wind.

***Gripe water:*** A remedy containing any of the following herbs: cardamom, chamomile, cinnamon, clove, dill, fennel, ginger, lemon balm, licorice, peppermint, or yarrow, gripe water is thought to relieve symptoms of wind and indigestion though P. Lucassen (2010) found no real evidence to suggest that it has any effect on colic. Nor is it suitable for young babies as it is not entirely risk-free. If you choose to give your baby gripe water, and many parents do, then use one that doesn't contain sugar or alcohol. Be aware, too, of the product's country of origin; D. Sas et al (2004) have found reports of contaminated bottles causing septic shock.

***Pain-relieving medications:*** Certain medications can prevent or slow down the involuntary movements of the smooth muscle found in the stomach and intestines. They work by limiting peristalsis, which is thought to be overactive in a colicky baby. Two of these – dicyclomine hydrochloride and cimetropium bromide – have been shown to be effective in the treatment of colic, but are not a cure. Dicyclomine hydrochloride, however, has a few rare but very serious side effects such as breathing difficulties, seizures and coma, so is not recommended by NICE. Cimetropium bromide, which is widely used in Italy, has the less serious side effect of sleepiness (Savino, F, et al, 2002) although there is still some debate as to its effectiveness. Neither has been effective for all colicky babies.

A herbal remedy made of a mixture of three herbs (extracts of chamomile, fennel and lemon balm sold as 'Collimil') has been shown to work in a similar fashion to cimetropium bromide (mentioned above) by slowing intestinal peristalsis. Colicky babies cried for less time after receiving the extract twice a day for one week (Savino, F, et al, 2005). There are no apparent side effects.

***Changing a baby's milk:*** If there is a family history of allergies and your baby is suffering symptoms of colic, it's possible that a temporary allergy to milk proteins is causing the problem. Some babies, who are not necessarily allergic to cow's milk, may still have some trouble breaking down the proteins found in it. This does not mean that your baby is permanently allergic to cow's milk because such intolerance can be short-lived.

If you are breastfeeding, you could try (under guidance from a GP or medical professional) eliminating known allergens from your diet, such as dairy, nuts, wheat and soya. These would need to be eliminated one at a time so the culprit could be identified. Studies conducted on breastfed babies whose mothers reduced their intake of allergens, showed a lessening of colic symptoms (Hill, DJ, et al, 2005).

If your baby is formula fed, there are partially hydrolysed- (broken down) protein formulas available. The inability of some babies to break down proteins could account for some cases of colic. Fully hydrolysed-protein formulas are only available by prescription and will be needed if your baby has a true cow's milk intolerance. Results are fairly inconclusive regarding the effect that this type of formula has on crying time or on colic. Some research that shows switching to a hypoallergenic formula can alleviate symptoms for some babies.

A temporary lactose intolerance in a baby is a result of a deficiency of the enzyme lactase, which leads to an inability to properly break down the lactose in milk. It could result in colic-type symptoms. Most babies grow out of it. Lactose-free formula milks are available, as are lactase drops, which aid the breaking down of lactose in either formula or breast milk. One study found that its use produced a significantly positive effect on colic symptoms, but it was not useful for all babies, again implying that there is probably more than one cause for colic (Kanabar, D, et al, 2001).

Soya milk formulas are not recommended for babies with an intolerance to cow's milk because these babies become more at risk of developing an allergy to the soya protein. Morever there is also little evidence that soya milk has any effect on colic (Bocquet, A, et al, 2001).

No change should be made to formula milk without medical advice, as changes could do more harm than good.

***Anti-colic bottles:*** These 'work' by preventing excess air from mixing with the baby's milk thus reducing the amount of air a baby takes in while bottle feeding. The evidence suggests, however, that a build up of wind is not usually the cause of colic and breastfed and bottle-fed babies suffer equally. So whilst such bottles may be very effective at limiting wind in a newborn, they may not have any effect on colic. There is, however, plenty of anecdotal evidence from parents saying that anti-colic bottles are effective at reducing their baby's symptoms, although it's unproven whether these babies truly have colic. The bottles tend to be expensive, but if you're at the end of your tether, they may be worth a try.

***Probiotics:*** Liquids and pills that contain 'friendly bacteria' – those that decrease harmful bacteria – work by optimising the existing 'friendly' bacteria of the digestive system; studies show they produce promising results in treating colic. Specifically, F. Savino et al (2007) showed that Lactobacilus reuteri DSM 17931 decreases the crying time of colic by up to 50% and works within seven days. So far, it has been shown to be safe for babies. Further research is being conducted and hopefully probiotics will soon be proclaimed as a recommended, safe and effective treatment for colic.

***Glucose or sucrose:*** M. Akcam et al (2006) looked at the effects of giving a glucose solution (30% glucose, 70% cooled boiled water) to babies everyday for 4 days and found that parents reported a moderate decrease in symptoms. Glucose is cheap and safe for a baby but when used in excess has been linked to tooth decay, even before teeth have appeared. Another study, which looked at the effects of a giving a 45% sucrose solution before a crying bout was fully underway, showed that it had some calming affects on a colicky baby (Barr, RG, et al, 1999). If you want to give some to your baby, make sure everything is totally sterile.

***Massage and manipulation:*** Although not proven to treat colic, massage may have indirect benefits; it provides sensory stimulation, which can pacify a distressed baby, and it can improve parent-infant interactions adversely affected by colic.

Manipulative therapy is based on the concept that a baby's well-being flows from the smooth functioning of his muscles, skeleton, ligaments and connective tissues. Techniques include spine and head palpation and must be carried out by chiropractor or osteopath, which limits its usability for a typical evening bout of colic and its effectiveness has not been proven.

Studies have also found no evidence that rocking or vibration (whether when holding a baby in one's arms or when pushing him in a pram or driving him in a car or from a device) is effective at relieving symptoms.

Carrying a colicky baby has also not been shown to decrease his crying time although it does decrease the crying bouts of non-colicky infants. I. St. James-Robert et al (2006) looked at the effects of parent-baby contact on crying and found that babies who had more contact with their parents (up to 15 hours a day) cried for less time. However, they found no significant difference with colicky babies. The chances of a baby suffering colic remained the same no matter how often he was held . What it does show is that puting a colicky baby in his cot and taking a break won't affect the colic. The colic won't last longer because

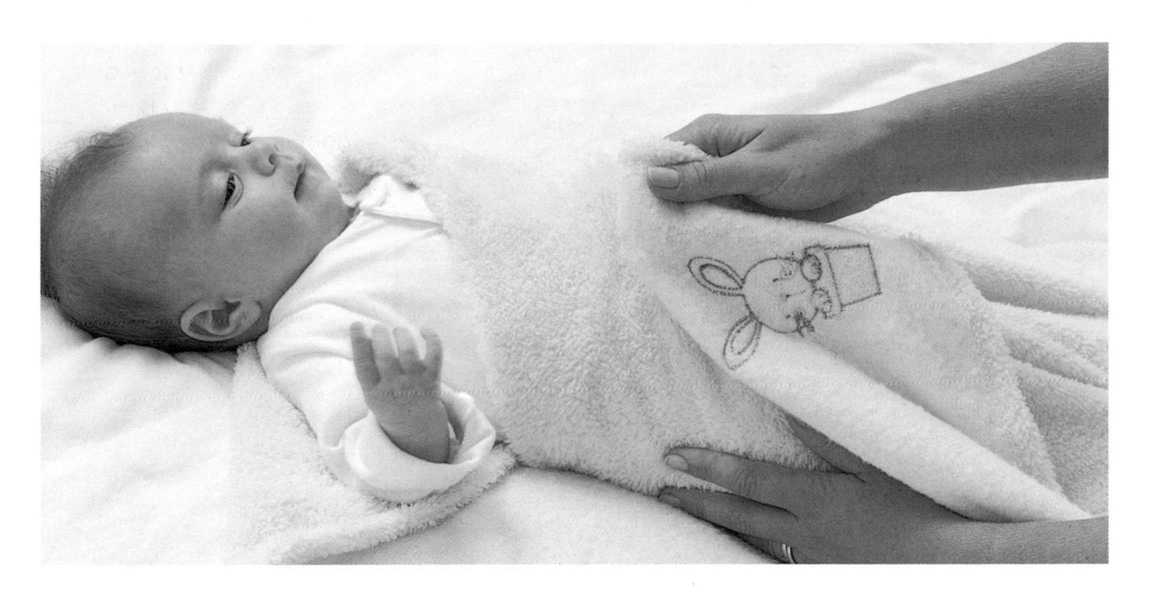

***Swaddling***
*Wrapping a baby securely in a light covering can prevent her being 'disturbed' by movements of her arms and legs.*

you had a break, and your baby won't suffer more by being on his own.

***Avoiding overstimulation:*** Some babies appear to be more responsive than others, which may make them more vulnerable to loud noises and/or a fast-paced environment with the result that they cry inconsolably (St James-Robert, I, et al, 2003). A study that compared colic sufferers with non-colicky infants found no difference in cortisol (a hormone released by the adrenal glands when the body is under stress) levels or other stress-related symptoms. This would imply that colicky babies do not necessarily react to a stimulating environment (White, BP, et al, 2000).

## 5'S' approach

The following calming techniques are effective if done correctly and used in combination.

1. **Swaddling** done so that the hips can be flexed and it doesn't cause overheating.
2. **Side or stomach lying** but not for when a baby is sleeping, because of the risk of SIDS; babies must always be put to bed on their backs.
3. **Shhh (shushing) sound** made loud next to baby's ear.
4. **Swinging** baby back and forth and very slightly up and down (jiggling)*
5. **Suckling** on breast, clean finger or dummy.

** There is a clear distinction between safely swinging or gently jiggling a very young baby and other more violent activity. 'Shaken Baby Impact Syndrome', a form of child abuse, can result in permanent brain injury or broken bones and occurs when distressed and frustrated parents handle an unsoothable baby roughly. If in doubt about what is safe, ask your GP or health visitor for advice; get help immediately if at any point you are in danger of shaking your baby.*

## Recommendations from the research

The first thing to do if your baby cries a lot is to get him properly diagnosed so as to rule out any other causes. Seek help and support from your health visitor and/or GP and family, friends and neighbours – immediately if you feel you cannot cope. Colic is caused by different things in different babies so if something hasn't worked for your friend's baby, that doesn't mean it won't work for yours. You could try simethicone, completely draining one breast before moving onto the next or an anti-colic bottle to rule out excess wind being the problem. A herbal remedy, glucose and probiotics could be tried with your doctor's advice. If symptoms persist, changes to your baby's milk or to your diet, if breastfeeding, could be explored with your doctor. Perform the '5 S's' – perhaps learn baby massage – and take a break during colicky bouts so you are rested when the next bout manifests. Above all, don't try to go through it alone.

**Other helpful hints**

When your baby cries, make sure he is not hungry, has a wet nappy or is cold or hot. Playing music and giving him attention such as eye contact, talking or walking may help soothe your baby as can the 5 S's (see the box, opposite). If you place your baby near the washing machine or the vacuum cleaner, the white noise effect may be beneficial.

Colic is not a condition that can conceivably be faced by one parent alone; both parents should face this together or if you're a single parent, getting help from family and friends is very much justified (Kheir, AE, 2012). Bear in mind, it's much worse for you than it is for your baby!

# Co-sleeping

Up to 65% of parents share their bed with their baby sometime during the first month of life. Parents, however, often feel the need to be covert about their sleeping situation, fearing disapproval from health professionals, the majority of whom discourage sharing a bed because of the risks involved.

Historically, co-sleeping was widely practised. Mothers kept their young close by throughout the night for warmth, protection and frequent breastfeeding. Our ancestors would probably have placed their babies on their backs to sleep because it made breastfeeding easier during the night and since 2005, we know, thanks to J.J. McKenna et al, that back sleeping is by far the safest position if a baby is to avoid SIDS (Sudden Infant Death Syndrome). Now that mothers and infants are separated at night, back-sleeping has to be actively promoted. Could the same logic (it's what our ancestors did), therefore, extend to the notion that co-sleeping is also safer for a baby, perhaps even for reasons that we do not yet fully understand?

## CURRENT RECOMMENDATIONS

The NHS and National Institute for Health and Care Excellence (NICE) advise parents that the safest place for their baby is in a cot next to their bed for the first 6 months of life. If parents feed their baby or desire a cuddle, they should aim to put the baby back into the cot for sleeping. They also state that there are risks to bed sharing, and that if parents choose to bed share with their baby, they should be aware of the risks.

## WHAT THE RESEARCH SAYS

### The risk of SIDS

Research has conclusively found that there is a greater risk of SIDS when baby shares a bed with mum and/or dad. However, although 'bed-sharing' is labelled as a variable by itself, it is not just one variable as it can take many forms. Some of the multiple aspects of bed-sharing dramatically increase the risk of SIDS, but others decrease it, so it is not entirely appropriate to say that bed-sharing causes infant deaths. What is notable is that a baby sleeping in a different room to mum and dad is more than twice as likely to die from SIDS than a bed-sharing baby.

What is risky about co-sleeping is whether mum or dad smoke (during pregnancy or afterwards), drink alcohol or take drugs. Other risk factors include maternal exhaustion, the position baby sleeps in – those who share a bed have been found to be more frequently lying on their tummies or sides, which greatly increases the risk – or if parent and baby are sleeping on the sofa. There's also a much greater risk for babies who sleep with a pillow and with someone other than their mum (Mace, S, 2006). The risk of SIDS is most pronounced between birth and 12 weeks of age but diminishes quickly after that. If a baby tends to normally sleep by himself there is a greater risk of SIDS if only sharing for one night (Shaefer, SJ, 2012). There is also a greater risk of SIDS if a baby is premature and/or has a low birth weight. Babies who are often poorly and fail to thrive could also be at greater risk.

In addition to SIDS, babies or toddlers can die from other causes if sharing a bed. S. Nakamura, et al (1999) listed the most common as being suffocated by mum, dad or a sibling lying on them, or from using a waterbed, or by becoming trapped between the mattress and another object, or by their head becoming trapped within bed railings.

**Breathing problems**

S.A. Baddock et al (2012) found that babies who slept with their parents were more likely to stop breathing briefly and, as a result, to have slightly lower oxygen levels in their blood although it was very rare for a baby to stop breathing for longer than 15 seconds. They also found that these babies were more likely to breathe in previously exhaled air, which had a lower oxygen content, usually because a sheet or bedding was placed over the baby's head. Whilst the researchers noted that most babies were far from any risks connected with SIDS, they did point out that repeated scenarios like that described above may have severe consequences for vulnerable babies.

**Sleeping behaviour**

A Russian study found that babies who slept with their mothers were more likely to sleep less, wake more than twice a night and need more help to get back to sleep. Babies who

### Preventing SIDS while sharing a bed

* Always put your baby on his or her back to sleep even when in the bed with you.
* Avoid sharing your bed if you smoked whilst pregnant, are currently smoking or if anyone in the house smokes.
* Never drink alcohol and share a bed with a baby (neither mum or dad).
* If either parent is on medication, or has taken drugs or is ill. it's not a good idea to share a bed as you will not be as easily roused.
* Breastfeed; if bottle feeding you won't be as attuned to your baby's signals.
* Don't bed share if you're exhausted.
* Consider very carefully the environment your baby will sleep in – never on a sofa or waterbed; always on a firm mattress; no pillows, duvets, blankets (using only sheets is a good idea while baby is young) near baby or anything that could cause strangulation (toys, mum's long hair etc.), there are no railings on the bed or places in the bed where your baby can become trapped or fall out. Make sure your baby is dressed appropriately.
* Older siblings should not sleep with young babies.
* It's not advisable to share a bed with a baby who was premature, had a low birth weight or is frequently unwell.

shared a bed were also more likely to be noisy breathers. Another study looking at sleep rhythms of young babies (Kelmanson, IA, 2010),, also found that babies were more easily roused during sleep when sharing a bed.

A sleep cycle is divided into five stages; stages 1–4 and REM. Stages 1 and 2 represent lighter sleep, stages 3 and 4 represent heavier sleep and REM (rapid eye movement) is dreaming sleep. Babies who share a bed with parents have longer 1 and 2 stages and shorter 3 and 4 stages. The amount of REM sleep is unaffected. S. Mosko et al (1997) has posited that if a baby spends more time in stages where he is able to be easily roused, this may reduce his risk of SIDS if all other risks are removed, but no research paper has found significant evidence of this.

The same study showed similar results for mums; the amount of sleep a mum gets while sharing a bed with her baby was almost identical to her sleeping alone although the sleep rhythm was somewhat altered (more 1and 2 stages and less 3 and 4 stages with REM sleep being the same). Mums who co-slept were more frequently roused but they also went back to sleep more easily. That both mother and baby are more easily roused highlights that mums are attuned to their infants and vice versa, and in the case of a life-threatening event, this heightened sensitivity could more easily be called upon. Another example of mothers and their babies being attuned to each other includes the observation that a breastfed baby sleeping on her back will almost always face her mother.

Toddlers who co-sleep wake up during the night more often than those who sleep alone but interestingly, parents who share a

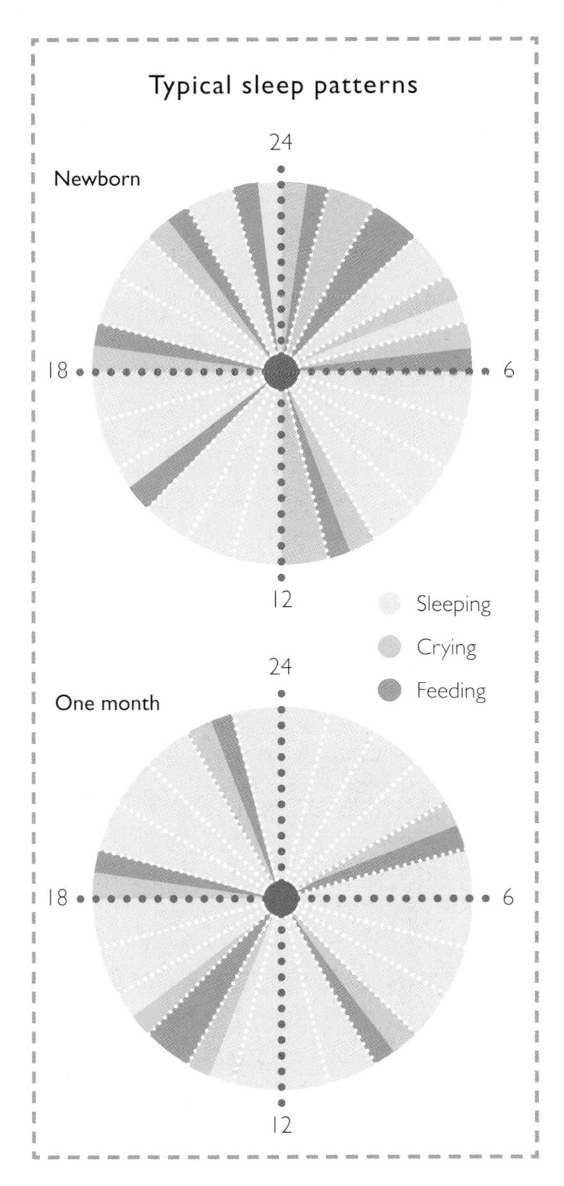


bed with a waking toddler tend be less disrupted or frustrated with such occurrences than parents whose toddler sleeps in the next room (or has recently begun sleeping in the parental bed.

**Breastfeeding**

Breastfeeding has been found to last longer for those mums and babies who share a bed but whether one actually directly causes the other is unknown (Blair, PS, et al, 2010). I.S. Santos et al (2009), found that babies who share a bed with their mums until three months of age are still being successfully breastfed at one year old. This is also one of the main reasons why mothers choose to co-sleep; it makes feeding easier during the night. Young babies need to breastfeed often and bed-sharing more easily facilitates this. The longer a baby is breastfed, the better it is for a baby, so the advantages of breastfeeding can also be extended to bed-sharing since the latter affects the former (Ball, HL, et al, 2003).

**Bonding**

The vast majority of cultures (outside the UK and USA) show a tendency towards co-sleeping; many view the independent sleeping toddler with pity. There is very little research into whether bed-sharing affects bonding with either parent. One piece of research, however, found that co-sleeping parents allowed their children greater independence, which could be a result of parental satisfaction with night-time bonding.

**Co-sleeping and the grown-up baby**

R. G. Barajas et al (2012) discovered that there doesn't appear to be any difference in cognitive ability amongst children (between ages 1-5 years) who share a bed and those who don't. But the theory that sharing a bed leads to toddlers becoming more dependent on their mothers is somewhat backed up by research. Babies who sleep in their own room are better at getting themselves to sleep and at sleeping through the night than babies who sleep in their parents' bed. It takes co-sleepers on average, over a year, to reach the same level of undisturbed sleeping. However, toddlers (aged 1-3) who sleep with mum and dad are better at dressing themselves and are more independent and successful when it comes to making friends.

That being said, it is not a good idea to allow a toddler to co-sleep if he hasn't done so previously. Parents who do this find the frequent night waking very disturbing (Keller, MA and Goldberg, WA, 2004).

## Recommendations from the research

The risk of SIDS should be taken seriously because research suggests that to a large extent it is preventable. Mums and dads should be well educated in reducing the risks whether their baby sleeps in a cot or in the parental bed. That said, and appropriate precautions taken, it really is a choice to be made by individual parents. If you are breastfeeding there are obvious benefits. If you are aware of the risks and have taken all the necessary precautions, there's no need to be covert about sharing a bed with baby, after all, most of us do, at some stage or another.

# Baby-led versus parent-led routine

There has been much debate on the topic of who 'dictates' the pace of daily life and which method is better for both baby and parents. Accompanying this debate is a continuum of slightly differing routines, from the extremes of proximal care (whereby parents are encouraged to hold their babies for much of the day, breastfeed frequently, co-sleep and respond immediately to their baby's cries – in other words, baby is the initiator) to the more rigid routines advocated in *The New Contented Little Baby Book*, whereby parents are advised to feed every three to four hours, delay responses to crying, not hold their baby for prolonged periods and to make sure baby sleeps in her cot at night.

There are passionate advocates for both routines and the many that lie somewhere in the middle, and often objectivity is difficult to find. It's also worth noting that it may not be appropriate to only adopt one routine for the entirety of babyhood; a combination of different approaches may turn out to be more beneficial for both parents and baby.

The research is, unfortunately, fairly generalised, as only clearly defined factors can be investigated. For example, what does 'contented' actually mean? Researchers, have predominantly focused on crying – the amount and age to which it persists, and sleeping – the sought after 'sleeping through the night' achievement. They also compare routines that lie within the middle ground as parents who practise the extremes are difficult to find.

## CURRENT RECOMMENDATIONS

The World Health Organisation recommends that breastfeeding be on demand – as and when baby requires (proximal care). The NHS website provides strategies for helping a baby distinguish between day and night (to aid sleeping through). Beyond this, parents are left to decide which routine to implement.

## WHAT THE RESEARCH SAYS

### Feeding

Feeding on demand is defined as there being no restrictions on the frequency or length of a baby's feed and there is a lot of support for this type of baby-led routine. Although generally applied to breastfeeding, it is possible to bottlefeed on demand (see box, page 108).

Breastfeeding on demand has been shown to encourage exclusive breast-feeding, and boosts the amount of time for which breastfeeding continues. However, there is still pressure to conform to a more parent-led scheduled routine. Today we know that a baby's need for milk varies throughout the day, and that breastfed babies require more feeds than bottle-fed babies. This is due to breast milk containing more easily digested whey protein compared to formula (see page 55). Mothers who breastfeed are able to respond to their baby's changing needs, which is known as a demand-supply feedback.

This focussing on a baby's needs is great for her what about mum? Demand feeding can make some mums feel like a milk production line, and if a baby needs to be fed often during the night and makes no headway towards a full night's sleep, it can result in her mum giving up breastfeeding early (before the recommended 12 months).

Scheduled breastfeeding, on the other hand, is when the feeds are predetermined – usually every 3–4 hours. Research shows that mums who follow this more parent-led routine also tend to give up on breastfeeding sooner, probably from a feeling of not having enough milk, as the crucial demand-supply feedback may be affected by feeding at set times.

**Crying**

There is some controversy over whether a baby-led or parent-led routine results in less crying. I. St. James-Roberts et al (2006) showed that a baby-centred routine, during which parents race to ease a baby's apparent suffering at the first whimper, results in the baby crying less when compared to a more parent-led routine. The same researchers found (2007) that a more structured routine, early on, results in more crying and fussing. However, they also discovered that a baby-centred or less structured routine was associated with more nighttime waking and more crying when a baby reached 12 weeks of age whereas, at this point, crying amongst babies used to parent-led routines started to decrease.

A variation of proximal care that is a step closer to a parent-led routine was found to be just as effective at decreasing crying time as a totally baby-led routine. In particular, holding time was reduced when the baby was awake (although still a substantial amount – 584 minutes per 24 hours compared to 989 minutes for the proximal care group) – but there was also some holding time whilst the baby slept (211 minutes compared to 571), while the parent-led routine has neither. In their 2006 study, I. St. James-Robert et al found that the amount of time the baby was allowed to fuss or cry was relatively similar in both the proximal care and the variation. An earlier study of theirs (1995) found that carrying a baby more often, particularly when a baby is happy and not crying, doesn't decrease overall crying time.

Up until 3 months of age, M. Alvarez (2004) demonstrated that variations of the proximal care strategy – frequent feeding, speedy response to crying and lots of time holding the baby – have been shown to be most effective at decreasing crying time;

**Bottlefeeding on demand**

You can make bottlefeeding more like breastfeeding with a little bit of effort.

* Learn to 'read' the signals when your baby is hungry – movements of of head, hands, arms and legs becoming more agitated the longer she waits to be fed;
* Hold your baby as close to your body as you can, and if possible with her naked skin next to yours;
* Don't force her to finish a bottle or feel disappointed when she leaves milk over. Follow her appetite;
* Create a routine so that your baby is bathed, sleeps and feeds at regular times.

beyond this time if structure is not incorporated, crying and not sleeping will usually continue for longer.

**Sleeping through the night**

Problems with infant crying and sleep affect about 30% of families leading to depression, an early end to breastfeeding and behaviour problems when children are older. A key risk factor delaying a baby's ability to sleep through the night (which researchers deem unbroken sleep of 5 hours although most parents would wish for longer) is frequent demand-led feeding day and night (more than 11 feeds within a 24-hour period). This is true whether babies are breast- or bottle-fed. But according to M. Nikolopoulou et al (2003), if nighttime feeds are delayed and less attention is given at night, babies can become better sleepers when 12 weeks old.

Nighttime awakenings are to be expected until 3 months of age, when two-thirds of babies are reported to sleep through the night. Beyond this time, the most important factor for babies waking in the night was whether the parents stayed with them until they were asleep (Touchette, E, et al, 2005). When parents actively encourage their babies to fall asleep by themselves, there are fewer interruptions during the night and less sleep-related problems later on (Anders, TF, et al, 1999). Babies who are put in their cots awake are also less likely to need parental support when they wake at night.

L. A. Adams et al (1989) found the most effective treatments for teaching a baby older than 6 months to sleep through the night are 'controlled crying' and a 'positive' bedtime routine. Controlled crying is discussed in more detail on page 80, and it is considered an important although not

crucial aspect of the parent-led routine. A positive bedtime routine (shown to be effective at reducing bedtime tantrums in 36 month olds) consists of up to 7 'enjoyable' activities such as bath, stories, milk, etc. These are first offered when a child starts to get sleepy but as time goes by, are introduced earlier – 30 minutes sooner each week, for example.

Parents may be able to prevent the use of controlled crying, which is distressing for some, by incorporating some structure into nighttime awakenings. Instead of feeding, they can change nappies or walk around with their infants during the night and not reduce the overall amount of milk taken in a 24-hour period (Renfrew, MJ,, et al, 2000). However, J. Hayama et al (2008) found that changing nappies immediately throughout the night leads to increased sleep problems.

If a baby usually wakes during the night, it can be beneficial to rouse her 15–60 minutes before her usual waking time to feed and put her back to sleep. You can then gradually reduce the scheduled awakenings. V.I. Rickert, et al (1988) showed this is effective from 8 weeks of age when babies were seen to awake and cry less.

Introducing a nighttime feed between 10 p.m. and midnight is also effective at reducing overall waking during the night. E. Touchetteet al ( 2005) demonstrated that the longer nighttime awakenings persist, the longer the problem will last. If a toddler aged 17 months is not sleeping through the night, there is a 30% chance she won't be at age 29 months.

**Psychological distress and its effects**

Parents whose babies are going through challenging phases related to sleep and crying are more likely to suffer from stress, anxiety and depression. Where parents have suffered from depression, their children are 40% more likely to suffer from depression when they, too, are adults. Furthermore, children who have a psychologically unwell mother are more likely to experience behavioural and emotional problems. If parental health is affected by a particular routine, it is vital to try an alternative approach – one where parental concerns are more to the forefront – to reduce the chances of long-term psychological problems.

## Recommendations from the research

If less crying and waking during the night are your goals, then some variation of proximal care is best for the first three months but after that, a more parent-led routine should start to kick in. Nighttime routines, including bedtime activities (bath, soothing music, etc.) and introducing the concept of day and night – whereby contact and communication is limited at night, nappies are only changed if absolutely necessary, and the times between feeds are regularly lengthened – can be introduced sooner than at age three months Some success has been achieved by introducing a feed between 10 p.m. to 12 a.m. The concepts of a bedtime routine and a child learning to fall asleep without help are very important for an unbroken night's sleep.

Extremes of routine tend to be less effective for successful breastfeeding.

All said and done, if a routine, whether baby- or parent-led, is causing anxiety or stress, a solution should be sought as quickly as possible because a psychologically unwell parent can have lasting effects on his or her infant.

# Car seats: ISOFIX versus standard-fix

A car seat is an essential buy for an infant and there is a plethora of types, models and mechanisms to choose from, not to mention colours. Safety is high on every parent's agenda but how much safer is an ISOFIX car seat and is it worth the extra cost? And, what exactly is ISOFIX anyway.....?

ISOFIX stands for 'International Standards Organisation Fix'. The name is derived from the International Organisation for Standardisation or ISO (from the Greek isos meaning 'equal'), set up to facilitate the international coordination of industrial standards. It is a network of national bodies; each country has its own standards organisation that is a member of the ISO.

Whereas 'standard' car seats secure using an adult seat belt (and then the baby/child is strapped into the seat with a harness), ISOFIX ones plug into two fixing points at the rear of the backseat, attaching rigidly to the car chassis. In the case of infant carriers, this is via a base that is semi-permanently fixed in the car; you simply clip the carrier on and off the base. A harness secures your baby to the carrier itself.

As well as the two rear seat latches, some ISOFIX systems use a third fixing point called a top tether (but not all cars have these), or have a rigid support leg to enhance stability.

Without ISOFIX points you cannot use an ISOFIX car seat. However, not all car seats with ISOFIX points are approved for use with all ISOFIX infant car seats. This is why it is best to get professional advice before buying and why you may also need a tethering system.

## CURRENT RECOMMENDATIONS

All children in the U.K. must use the correct car seat until they are 135cm tall or turn 12 years old (although there are exceptions). Correct car seats depend on weight (see table). Since November 2012, all new car models had to be fitted with ISOFIX and from November 2014 all new cars must have ISOFIX. There are no current recommendations against using standard-fix seats but RoSPA (Royal Society for the Prevention of Accidents) on behalf of the Department for Transport highlights that ISOFIX is a safer and easier method than seatbelts for attaching an infant car seat. Children under 13kg must use a rear-facing infant car seat (secured by seatbelt or ISOFIX) and the safest place is on the back seat of the car, preferably in the centre.

## WHAT THE RESEARCH SAYS

### Safety in crash conditions

During a crash, a child's car seat can be thrust forwards but also, depending on the direction of the impact, for example from the side, be thrown sideways or even start to rotate. The more a car seat is able to move, the greater the chance of injury to the baby or child inside. ISOFIX decreases the amount of movement a car seat sustains in crash conditions compared to one being held with a normal car seatbelt, which usually results in fewer injuries. In particular, L. E. Bilston et al (2005) showed that ISOFIX lessens the amount of head movement and therefore decrease the risk of head injury.

In a frontal crash, an adult seat belt used to attach a car seat can be just as effective as ISOFIX as long as the belt is pulled tight. However, a belt's effectiveness will be compromised if there is any slack on the belt, or if it's not pulled tightly. If a crash includes a side impact, R. Lowne et al (1998) showed that ISOFIX performs better than a standard-fix car seat.

## Fitting

Numerous studies have shown that two thirds of standard-fix infant car seats are not attached correctly using the car seatbelt (up to 83% in one estimate) primarily because the seat belt is not secured and tested* every time a baby is put into one (in some cases, several times a day).

*Testing requires you to determine if there is slack or twisting on the seatbelt, which will cause greater movement should a crash occur. An ISOFIX base for a rear-facing car seat, however, is left attached in the car and the baby car seat simply clicked into place for journeys. No testing needed and less room for error.

The incorrect fitting of an infant car seat using a seat belt, not surprisingly, results in a much greater chance of injury to the child (Kapoor, T, et al, 2011). Even when parents are given detailed instructions, many are still unable to fit one correctly.

On the other hand, studies show that up to 96% of ISOFIX car seats are properly fitted; they were designed to be easier to put in place and are therefore safer because they are less likely to be fitted incorrectly. An added bonus with ISOFIX is that even when a rear-facing infant car seat is removed regularly, the base remains in place, unlike a standard-fix seat, which needs to be re-attached each time it is removed. Should you need to use your ISOFIX car seat in a different car, you will be less likely to make an error in attaching it compared to using a seatbelt.

## Varying standards

Car models come in many different designs and so too do their seats and length of seat belts. This makes it difficult to design a child car seat that fits all cars and there is more room for error when using a seatbelt to attach a seat. ISOFIX ensures that despite any design differences, your child's safety will be preserved because the method of attaching a car seat is always the same in every car with ISOFIX. That being said, however, not all ISOFIX infant car seats will fit all ISOFIX-enabled cars. For this reason it is very important to get professional advice.

## Cost

Currently, a group 1 ISOFIX rear-facing car seat costs, on average, about £200 more than a standard-fix seat. Some models, however, use an ISOFIX base that fits a group 0+ and the group 1 seat, which could work out cheaper.

## Recommendations from the research

ISOFIX is definitely a safer choice for baby. It comes out top on all tests for allowing the least amount of movement, particularly on sideways crashes. A standard-fix seatbelt attachment can also be safe if used correctly but it is easier to get this wrong. Making sure that you secure it properly can make it almost as safe as an ISOFIX (except in sideways crashes). Unfortunately, a large percentage of people fail to attach it correctly.

| Type of Seat | Group Number | Weight Range | Approx. Age |
|---|---|---|---|
| Rear-facing baby seats | 0+ | up to 13kg (29 lbs) | Birth to 12-15 months |
| Forward or rear-facing * baby seats | 1 | 9-18kg (20-40 lbs) | 9 months to 4 years |
| Booster seats | 2 & 3 | 15-36kg (33-79 lbs) | 4-12 years |
| Booster cushions | Some are approved for group 2 and 3, others for just group 3 | Over 25kg until 12 years old or 135cm in height | 6+ |

* Many research articles show that keeping a baby/toddler in a rear-facing car seat for as long as possible is safest. It is common in some Scandinavian countries for a child to stay rear-facing until he or she is four years old. There are a couple of group 1 rear-facing car seats available in the UK but not all cars can safely fit them; they must be fitted by professionals. The RoSPA website has more information about rear-facing car seats including a link to where you can buy them in the UK: www.childcarseats.org.uk.
* To check whether your car has ISOFIX fittings, feel between the rear seat back and base (or get a car seat retailer to look for you), check the manual, or ask the car dealer, if you bought the car from an official dealership.
* Be aware that just because your car has ISOFIX fixings, it doesn't mean that all ISOFIX seats will fit; you'll still need to check the compatibility of a particular model for your car.
* Check your car seat has an 'E' label to make sure it meets the UN standard regulation 44.03.

# Introducing solids

When to start weaning depends on social pressures, guidance from health professionals, medical histories and the individual baby. There are also many sources of information. Most midwives and health professionals follow Department of Health guidelines but overwhelmingly, the advice from parents and friends tends to be that early weaning is both acceptable and advisable ("You need to fill up her tumtum if you want your baby to sleep all night…").

### Age in weeks and months

I remember being rather confused with knowing the actual age of my newborn – how exactly did weeks relate to months! Was a month four weeks?! Health professionals and books jump from one to the other quite fluently with little explanation. It took me a while to understand that when they say weeks, they mean weeks and when they say months, they mean calendar months. That may seem obvious to you but for those of you who were wondering the following may be helpful:

| Age in months | Age in weeks |
|---|---|
| 1 | 4 |
| 2 | 9 |
| 3 | 13 |
| 4 | 17 |
| 5 | 22 |
| 6 | 26 |

After all, they weaned their little one earlier than six months. So, what is best for your baby? The following is a summary of the many different views on the question, which you'll need to make an informed decision.

## CURRENT RECOMMENDATIONS

In 2001 the World Health Organisation (WHO) changed its previous guidance to start weaning at age 4–6 months to a full 6 months of exclusive breastfeeding. The UK Department of Health subsequently adopted the same guidance after considering the evidence. However, both recognised that each baby should be managed individually to avoid insufficient growth and its related conditions. Weaning is defined as the introduction to any substances other than formula or breast milk, with the exception of vitamins, minerals and medications.

More recently, European Food Safety Authority (EFSA) has stated there are no disadvantages to weaning a baby between the ages of 4 and 6 months, and that gluten should be introduced no later than at 6 months (EFSA, 2009). Many other European countries and the USA also choose not to adhere fully to the WHO guidelines. This opinion is not supported, however, by the U.K. Department of Health's Scientific Advisory Committee on Nutrition (SACN) or the Committee on Toxicity (COT), who are currently conducting another review on the subject. And so the UK advice remains fixed at the 6-month mark.

| Age of weaning | % of mothers introducing solids |
| --- | --- |
| By 3 months | 10% |
| 3-4 months | 42% |
| 4-6 months | 46% |
| After 6 months | 2% |

*Table of weaning ages – taken from the Infant Feeding Survey 2005*

## WHAT THE RESEARCH SAYS

### Protection from infections

The longer a baby is breastfed, the greater the protection against gastrointestinal and respiratory infections such as gastroenteritis and middle ear infections (Kramer, MS, and Kakuma, R, 2012). However, in terms of weaning, it doesn't appear to make much difference whether a baby is weaned at 4 or 6 months.

M. A. Quigley et al (2009) looked at babies who had to be hospitalised because of gastroenteritis or respiratory tract infections. They found that starting solids before 6 months had no impact while drinking formula milk was a much larger factor. However in 2006, C. J. Chantry et al researched the risk of children developing recurring middle ear infections or pneumonia and found that those who had been exclusively breastfed for a full 6 months as opposed to 4–6 months were better protected.

Although not all studies agree, J. S. Forsyth et al (1993) reviewing those looking at babies who had been introduced to solids at around 2–3 months of age, found that they were more likely to suffer from respiratory illnesses and possibly gastrointestinal infections.

### Dental caries

There is very little research carried out on the impact of early weaning on tooth decay. However, one large investigation found no link between early weaning (3 months) and a risk of caries in childhood (Kramer, MS et al, 2007).

### Energy deficit

There is a theory that the reason so few women exclusively breastfeed for the recommended six months (1%) is because the energy needs of a baby of that age are not met by breast milk. One estimate is that breast milk is on average about 100 kcal short of what is required by the average 6 month old (Reilly, JJ and Wells, JCK, 2005). This will not be true of all babies, but if your baby starts losing weight around this age, discuss with your health visitor. If you feel your baby is hungry even after a feed, it may be worth getting her weighed.

### Digestive system and renal function immaturity

A report by the EFSA in 2009 stated that a newborn's digestive system is not mature enough to cope with many different types of food; in fact, a child does not possess a full quota of enzymes for breaking down food until she is about three years old. A baby will have sufficient enzymes to break down food by about 4 months, so food should not be introduced before this time.

One function of the renal system is to enable the kidneys to remove salts from the blood stream and to dispose of them in the urine. According to the EFSA report of 2009, a newborn's renal system is very immature and breast milk (and to a lesser degree formula milk) contains very low amounts of salts that need to be removed. Until at least 4 months of age, a baby's renal system is not sufficiently mature to be able to cope with the higher naturally occurring salts found in solid foods.

**Food preferences**

The optimum age for introducing different tastes and flavours to a baby to ensure she grows up to be a 'good eater', content to sample a wide range of foods, is yet unknown. However, if you delay the introduction of new foods and textures or if you serve puree for too long, this can impact on your child's willingness to eat different foods.

How well your young child accepts healthy foods, such as bitter-tasting leafy green vegetables, is often an issue in relation to weaning. Parents are sometimes fearful that if they wait too long, their little one will not be so open to new tastes.

C. Schwartz et al (2012) found evidence that prolonged breastfeeding can positively affect how well a child receives new tastes at 6 months. Elements of the foods a breastfeeding mum eats pass into her breast milk so, from a young age, her breastfed baby will experience different tastes. However, these effects are no longer evident at 12 months, indicating that parents must continue to offer a variety of foods if they

wish their toddler to continue to enjoy eating them, or at the very least avoid mealtime tantrums.

There is little research carried out on whether introducing certain foods earlier than 6 months will have beneficial consequences later on. J. A. Mennella et al (2004) found babies were more accepting of a not-so-great tasting formula milk at 2 months than they were at 7 months, but they were better at accepting it later on if they had taken it at a younger age. As for other foods, S. Nicklaus (2011) found evidence that introducing fruit during weaning (at whatever age) means a baby is more likely to enjoy fruit as a toddler, but introducing vegetables during weaning is not necessarily the same. In other words, the early introduction of fruit has a positive effect but that of vegetables may not have any effect on future acceptance.

R. J. Cohen et al (1995) compared weaning at 6 months with weaning between 4 and 6 months to see which group of babies were happier to accept a greater variety of foods. They found that the babies who were weaned between 4 and 6 months were slightly more accepting of foods, such as potatoes and vegetables, at 9 months but that by 12 months, any differences had disappeared. They concluded that it does not seem to matter to later acceptance of food if a baby is introduced to new foods at 4 or 6 months.

**Allergies and autoimmune disease**

***Gluten intolerance and coeliac disease:*** If a child is at risk of developing coeliac disease (a close relative has it), S. Guandalini (2007) determined the best protection may come from introducing small amounts of gluten (breakfast cereal) between 4 and 6 months whilst breastfeeding for a further 3 months. J. M. Norris et al (2005) looked at children at risk of developing type 1 diabetes and found that they had an increased chance of developing auto-antibodies (antibodies that attack the body's own proteins) if they were first given gluten younger than 3 months as well as after 7 months of age. Moreover, if a child is given lots of gluten from a young age, she is more likely to develop coeliac disease, and continued breastfeeding seems to give no protection against this. Other research, however, suggests that breastfeeding whilst introducing gluten can also help delay the incidence of coeliacs.

That breastfeeding protects against the development of coeliac disease is well documented (see page 00), as is only giving small amounts of gluten in the first few months; however, the timing of when gluten is introduced is still debated (Ludvigsson JF and Fasano A, 2012). H. Szajewska et al (2012) suggested that in the absence of clear evidence, the optimal time for introducing gluten is not earlier than 4 months but also not later than 7 months. Unfortunately, the evidence for this advice is largely based on observations and rigorous, conclusive research is yet to yield any firmer results. It may be that when higher quality investigations are carried out, we may see the official advice change.

***Eczema:*** Many older studies have found a link between early weaning (3 months compared to 6 months) and the risk of eczema. However, the majority of more recent studies have found no link at all, even for high-risk babies (those whose mum, dad or any siblings suffer from eczema, asthma or hay-fever). Limiting the risk of eczema is more to do with whether a baby is breastfed rather than when solids are first introduced. Breast milk's protective effect is most pronounced if weaning happens between four and six months. It does not appear to matter whether introducing foods considered to be high allergens (eggs, nuts, fish, cow's milk) is delayed beyond 6 months as is often advised (Filipiak, B et al, 2007). In fact, current thinking is that foods most likely to cause an allergy may be best introduced before 6 months whilst breastfeeding is ongoing (Grimshaw, KEC, et al, 2009).

***Asthma:*** A. C. Wilson et al (1998) showed that introducing solids at an early age (before 15 weeks) increases the chances of the baby developing a wheeze in later life (measured at age 7).

***Diabetes, type 1:*** For this autoimmune disease in which the body's immune system attacks the cells that make insulin, there is a connection between it and introducing gluten and coeliac disease. Autoantibodies, which attack the body's own proteins such as with coeliac disease and diabetes, are more common in high-risk children (those whose mothers have either coeliac disease or diabetes) introduced to gluten early (before three months) or late (after seven months) during weaning. Other factors contributing to increased levels of autoantibodies are short breastfeeding (less than two months) and an early introduction to cow's milk (Wahlberg J, et al, 2006).

***Other allergies:*** The advice to date for introducing known allergens, for example, eggs, is to delay introduction but research is ongoing into the time when allergens can be safely introduced into a baby's diet and that that time is most likely to fall within the four to six month bracket.

**General health**

***Obesity and BMI:*** The effect of early weaning on obesity or BMI later in life is still very much under discussion. K. C. Mehta et al (1998) ound no effect at all on obesity in childhood when weaning happened before 4 months, while S. Chomtho et al (2008) found a link. A. C. Wilson et al (1998) comparing babies weaned before 15 weeks (about 3½ months) with babies who were weaned later, found no weight difference when the babies were 2 years old but did find a difference when they were 7. L. Schack-Nielsen et al (2010) found no link between obesity and early weaning as children, but found one in adulthood (at 42 years of age).

***Blood pressure and cholesterol levels:*** It is well understood that prolonged exclusive breastfeeding lowers blood pressure and cholesterol levels in later life (Robinson, S and Fall, C, 2012). Delaying weaning until 6 months means that the protective effects of breast milk, therefore, have a chance to reach their full potential.

***Diabetes, type 2:*** We know that lower insulin and glucose levels found in breastfed infants are also seen in adulthood. This suggests that the body is programmed from a very early age (Owen, CG, et al, 2006). Early weaning, which is likely to boost insulin levels, may programme the body in a similar way. However, there is no evidence

### Weaning signals

The common signs that your baby is ready to begin eating solids are if she puts toys in her mouth, chews her fists, takes an interest in other people eating and seems hungry even when extra milk has been offered. These developmental readiness signs are quite common in babies aged between 4 and 6 months.

Another common reason for weaning early is if a baby has started waking frequently during the night. Unfortunately, the research does not support the notion that all babies who wake will go back to sleeping well after food has been introduced (Yilmaz, G, et al, 2002). There is the possibility that about of a third of babies will sleep better but there is frequently a delay of a few weeks before the effects of weaning on sleeping can be seen (Robertson, RM, 1974). A quick look through online forums of baby sleep will confirm that for the vast majority of waking infants, weaning does little good.

to suggest that early weaning (of any foods) has an effect on babies developing diabetes in later life.

***Growth and body composition:*** The effect of weaning early (from three months) or late (after six months) on weight gain and length is not entirely conclusive. Some research has found little connection (Mehta et al, 1998) whilst others have found that weaning before four months is associated with weight gain, which could impact on future health, resulting in obesity, type 2 diabetes and heart disease in adulthood (Leunissen, RW, et al, 2009). However, L. Salmenperä et al (1985) found some evidence to suggest that if food is not introduced until after six months, there's an increased risk of an exclusively breastfed baby being undernourished and his/her weight gain and length being stunted.

***IQ:*** It doesn't appear to make a lot of difference whether you exclusively breastfeed for 3–4 months then partially breastfeed until 6 months or if you exclusively breastfeed for the full 6 months (Kramer, MS & Kakuma, R, 2012). See also page 58.

***Nutrient deficiencies:*** Breast milk does not contain a great deal of iron, regardless of the mum's diet, and a baby is dependent on his/her own iron stores for the first few months of life. Formula milk is enriched with iron so anaemia is very unlikely in bottle-fed babies. In the UK and other developed countries, a breastfed baby is also unlikely to suffer from iron deficiency, provided mum's iron levels were normal at the time she gave birth and her baby's birth weight was above 2.5kg (5lbs 8oz); baby boys weighing between 2.5–3 kg (5lb 8oz–6lbs 10oz)§ at birth have a slightly higher risk. Only about 2% of babies will develop anaemia, although screening for iron deficiency does not happen routinely in the UK (Fewtrell, M, et al, 2011). There is no real evidence that weaning at 6 months has any effect on iron levels in baby. If solids are not introduced at around six months, however, a breastfed baby will be at an increased risk of becoming anaemic (Butte, NF, et al, 2002).

Levels of zinc in breast milk also decrease during the first few months and the exclusively breastfed baby will be at risk of deficiency if solids, in particular meats, are not introduced by six months (Krebs, NF, & Hambidge, KM, 2007).

Levels of vitamin D, particularly in the autumn and winter, are likely to be insufficient in breast milk unless a mother takes a supplement. Formula milk already contains added vitamin D.

Vitamin K is also quite low in breast milk but unlike iron, a baby's stores are also low. This is why newborn babies are given vitamin K during the first few days of life.

Other nutrients such as vitamin A, B vitamins and long-chain polyunsaturated fatty acids are all found in breast milk but depend on mum's healthy diet for stores to remain adequate. Deficiencies in these nutrients are unlikely to occur and therefore won't be affected by waiting to wean until 6 months.

## Recommendations from the research

It is very clear from the research that weaning should not take place before 4 months. Never offer any food, even mixed with formula in a bottle, to a baby less than 4 months old. However, whether you wean at 6 months or at 4 to 6 months depends largely on whether you are breastfeeding and your baby's appetite.

If you are breastfeeding, weaning at 6 months is recommended for your baby's present and future health mainly because giving breast milk for a full 6 months has benefits. There are no particularly beneficial reasons to wean before 6 months and little risk with waiting until that time. However, there is also no real risk if weaning happens between 4 and 6 months. You should, however, introduce gluten before 7 months whilst continuing to breastfeed.

If you are bottle-feeding, the recommendation is still to wait for a full 6 months but formula milk does not have the same benefits as breast milk. No harm is likely if weaning happens between 4 and 6 months. Aim for the full 6 months but if you feel your baby really needs something extra, feel free to oblige, although don't be fooled by night waking.

# Baby-led weaning

Weaning is a key milestone and an important time for parents and baby. It is the start of meals that will affect a baby for the rest of her life. Decisions made now will impact on later health so it's important for every parent to 'get it right'.

Baby-led weaning (BLW) is, as the name suggests, when a baby ready for solid food is allowed to reach out, grab an item and eat it by herself. No spoons, no puree, and parents on hand only if needed. Parents provide a variety of food in easy-to-grab sized pieces, and the baby is given the opportunity to 'maul' it (since she usually doesn't have teeth) until it is broken up or soggy enough to swallow. BLW is not intended for babies younger than 6 months, who usually cannot sit up unaided, and a key aspect is that the baby sits with the whole family at mealtimes and always eats in the company of others.

Puree is often the first food introduced to a young baby because, historically, weaning occurred before six months. Today, the Department of Health's recommendation is that weaning not start until 6 months, by which time, BLW advocates state, the puree stage is already past.

Exclusive breastfeeding on demand for the first 6 months is arguably related to BLW. For their first 6 months, babies dictate when and how much food they take in the form of a breast-milk feed. With BLW, this

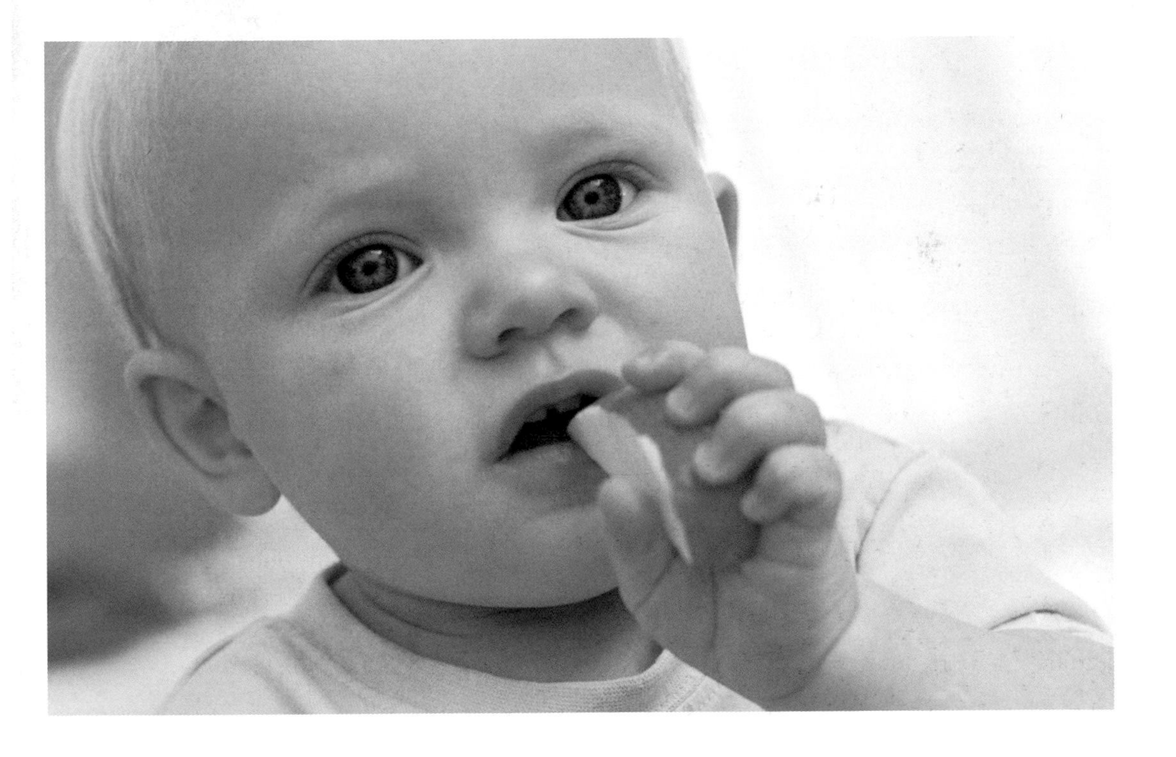

on-demand style feeding continues with the introduction of different foods. Baby chooses which foods and how much to eat.

## CURRENT RECOMMENDATIONS

The Department of Health currently recommends offering mashed fruit and vegetables and cereals mixed with baby's usual milk as first weaning foods. These should be given with a spoon and introduced slowly from 6 months of age. It also recommends providing a baby with some finger-sized soft fruit or vegetables or other finger foods, which a baby can pick up and practise 'chewing' on. Once these foods are accepted, parents can move onto mashed up meats, lentils and dairy products. Amounts, textures of foods and frequency of meals should all be increased as a baby develops over the next 6 months. Baby-led weaning is not yet officially recommended.

## WHAT THE RESEARCH SHOWS

### Less stress for parents

A. Brown and M. Lee (2011) found that parents following the baby-led weaning approach tend to be less concerned about the weight gained by their little one, feel less need to monitor what their child is eating and generally feel less pressure about weaning in general. Mums who follow the BLW approach also report less anxiety.

Food controlling parents can have a negative effect on toddlers and how well they eat but these researchers found that BLW parents were less controlling of how much or how well their babies ate. This makes sense as BLW is all about allowing a baby to do it herself. However, it is not clear if BLW makes parents less controlling or whether more laid-back parents tend to opt for it as it better suits their disposition. A parent who likes to be in control may be more likely to choose a puree approach.

### Development is key to success

Being able to feed oneself is the backbone of baby-led weaning. But can all babies manage this task at weaning age? The answer depends on when you choose to wean. Just under half of babies can reach out, grab and eat food before the age of 6 months, which means the rest will develop this ability somewhere between 6 and 8 months. If a baby develops this skill a little late (in comparison to other babies), C. M. Wright et al (2011) found the baby-led weaning approach could cause that child to become undernourished. Babies who, for whatever reason, may be developmentally delayed may not manage the BLW approach until they reach certain milestones, for instance, being able to sit up unaided.

The method could also cause complications if gluten is introduced too early; research suggests there is an optimum window for introducing gluten into the diet in order to reduce the chances of developing coeliac disease (see page 120 for more information).

### Earlier acceptance of lumpy food

The traditional weaning approach is to introduce food that is of a relatively liquid consistency. When a baby copes well with this consistency, further foods are offered that are coarser and less pureed until a child is given food containing lumps that can be effectively mashed with a fork.

BLW involves the initial introduction of whole foods, such as a chip-sized piece of carrot or toast, as well as things like thick soup. This means that a BLW baby will be given greater food options and a range of textures simultaneously, rather than waiting for them to be presented.

Problems can occur with the traditional puree route if lumpy food is not introduced early enough. K. Northstone et al (2001) found that if lumpy food is not introduced until ten months, a toddler is much less likely to be eating family food by age 15 months. In contrast, babies who are introduced to lumpy food between the ages of 6 and 9 months – as they more frequently are with BLW – are much happier to be eating what the family eats by toddlerhood. Being given food with a range of consistencies at the start of weaning generally results in a baby being more accepting of the usual array of different textures available in most family meals when she is older.

**Diet diversity**

H. Rowan and C. Harris, C (2012) found that parents following the BLW approach introduce similar foods into their baby's diet as those following the puree approach. The main difference is that BLW parents offer their babies most of the same foods they eat at mealtimes.

**Iron needs may not be met**

An exclusively breastfed baby's iron stores may begin to run low after 6 months, which is one of the reasons weaning is recommended at this age. Conventionally, parents will offer spoon-fed iron-fortified cereals to a baby as an early weaning food. The BLW baby is likely, therefore, to miss out on iron unless her parents provide her with finger foods containing the mineral such as cooked beef or other red meat strips. (This iron is actually better absorbed by a baby.) Parents also need to introduce iron-rich foods quite early on in order to ensure their baby does not become deficient. Although parents are frequently wary of introducing meat as an early weaning food, BLW requires it. Unfortunately, there has been no research on whether BLW babies are meeting their nutritional requirements.

**Food intake difficult to assess**

Parents are often concerned about how much milk a baby takes in, which is understandable as weight gain is crucial for health and development and is closely monitored during the first few months. It's therefore natural that this concern extends to food intake during weaning. However, if a baby is given too much puree before four months, she will tend to drink less milk. Milk contains a vast selection of vitamins and minerals, which is only replicated when a full, varied, diet is introduced. It is therefore conceivable that too quick an introduction to puree could result in a baby lacking certain nutrients, which could impact on her weight gain. The BLW baby, on the other hand, does not appear to actually consume a great deal of food, particularly as she is supposed to 'play' with food until her first birthday.

A recent study (Townsend, E and Pitchford, NJ, 2012) reported that babies who were fed puree were more likely to have higher BMIs than babies who fed themselves. Though much publicised, the conclusions were criticised by other researchers because the numbers involved were too low. The researchers also found that there were slightly more underweight BLW babies when compared to those eating pureed foods. Other research, however, found no difference in weight or length between the two (Brown, A and Lee, M, 2011). Getting clear answers from research is difficult as it often relies on parental questionnaires and because parents often use both methods, particularly if they wean before six months (which the vast majority do). A. Brown's and M. Lee's definition of BLW alllowed for up to 10% of food to be given as puree or with a spoon.

Unfortunately, there is simply not enough good research to make conclusions about whether a BLW baby is getting too much or too little food. There is also a worry about whether a BLW baby gets sufficient food is when she becomes unwell. Illness is common during the first 2 years of life, and if poorly, a child will need spoon feeding to ensure her energy needs are met (Cameron, SL, et al, 2012).

**Risk of choking**

When a piece of food becomes lodged in her throat – either partially or totally blocking the passage of air – a baby will cough and generally be capable of bringing up the food. A baby's gag reflex, which is initiated when food that is not ready to be swallowed but has travelled to the back of the mouth is pushed to the front for further 'chewing', is much further forward than a toddler's or adult's so gagging happens more often. Although its choking that can cause harm while gagging is a safety mechanism that prevents it, both situations are distressing for parents.

Whilst gagging is generally agreed to be more common with BLW, a recent study (Cameron, SL, et al, 2012) found that 30% of BLW parents reported a choking incident – although all the babies were able to cough up the food without parental help. Raw apple was the most common perpetrator. There are specific rules to follow when attempting BLW so as to limit any risks and parents who use it are well advised to learn first aid treatment for a choking baby.

## Recommendations from the research

The research has not highlighted any reasons that BLW should not be employed. In fact, if it encourages parents to delay weaning until six months, which is what is recommended, then it may have health advantages for baby. It may also help alleviate stress for parents and ensure baby is happy to accept a variety of textures in later life. However, the research is still, relatively speaking, in its infancy and there may be further advantages, such as decreased BMI, for baby.

# Jarred baby food

Since the 1930s, the pendulum of popularity has continued to swing between whether to go with commercial baby food or a parent's own homemade version. Today, with our busy schedules and the high quality of commercial foods available, many parents still opt for the bought variety as a fallback. There is much encouragement of late, however, for parents to serve homemade food because it's supposed to be healthier, but is that really the case?

## CURRENT RECOMMENDATIONS

The NHS website encourages parents to serve homemade food and to save jarred (or other commercially prepared food) for when out and about. With homemade food there's more control over things like added sugar and salt and it offers a wider variety of textures.

## WHAT THE RESEARCH SAYS

### Nutrients

Commercial baby foods are currently required to display the amounts of macronutrients (protein, fat and carbohydrates) on their labels, as well as salt and sugar content, but they are not required to display all the micronutrients (vitamins and minerals).

A study that compared the nutrient levels (macro and micro) of commercial and homemade baby food puree found that there was no significant difference. Both contained more or less equal nutrients over a four-day period. There was also no difference in the weight of the babies eating them. Interestingly, the researchers found that baby's who had a mix of both diets, commercial and homemade, were actually receiving greater amounts of nutrients. They didn't really know why this was but guessed that it may be a sign that these babies were introduced to food that had been prepared for the whole family (Yeung, DL, et al, 1982).

According to research carried out by E.L. Ferguson and N. Darmon (2007), ensuring a baby gets all her necessary nutrients from a homemade diet can be difficult and, in fact, a balanced diet is more easily attainable from commercial baby foods. N. Zand et al (2012) found that commercial baby foods show less variation in nutrient content compared to homemade meals and that the percentages of macronutrients listed on the jar are generally accuratel.

Commercial baby foods, however, have higher fat and protein contents than homemade foods and also tend to offer a better balance of the three macronutrients: fat, protein and carbohydrate (van den Boom, S, et al, 1997), so if toddlers eat additional snacks in combination with commercial baby foods, the recommended daily allowance for fat may be exceeded (Zand, N, et al, 2012).

In terms of micronutrients, Zand and his fellow researchers (2012), found that there was insufficient amounts of vitamins $B_6$ and $B_2$ in commercial baby food in the UK (although they noted that the amount of these vitamins actually being consumed by 6-9 month olds was sufficient due to their consumption of formula milk).

In another recent study, Zand and his fellow researchers (2012) looked at 20 different nutrients in popular jarred baby food aimed at 6-9 month olds. They found that, on average, the baby foods had insufficient levels of calcium and some had excessive amounts of potassium. This is probably due to the move away from sodium salts towards alternatives, such as potassium chloride, for flavour. Magnesium was also found to be higher than needed; this is thought to be due to the use of fertiliser in growing vegetables. Different brands also differed a great deal in terms of their nutrient content. The European Union Scientific Committee on Food (EUSCF) directs the level of nutrients required in baby food but, as yet, nutrients are not required to be displayed on the containers. According to the researchers, this may be because the manufacturers cannot guarantee the minimum levels of certain nutrients like potassium and magnesium.

Interestingly, children of a lower socioeconomic status who ate more commercial baby food had, as a result, more fruit and vegetables in their diet and thus healthier meals than those who ate meals prepared at home (Hurley KM, and Black MM, 2010).

### Hygiene

There is an argument that commercial baby food is less likely to cause infections because it is manufactured in sterile conditions. Homemade baby food preparation can, of course, attain similar levels of hygiene, as long as mum and dad follow the recommended guidelines.

### Cost

It is cheaper to make your own baby food. A modest estimate is that during the first year of weaning (from 6–18 months), parents will spend just under £250 on baby food, however, if making homemade and buying what's in season, parents could spend as little as £45. The cost of food in general has risen generally in the UK over the last 5 years but baby food, in particular, has risen more sharply, so there may even be greater savings.

### Brand awareness

There are a number of competing baby food brands, which vary not only in cost but also in water content and, therefore, nutrient content. Parents need to be particularly careful that they avoid baby food that has added salt or sugar, which has been largely removed by public pressure within the last decade, but still exists in about 15% of available baby food (Bennett, AE, et al, 2012).

### Acrylamide and furan

These are substances produced during the normal cooking of starchy foods (except for boiling or microwaving) and are particularly likely to be produced when food is cooked to a high temperature such as with processed foods. Crisps, bread and chips tend to be relatively high in acrylamide while furan can be found in coffee, and

canned and jarred foods. Both substances have been shown to cause cancer in animals and could, therefore, have a similar effect in humans if levels are high enough. The Foods Standards Agency (FSA) has recently reported that the levels of acrylamide and furan are increasing in cereal-based baby foods although levels are still relatively low and pose no immediate danger. However, they are being monitored closely. Parents making homemade baby food can more easily avoid the high temperature cooking that produces these substances and opt to boil or microwave food instead.

Another area of concern is the transfer of mineral oil saturated hydrocarbons (MOSH) from recycled food packages. Many different types of MOSH chemicals are used by the food industry either directly or indirectly, and many foods become contaminated with them. These chemicals are easily absorbed by the body but are not broken down. Animal studies have shown that they can accumulate in certain organs, such as the liver, heart and lymph nodes, and could cause them to become damaged or diseased.

A recent Italian study was concerned with levels of MOSH chemicals in baby food (Mondello, L, et al, 2012); it found them in all the samples tested. The researchers concluded the contamination was most likely caused by the use of vegetable oil, sugar and commercial corn starch. It is not known how MOSH chemicals enter the plant from which the vegetable oil is produced, but it is thought that particulates are absorbed as the plant grows and during the manufacturing process. As yet, there is no official safe limit for these chemicals and far more research needs to be conducted in order to determine what the detrimental effects could be. Homemade food is more likely to be prepared without sugar and corn starch and will also use a greater variety of foodstuffs so levels of MOSH chemicals are likely to be lower.

## Recommendations from the research

Bought baby food can provide appropriate nutrient levels and a balanced diet but recent research suggests there may be nutrients, which are not sufficiently high. The number of different varieties on sale can make it complicated and stressful to ensure a baby gets the best nutrients, and is able to avoid substances that she doesn't need, such as extra salt and sugar as well as potentially harmful substances. Hygiene and ease of use whilst out are other positive considerations.

Making your own baby food, however, would appear to be the better choice, but only if you aim for a good variety in order to produce a balanced diet. Homemade food is also cheaper and should contain less furan, acrylamide and MOSH chemicals.

However, for days when you're a little too busy or on the go, you can guiltlessly reach for the bought stuff, as long as you've bought a good quality product.

# Go organic?

Every parent wants the best for their little one and that includes the foods she eats. Even if you don't choose organic for yourself, you may be tempted to offer it as a first food, or at least until your baby eats what you eat. After all, first foods are important and babies really don't eat that much. Commercially produced organic baby foods are readily available and quite often only marginally more expensive. But is it important to offer organic food to a baby, and if so, for how long?

Organic food is that produced without using artificial fertilisers, antibiotics, growth hormones and most pesticides, and in such a way as to not interfere with the natural surroundings and with greater care of animals. There are stringent regulations that must be met in order for a food to be labelled 'organic'.

## WHAT THE RESEARCH SAYS

### Nutrient quality

A large and recent review investigated organic food and found no real difference in health benefits compared with non-organic foods (Smith-Spangler, C, et al, 2012). An earlier review by The Foods Standards Agency came to a similar conclusion (FSA, 2009). Both found little or no difference for vitamin C, calcium and iron. Organic crops

were found to contain higher levels of phosphorous, while non-organic crops contained more nitrogen (likely due to the fertiliser used and not considered to be a health benefit). Organic livestock was also found to contain similar amounts of nutrients to non-organic.

However, small individual studies have found some differences. One study that compared organically grown grapefruit with inorganic fruit found that the organic fruit had slightly higher levels of vitamin C (Lester, GE, et al, 2007). Other studies found that organic peaches had higher levels of vitamin C and citric acid (Carbonaro, M, et al, 2002), and that there were increased levels of nutrients in organically grown aubergine (Raigón, MD, et al, 2010).

Antioxidants are substances that prevent damage to cells from free radicals. Free radicals are produced by normal body functioning and environmental pollution, such as smoke from cigarettes. Free radical damage is thought to contribute to a host of ailments including heart disease and cancer. One 3-year study that measured the levels of antioxidants in apples found that there were slightly more antioxidants in organic apples but that levels changed dramatically over the years; in some years, there was a noticeable difference in levels between organic and nonorganic, whilst in other years, there was little difference. This demonstrates that whether or not organic food is high in antioxidants depends on the climate (Stracke, BA, et al, 2009).

Another study found no real difference in levels or nutrients or antioxidants when comparing organic versus non-organic tomatoes. It did, however, report that different varieties produced different amounts of nutrients, indicating that a varied diet may be more beneficial than eating organic (Juroszek, P, et al, 2009).

**Allergies**

There is no evidence that organic food limits the chances of a baby developing eczema (Shams, K, et al, 2011) or a wheeze.

**Patulin**

This is a naturally occurring chemical found in apples and therefore apple juice or puree. In high doses, it has detrimental effects on a developing baby and the immune system but tests have only been carried out on animals. An Italian study comparing patulin levels in organic and non-organic juices found that while organic juices had a lot more patulin, the levels were still considered to be well within safe limits (Piemontese, L, et al, 2005).

**Pesticides**

Repeated low level exposure to pesticides during fetal development and early life has been linked with neurodevelopment problems (Eskenazi, B, et al, 1999) and cancers (Meinert, R, et al, 2000), but these studies only focussed on children in farming communities where exposure to pesticides would have been regular and potentially higher than for other children. Compared to non-organic foods, the level of pesticides in organic foods is significantly less (Smith-Spangler, C, et al, 2012) but pesticide levels and types are closely monitored in non-organic food intended for babies and levels are kept very low which means they are unlikely to cause any harm.

However, scientists are still unsure whether there are any long-term effects from the chemicals used in producing non-organic foods. While chemicals are tested regularly and removed or levels adjusted if any evidence of harm is uncovered, pesticides tend to be tested independently of each other and only short-term effects have been noted. There are hundreds of different pesticides used in the growing of our food and it's likely that a toddler eating a varied diet may well encounter a number of different residues. There is a possibility that this cocktail of residues could interact with each other and cause harm to the body in the longer-term. There are few investigations looking at the possible interactions of pesticides within the body and the long-term effects.

**Organic milk**

K. A. Ellis et al (2006), found that organic milk was higher in beneficial omega-3 fatty acids, a type of polyunsaturated fatty acids (connected with a range of health benefits) than normal milk but this depends on the time of year and a few other factors such as breed of cow. Overall, the differences in types of milk are small and there would be little impact on health (O'Donnell, AM, et al, 2010). But organic milk is also used for a variety of dairy products such as yoghurt and cheese, so that they, too, would have higher levels of beneficial omega-3 fatty acids (Palupi, E, et al, 2012).

There is also an added complication that while iodine is added to non-organic cattle feed it is not added to organic cattle feed so is not present in high levels in organic milk. There are many dietary sources of iodine so a varied diet should ensure that baby stays healthy but this is one of the reasons why the UK is currently thought to be deficient in iodine. The effects are most noted in women because more iodine is needed during pregnancy for healthy brain development (Vanderpump, MP et al, 2011).

**Iron supplementation**

An important consideration if you're aiming to go organic is that organic baby food cannot be supplemented with iron, which means that if you weant your baby onto an organic baby cereal, she may not receive enough of this mineral. This is particularly true if you are breastfeeding and, as recommended, you don't introduce solids until she is 6 months old.

**Cost**

Organic food, particularly if you're preparing baby's food from scratch, is noticeably more expensive, particularly for meats. However, given the recent popularity of organic baby foods the cost of buying pre-prepared foods is largely similar to the cost of non-organic ones and organic, pre-pared foods are often more easily accessible.

**Antibiotic resistant bacteria**

Animals can be fed antimicrobial substances or be given antibiotics in order to improve their health and boost production. In conventional rearing, animals are often given these drugs as preventatives, which increases the likelihood of certain bacteria becoming antibiotic resistant so the World Health Organisation has called for fewer antibiotics to be used in this manner. The risk to humans is not clearly understood although it is possible that some bacteria could become resistant to antibiotics normally used to treat humans. Antibiotic-resistant bacteria are killed by the proper cooking of meats.

Organically raised animals are only given antibiotics if they become ill and need them. Due to the nature of organic farming, fewer animals are in confined spaces so the spread of disease tends to be less and antibiotics are not required as frequently. As a result, organic meat and eggs contain fewer antibiotic-resistant bacteria (Smith-Spangler, C, et al, 2012).

E. Alvarez- Fernández et al (2012) found that the eggs from caged, barn-raised and free ranging hens were higher in antimicrobial resistant bacteria than organic or domestically produced eggs.

## Recommendations from the research

Buying organic is not necessary for your own or your baby's health. A healthy diet can be achieved easily and more cheaply through eating a variety of foods. Organic food, however, is certainly not detrimental to baby. If opting for organic milk (after one year) or organic baby cereals, make sure your baby's intake of iodine and iron are sufficient, particularly if breastfeeding. Organic baby food will contain fewer pesticides than non-organic foods, but the latter are monitored to ensure levels are safe, although little is known about long-term exposure. There are other reasons why buying organic is a good idea, too. Organic farming has been shown to increase biodiversity, produce fewer antibiotic resistant bacteria, better animal welfare and, many people will argue, better taste.

# Handling a fussy eater

The early months of introducing food can often be a fun time for both parents and baby. A baby is keen to try new things, explore and experiment, and healthy foods can be easily spooned into her smiley mouth. But as she grows older and her ability to make decisions becomes more pronounced, the spoon becomes less readily accepted. A once explorative baby transforms into a picky toddler.

A tendency to avoid new foods or food neophobia, is common amongst toddlers and young children, and most prevalent around age 20 to 30 months. It can last until about ten years of age but the vast majority of children grow out of it, although some still suffer as adults. The terms 'picky' or 'fussy eaters' are often used colloquially to describe food neophobia but some researchers would define it slightly differently. Food neophobia is a reluctance to try new foods, whereas picky eaters are also reluctant to eat familiar foods, such as vegetables. While the former decreases with

age (possibly because there are less new foods to try), pickiness remains quite stable throughout childhood. The two may be related, in that most children will experience food neophobia but the degree to which they suffer or how well parents handle the situation could determine whether they turn into picky eaters. Despite the subtle differences, these 2 types of eaters have much in common in terms of causes and treatments. Food neophobia tends to be limited to fruit, vegetables and proteins, such as meats; a liking for sweets is never affected!

**More about food neophobia**

Animals, such as rats and chimpanzees, exhibit food neophobia so it is likely part of our evolutionary heritage. For some species, an aversion to new foods could be protective as it prevents the eating of potentially poisonous foods.

Problems with food don't necessarily run in a family; 1 child can eat everything she is given, whilst her sibling can be extremely particular (Pliner, P and Loewen, ER, 1997). However, the degree of neophobia your child experiences is likely due to her genes rather than your parenting as demonstrated by research done by L. J. Cooke et al on twins (2007).

Humans differ in their ability to taste bitter flavours, usually derived from plants, due to the presence or absence of specific genes. A particular chemical known as 'PROP' (6-n-propylthiouracil) has been studied in some detail. About 30% of the population cannot detect it, some may taste it moderately, whilst others, often described as supertasters, taste it strongly. The ability to taste PROP and similar bitter substances may account for why some people are pickier than others with their food, particularly vegetables. If a young child is able to taste PROP, this might influence her degree of neophobia, which may be much worse in PROP tasters than in non-tasters (Tsuji, M, et al, 2012).

Even if food neophobia is genetically predetermined, the extent of the problem can be minimised (or maximised) by a parent's reaction to it. If a mum is reluctant to try new foods, her baby is likely to act similarly. Picky eaters, on the other hand, seem to be more affected by whether their parents eat a varied diet (Galloway, AT, et al, 2003).

That food neophobia generally arises in toddlerhood suggests that it might be linked to a developmental stage, perhaps of growing independence and a child's desire to assert herself, which is why tantrums also are more common at this age. And it tends to fall on a scale; some children are highly neophobic whilst others hardly at all, with all conditions in between. There does appear to be a link between the range of foods offered and the degree of neophobia; the more food offered from a young age, the lower the neophobia. A possible cause is a lack of variety in the weaning diet or even in a mum's diet during pregnancy or breastfeeding, as different tastes can transfer to baby whilst she is in the womb or through breast milk. Related to this is the idea that women who breastfeed tend to be slightly more relaxed about how much food their children have taken, as they have less control over this. Their babies determine how much they eat when breastfeeding and this could extend to their taking greater control over what they eat as toddlers.

Toddlers who had been breastfed take in more calories (Fisher, JO, et al, 2000).

There is also a possible link between food neophobia and levels of anxiety. Anxious children tend to be pickier than others.

It's also important to bear in mind that some food aversions can be learned. For example, if a child was ill after trying a particular food she may be put off trying it again. However, this is more likely with a small selection of foods, rather than an entire food group as is generally the case with food neophobia.

## CURRENT RECOMMENDATIONS

The NHS recommends that as long as your child is active and gaining weight and eating any foods from the four main food groups (dairy, carbohydrates, fruit/veg and proteins) – even if they're the same foods over and over again – there is no reason to be concerned. You should attempt to improve the variety in her diet by gradually re-offering disliked foods and modelling a healthy diet yourself. Be patient, give lots of praise but try not to use foods as rewards, have a family dinner at the table and, if possible, invite other children or adults round to model good eating habits.

## WHAT THE RESEARCH SAYS

Early and frequent exposure to foods is important. Providing a baby with a wide range of tastes, particularly fruit and vegetables, may make food neophobia less pronounced. Early exposure doesn't just mean when weaning; J. A. Mennella et al (2001) showed that babies who are exposed to tastes in the womb or whilst being breastfed are more likely to accept them later on.

If a child is repeatedly offered, but not forced to eat, an originally disliked food, she will begin to accept it – and this will be more effective than offering a reward (Wardle, J, et al, 2003).

Unfortunately, many parents tend to avoid cooking a disliked food (who wants to cook something that will be rejected), which can exacerbate food neophobia. A child needs to be offered, or at least see another person eating, a food between 10–15 times before she begins to accept it. Most beneficial is to continue offering a 'rejected' food (usually a vegetable) on consecutive days, possibly because a toddler will remember it and it'll become more and more familiar. Because toddlers have fairly short memories, if you allow a period of time to pass before reintroducing the item, you are likely to be back at square one. B. R. Carruth and J. D. Skinner (2000) found that failing to offer once disliked foods over and over again is a major contributing factor to a child's neophobia as is a mother's neophobia and her unwillingness to cook particular foods.

### Modelling good eating behaviour

Children eat better when they see an adult eating the same food. E. Addessi et al (2005) found that children aged between 2–5 years were more willing to eat a novel food if a nearby adult was eating a food of the same colour. The inclination to try new foods is also enhanced if siblings and peers do the same. Unfortunately, the opposite is also true. If you are a picky eater and tend to avoid trying a food before offering it to your toddler, you will encourage her neophobia.

Pre-school teachers who enthusiastically model the trying of different foods (new to the child) produce positive effects but their efforts can be undermined by the influence of fellow picky toddlers. Girls are more susceptible to this kind of peer pressure than boys (Hendy, HM and Raudenbush, B, 2000).

Snacking parents produce snacking children. R. Brown and J. Ogden (2004) found that if parents frequently eat crisps, chocolate or biscuits, it's likely their children will want to eat them too. Likewise, if healthier snacks are chosen then it's likely the children will opt for healthier ones too; children copy what they see). The influence of modelling by parents is so strong that simply by eating healthily yourself you may not need to employ any other strategies in getting your child to eat well.

Similarly, the same study showed, if you are dissatisfied with an aspect of your body image, for example your weight, this dissatisfaction can be transferred to your child. This is particularly true for mothers and their daughters. In our society there are many complex behaviours surrounding the eating of food that have little to do with hunger and satiety, which can be modelled for children. Parents, who have a tendency to eat for reasons unrelated to hunger, are more likely to have children who eat food because they feel cross or upset.

**Parental control**

The majority of parents will attempt to get their child to eat more food at the dinner table. The most successful strategies include prompts or reminders to eat, the reward of a dessert and praise for eating. Threats to remove playtime or similar won't improve a child's eating. Generally, if parents try hard enough, they are usually successful in getting their child to eat more although this may not be a good thing. Parents who insist on controlling when, what and how much their child eats can be detrimental in terms of excess weight gain.

Children are also sensitive to their environments, and are more likely to eat where the atmosphere is emotionally positive and threats and punishments are not routinely used as methods to attain a healthy diet.

**Trust the child**

Parents have a tendency to underestimate how effectively a child can regulate her own energy intake. L. L. Birch et al (1990) gave children highly calorific and low calorific drinks at snack time and found that the children who had the highly calorific drink ate less food afterwards and the children who drank the low calorific drink ate more food afterwards to compensate.

Parents might feel that during one particular meal their child hasn't had enough, but it is likely that over a 24-hour period, the child will meet her energy demands by herself (Birch, LL and Fisher, JO, 1998). Parents can inadvertently override their child's body cue that she is full up and teach her

## Recommendations from the research

Although common, neophobia is heavily influenced by parental actions. A varied diet during pregnancy and breastfeeding can help limit its effects, as can introducing at weaning a wide range of healthy foods. Repeatedly offering once rejected foods – up to 15 times – and not limiting food variety will also help to curb symptoms. If you model eating a healthy variety of foods, it is likely your child will eat the same. Avoid using rewards for eating certain foods and never threaten or punish. Trust your child's innate ability to regulate her own calorific intake.

to overeat. This could eventually contribute to overweight and other health related issues later on in life (Orrell-Valente, JK, et al, 2007).

**A balanced diet and rewards**

Some parents may be more concerned with what their child is eating rather than how much and may reward the eating of vegetables with dessert. Rewarding a child for eating a food does yield results, in that the child will most likely eat her vegetables (and the dessert) but it also creates a distinction between vegetables and a far more pleasing dessert. By creating this distinction, parents will cause a reduction in the preference for vegetables, or for whatever food they were trying to get their picky toddler to eat (Birch, LL, 1999). Likewise, keeping favourite foods on the top shelf and limiting a child's access will actually increase her liking for that food.

**Try not to worry**

Despite a child being a picky eater, B. R. Carruth and J. D. Skinner (2000) suggest that over time she will eat an adequate diet and grow normally. Compared to children who eat well, toddlers with neophobia tend to reach the same levels of RDA (recommended daily allowance) for energy and most nutrients. The exception, according to G. A. Falciglia et al (2000), was vitamin E. They also reported that neophobic children tended to eat higher amounts of saturated fat.

Generally, if you are eating healthily, then this is likely to rub off on your little one. However, if you have unhealthy foods available at home, or if your child's peers eat unhealthily or if she sees unhealthy foods advertised on the television, these can undermine your efforts (Taylor, JP, et al, 2005).

# Restricting the bottle

The vast majority of toddlers still use a bottle well past current recommendations for them to stop. Just how important is it to adhere to the guidelines and are there ways to protect your baby from injurious effects if he won't be parted from a beloved bottle?

## CURRENT RECOMMENDATIONS

The NHS recommends introducing a cup from 6 months of age in order that the bottle is dropped by about 1 year. It also recommends using free-flow cups which are better for baby's teeth and only putting milk or water in bottles.

## WHAT THE RESEARCH SAYS

### Tooth decay

Bottle drinking, or any method requiring a baby to suck to get liquid, means that it takes longer to get a mouthful and that each sucked mouthful of milk (or other sugary drinks such as juice) pools around a baby's teeth for longer than if he used a normal cup. This enables the bacteria in a baby's mouth to change the sugars in the milk into acids, which 'dissolve' tooth enamel and result in tooth decay. A preschool child who has at least one decayed tooth (and some do so before age 2), is said to be suffering from early childhood caries. At its worst, this totally preventable situation, can mean that a child might require restraining or even sedation to have the affected tooth pulled. It is fairly well recognised that early tooth decay is heavily exacerbated by bottle drinking during the night (Tiberia, MJ, et al, 2007 and Mohebbi, SZ, et al, 2008). Toddlers allowed to fall asleep whilst drinking from a bottle are more likely to have tooth decay than those who fall asleep after they've finished a bottle. Those who fall asleep after discarding a bottle are more likely to have tooth decay than those not given a bedtime bottle at all (Schwartz, SS, et al, 1993).

C. Febres et al (1997) also found evidence that tooth decay is more common if weaning from a bottle does not occur by 14 months. That being said, a lot of children who use a bottle well into childhood (up to age 3 or 4) never develop early tooth decay.

However, there have been links made between socioeconomic status and early tooth decay. This has led to a theory that tooth decay is more likely to be dictated by the mum's diet whilst she is pregnant and her baby's teeth are being formed. If she is deficient in vitamin D and calcium, this can lead to defects in her baby's tooth enamel, which make tooth decay more likely later on (Smith, PJ and Moffatt, ME, 1998).

According to M. I. Pavlov and C. Naulin-Ifi (1999), although childhood caries affect the milk teeth, decayed milk teeth can cause the adult teeth to become more at risk of decay and be poorly spaced within the mouth. Baby teeth 'save' room for adult teeth but if they have to be pulled out prematurely due to decay, the adult teeth may become overcrowded, which can lead to tooth misalignment requiring orthodonture.

### Iron deficiency

Toddlers who drink from a bottle tend to take in too much milk, which can lead to a decreased amount of other foods eaten. K. Bonuck and R. Kahn (2002) found that toddlers who were still having between 3 to 10 bottles of milk or juice a day had a greater risk of being iron deficient. Cow's milk is not a great source of iron and if a toddler is filling up on milk rather than nutritious iron-rich foods, he is not likely to be getting enough iron in the diet (Brotanek, JM, et al, 2005). The longer bottle feeding continues, the greater the chances of iron deficiency.

### The risk of obesity

R. A. Gooze et al (2011) demonstrated that a 2 year old who still drinks from a bottle may become an overweight 5 year old since bottle-feeding toddlers tend towards taking in too much milk. K. Bonuck et al (2004) found that with each subsequent month of bottle feeding, after the recommended 12 months, there was a slightly increased chance of a higher BMI.

### Other possible issues

If a child is allowed to carry his bottle around and help himself whenever he fancies some milk or juice, then the bottle begins to mimic a dummy. Using a dummy beyond 12 months brings a host of other potential complications such as recurring middle ear infections and misalignment of teeth (see page 92).

Moreover, accidents can be exacerbated if a bottle is in the mouth at the time of a fall. Cuts and bruises to the mouth are more likely and a tooth can be chipped or even dislodged, depending on the severity of the fall. S. A. Keim et al's analysis of children aged under 3 admitted to emergency rooms in the U.S. (2012) found that bottles caused a significant number of injuries – more than dummies and cups. Limiting the time that your child is able to walk around with a bottle (or cup for that matter) will decrease the risk him ending up in the A & E after a fall.

### The best cups

There are many different cups available but the key factor in choosing one is whether it allows liquid to flow freely or has a valve that requires your baby to suck to get liquid. Free-flow cups are referred to as 'feeder' cups and ones with a valve are 'sippy' cups. Because sippy cups mimic bottle use, all the negatives associated with the latter can be extended to them, particularly, if your toddler is allowed to sip milk or juice frequently throughout the day or night. One will, however, allow your toddler to have a drink without him spilling the contents everywhere – a significant advantage in certain conditions. Free-flow cups may not prevent spills so well but are better for teaching your child to drink from a normal cup.

There are also a couple of lidless toddler cups such as a 'Doidy' cup or 'Babycup'; again, there's a risk of spilling but less of tooth decay. While health experts usually advise against using sippy cups one might still be handy for car journeys!

## Recommendations from the research

The bottle-using toddler faces certain risks but there are precautions, which can minimise them.

Your toddler should only need a maximum of 2 bottles of milk a day (approximately 500ml). Drinking more milk than this could increase the chances of tooth decay, obesity in later life and becoming iron deficient. You need to limit the amount of milk he takes, as well as your toddler's access to the bottle.

If you allow your child to have regular sips from a bottle throughout the day, you should limit him to only water for these periods, offer juice only at meal times and milk when your toddler will drink the whole bottle. Never allow your child to take a bottle to bed.

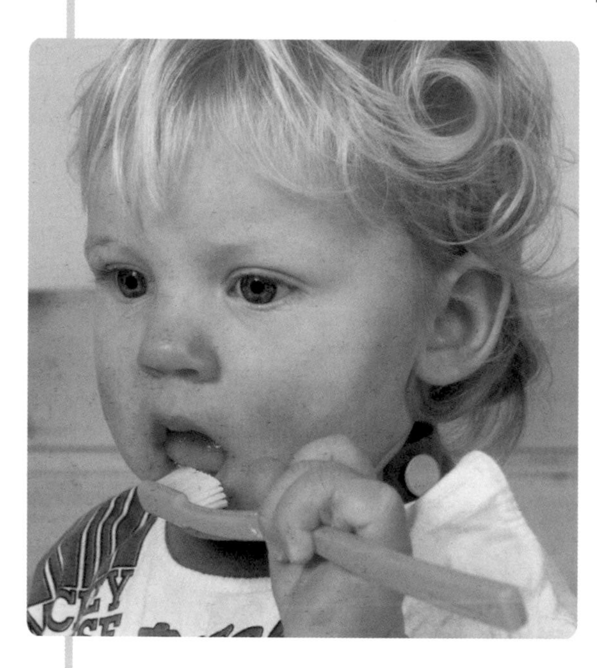

Of course, in order to eliminate any risks and for, theoretically, an easier time of it, follow the NHS guidelines and aim to be off the bottle by age 1 (or not introduce one at all). A 12 month old is easier to convince that change is good than an older toddler, so it may be easier to switch from bottles to cups at 12 months than it would be at 18 months or older but bear in mind that valved or sippy cups carry similar risks as bottles, so limit their usage.

Brush teeth daily, even if your child resists. Although most parents realise that tooth brushing is important, M. J. Tiberia et al (2007) have shown it can't overcome factors such as bottle use in preventing caries but not brushing due to battles will increase the risk of tooth decay.

J. Khadra-Eid et al (2012) have proved that a trip to the dentist is the best protection against tooth decay. Early signs of tooth decay can be detected and vital advice on protecting against further damage will be offered.

# Toddler milk

If your baby has been drinking formula milk, enriched with countless vitamins, minerals and other nutrients or you are weaning your baby from the breast, it can seem quite a jump to suddenly switch to cow's milk on her first birthday. OK, she is eating most foods by this stage, but are they really enough? How nutritious is cow's milk compared to formula? Are fortified milks for toddlers necessary or are we being duped by clever advertising? It can be tough to make the best choice for a baby, not to mention the effect on your wallet.

## CURRENT RECOMMENDATIONS

The NHS recommends that normal, full-fat pasteurised cow's milk should not be introduced as a child's main drink until she is at least 1 year old and half-fat milk can be offered from 2 years. Formula milk can continue to be offered beyond 1 year old, if desired. There are no specific guidelines on toddler milk. The Scientific Advisory Committee on Nutrition states that follow-on milks offer no nutritional advantages for healthy babies. There is evidence that the extra iron in these milks is not actually absorbed by baby and may hinder the absorption of other nutrients.

## WHAT THE RESEARCH SAYS

### Omega-3 supplementation

There has been much research conducted recently on the advantages of long-chain polyunsaturated fatty acids or omega-3 supplements for children, however, whilst some of it is promising, the benefits are far from conclusive. Despite this, much formula milk nowadays contains added omega-3 fatty acids.

Docosahexaenoic acid (DHA) is a type of fatty acids in which children are routinely deficient, as they often don't get enough from the diet. DHA is found naturally in oily fish and is needed for growth and brain development; deficiencies are associated with ADHD and L.A. Horrocks and Y.K. Yeo showed that supplementation improves eyesight in young children (1999). Toddler milk that has been supplemented with DHA is successful at boosting DHA levels in drinkers, and L.M. Minns et al demonstrated it had reduced respiratory illnesses, such as colds 2010). The advantages of omega-3 supplements may be inconclusive but when added to milk can certainly prevent deficiencies.

### Iron supplementation

Cow's milk is relatively low in iron while formula is supplemented with the mineral. This is why current guidelines recommend parents wait until their baby is a year old and eating solids before they switch (from formula or breast milk) to cow's milk as a main drink. Most babies will eat sufficiently well to gain enough iron from their diet. However, if a child has an inadequate diet, is a fussy eater or does not take enough iron-rich foods, she may not get enough. Or if a baby drinks too much cow's milk throughout the day, she may not be eating enough solid healthy food and, again, become iron deficient. Children who have cow's milk as their main drink are likely to have lower iron levels than those who drink formula (Thorisdottir, AV, et al, 2012) even if healthy. P. Soh et al reported that 66% of New Zealand toddlers were at risk of having insufficient iron levels (2002). Generally, however, even if a toddler has low iron stores it will not necessarily impact negatively on her health in terms of growth or development (Sachdev, H, et al, 2006).

Toddler milk has a much higher iron content than cow's milk so taking toddler fortified milk can ensure a child gets enough iron into her diet (Maldonado Lozano, J, et al, 2007). However, the optimum iron levels in formula milks are still relatively unknown. In 2012, B. Lozoff et al reviewed a study that started ten years ago and found that iron-fortified formula milks may not be as beneficial as first assumed. At the beginning of the study, some 6-month-old babies were randomly assigned either iron-fortified formula milk or regular formula milk. The researchers found that those that received the former scored less well on every developmental test, including spatial memory (the ability to remember where are things are, for example in a darkened room) and eye-hand coordination. They also found that those babies who had high levels of iron at 6 months and were then assigned to iron-fortified formula scored lowest on the tests 10 years later. Those with lower iron levels when they were assigned the iron-rich

formula did better in the tests. This implies that high iron levels at a young age may negatively affect a child's development later on. This was a single, relatively small study and its findings need to be further tested but it does make a case for aiming for iron-enriched foods as part of a varied diet rather than relying on fortified milks.

**Vitamin D**

The Department of Health recommends that children aged 6 months to 5 years receive a vitamin D supplement. However, formula milk already contains vitamin D so babies who are formula fed will not need vitamin D supplements until they are drinking less than 500ml or two full bottles worth per day. Roughly 500ml of milk per day is the maximum that a 1 year old should be having (see page 00), so a formula-fed toddler should be given a vitamin D supplement, which means there'll be no advantage in toddler milk compared to cow's milk (as long as the vitamin D supplement is taken).

**Other vitamins**

As well as vitamin D, the Department of Health recommends that children aged 6 months to 5 years, be supplemented with vitamins A and C so parents who follow this guideline can be assured that their child is getting enough of these vitamins without having to use toddler milk (even though it has much more vitamins C and A than cow's milk).

**Calcium**

There is more calcium in cow's milk than in formula milk. Once a baby begins to decrease the volume of milk that she drinks, it is important that she has a more calcium-rich source. By offering cow's milk, as opposed to toddler milk, a parent can maximise the amount of calcium a toddler receives.

**Added sweetener**

Most toddler milks contain vanilla to make them taste sweeter. Getting a young toddler used to drinking sweeter-tasting milk could make it harder for her to accept normal cow's milk eventually, or worse still, encourage a preference for sweetened drinks throughout childhood.

**The power of advertising**

It is illegal to advertise infant formula in the UK because it is designed to replace breast milk. However, it is legal here (but not in all other countries) to advertise follow-on (from 6 months) and toddler milks (from 1 year). By heavily advertising toddler milks, formula milk providers are also able to 'suggest' their infant formula as part of a product line, which encourages parents to buy it (Berry, NJ, et al, 2012). The Advertising Sales Authority has found that some toddler milk companies mislead parents in their advertising over how much iron toddlers need. Because only follow-on and toddler milks are allowed to be advertised, a parent can easily become confused over whether they are needed or not.

## Recommendations from the research

A mixed and healthy diet provides most toddlers with the majority of their necessary vitamins, minerals and fatty acids. For children who drink less than two bottles of formula a day (500ml), supplementation of vitamins A, D and C is recommended to ensure that anything lacking in or unattainable from the diet (such as vitamin D), is supplied. However, toddlerhood can be a time of picky eating, and omega-3 fatty acids and iron can sometimes be lacking in a limited diet. These are found in most toddler milks, although many children show no adverse effects from low iron levels. Calcium levels are higher in cow's milk than toddler milk and many of the latter are sweetened, which can make the eventual transition to cow's milk trickier and encourage a sweet tooth.

In summary, a healthy toddler should be taking about two bottles of cow's milk per day, a supplement of vitamins A, D and C, and eating a healthy, varied diet. There is no evidence, as yet, that cow's milk plus a multivitamin is less healthy than having toddler milk.

Picky eaters may benefit from a toddler formula, but depending on how much they are drinking, they might need a supplement of vitamins A, D and C, as well, which could be costly. If you're going to use a toddler milk, choose one that's recommended for use to complement cow's milk, i.e. one drink per day, as this may also help with the transition to cow's milk.

| Type of formula milk | Designated age | Description |
|---|---|---|
| Infant formula/ hungry milk | Suitable from birth to about 1 year | Everything a baby needs for the first year of life. Hungry milk has a slightly different ratio of proteins. |
| Follow-on milk | Suitable from six months in conjunction with a weaning diet | Contains slightly more protein, micronutrients and iron than infant formula. |
| Toddler milk/ growing-up milks | From age 1 year. Can be as an alternative to cow's milk or to complement cow's milk. | Contain higher amounts of some vitamins and minerals than other formulas or cow's milk. |

# Discipline: when to start

In order to rear socially healthy children and future adults, parents must begin to lay the groundwork early on for what is acceptable behaviour – and what is not. This can be a challenging time for parents, as doubts can arise as to whether the decisions they've made are good ones. Bad behaviour, such as kicking, biting and hair pulling can start at a very early age, but how old does a child have to be to know what he is doing is wrong? Or put another way, at what age should discipline begin? Although most people hear 'discipline' and think 'punishment', they are not the same thing. Discipline involves changing unwanted behaviour, and can be by positive reinforcement rather than punishment.

## CURRENT RECOMMENDATIONS

There are no real UK guidelines. A parent is permitted under the law to use 'reasonable punishment' which could include smacking. Physical punishment using an implement or leaves a mark is considered 'unreasonable'.

The American Academy of Pediatrics recommends using routines to limit unwanted behaviour from age 6 months and to employ distraction for unwanted behaviour and praise for desired behaviour from about 12 months. They recommend that if these strategies don't work that time-outs can begin to be introduced from 12 months.

## WHAT THE RESEARCH SAYS

### Age appropriate tactics

Changing unwanted behaviour can begin when you introduce a routine in response to your child's needs, such as hunger or sleepiness. For example, you will introduce routine when you put your well-fed baby, older than 3 months, down to sleep at the same time each evening in order to encourage himm to sleep through the night. Although babies begin to understand the concept of "No" at about 9 months old, it will take more time before they fully comprehend that it means you don't want

them doing something.

Between the ages of 1 and 2, children often lack the ability to understand verbal reasoning. At this time, they won't fully understand a direction not to do something or an explanation of why they are being punished. While your child should have some capacity to understand short statements, such as "No…hot!", sharing more complex thoughts, such as "Don't go near the cooker, if you touch it you may get burned…" may yield fairly limited results (Blum, NJ et al, 1995).

For young toddlers, distraction, praise for wanted behaviour and pre-emptive parenting, discussed below, may prove effective.

According to B.J. Howard (1991), time outs are a useful way to discipline older children and if carried out consistently, are one of the most effective methods of teaching consequences to a young child. Although there are variants, a time out usually consists of putting a child in an area (bedroom, stairway) where there is limited distraction (no toys or TV), giving a brief explanation of why he is there ("No biting"), and allowing the child to 'ponder' his misbehaviour by staying put for 1 minute per year of his age (a child aged 2 staying 2 minutes, etc.). Young toddlers may need to be gently held in order to keep them in the area where the time out is happening. At the end of the time out, an apology (kiss and cuddle) may be required or the parent may chose to actively forget the bad behaviour by moving immediately into a new activity. Results can be expected slowly over a period of time.

A time-out's effectiveness will be very limited for young toddlers but It may be a good idea to introduce the concept, perhaps without you leaving the area, in order to prepare your child for its inevitable future use. It is important to bear in mind, however, that when considering time-outs, young toddlers are more susceptible to fears of being abandoned than older children, and a parent's absence as a form of discipline may cause more pain than the poor behaviour warrants (Canadian Pediatric Society, 2004).

The terrible 2s are so-named because of the tantrums young children throw as they begin to become more independent and learn to assert themselves. A child of this age has more capacity to understand simple verbal reasoning but still has a limited ability to control his behaviour. So even if your child understands that he needs to put those sweets back on the shelf because you're not going to buy them, he may not have the ability to control the accompanying emotions and may throw a fit. In these circumstances, the Canadian Pediatric Society advises that you try to remove your child from the sweet aisle and hold him gently until his anger passes, after which you can give a short verbal instruction and maybe an example of how to behave better. If you can ignore the tantrum, even better, as negative behaviours can be attention seeking (a child may behave badly because it is a sure-fire way to get some attention from a busy, distracted parent). In order to avoid future tantrums, try to avoid trigger factors (those that caused the tantrum in the first place). For example, when you next go shopping, take along a snack that your child enjoys in case he spies the sweeties.

Children older than 3 have the ability to not only understand your verbal instructions but also have better control of their own behaviour, although they still need supervision. Time-outs or the removal of items used inappropriately, for example crayons used to draw on the wall, should work effectively with this age group.

**Spanking and harmful discipline**

No form of discipline should cause long-lasting harm. A child's self-esteem should not be negatively affected nor should he be made to feel that he has been abandoned. The many pieces of research done on the act of spanking and other physical punishments show they are associated with negative long-term outcomes such as aggression, delinquency and substance abuse (Knox, M, 2010 and MacMillan, HL et al, 1999), even though they may show signs of immediate effectiveness.

The effectiveness of frequent screaming or yelling becomes limited as a form of discipline as a child grows accustomed to them. They also can also cause long-term harm, particularly poor mental health in adulthood when associated with emotional abuse (Simeon, D, et al, 2001).

**Preemptive parenting**

This is a range of behaviours or strategies that a parent can employ that is designed to ward off misbehaviour by a child. Things like explaining a task very clearly, turning tidy-up time or other similar tasks into a game making them fun, removing objects that children are not allowed to play with, anticipating boredom, tiredness and hunger and simply using a bright voice rather than a frustrated or defeated one. Many parents will do some of these things instinctively. It has been shown to result in less aggressive behaviour, increased rule-following and more pleasant shopping behaviour. C. B. Dowling et al (2011) having conducted some good-quality research on this type of parenting, have shown that it doesn't necessarily result in less misbehaviour but it does result in more praise for the child and less lax parenting (see below).

**The negative effects of improper discipline**

Lax parents fail to follow through with discipline. They may make frequent threats but never actually carry them out. Over-reactive parents discipline can be unnecessarily harsh and can contribute to their child having mental health problems in later life, such as eating too much or too little, abusing substances or employing other methods of self-harm as well as acting aggressively or bullying (Bayer, JK et al, 2012). Harsh discipline can also cause a child to become more defiant. M. Byford et al (2012) have shown that parents who have a tendency to use harsh or coercive disciplining, including threats, may also impact negatively on their child's cognitive development.

I. Tung et al in the same year, showed that

## Some guidelines for parents

The following "Effective Discipline for Children" is taken from an article in 'Paediatrics and Child Health', 2004 January; 9(1): 37–41, which was primarily aimed at GPs giving advice to parents.

* Reinforce desirable behaviour. Praise positive behaviour and "catch children being good".
* Avoid nagging and making threats without consequences. The latter may even encourage the undesired behaviour.
* Apply rules consistently.
* Ignore unimportant and irrelevant behaviour, e.g., swinging legs while sitting.
* Set reasonable and consistent limits. Consequences need to be realistic. For example, grounding for a month may not be feasible.
* State acceptable and appropriate behaviour that is attainable.
* Prioritise rules. Give top priority to safety, then to correcting behaviour that harms people and property, and then to behaviour such as whining, temper tantrums and interrupting. Concentrate on two or three rules at first.
* Know and accept age-appropriate behaviour. Accidentally spilling a glass of water is normal behaviour for a toddler. It is not willful defiance. On the other hand, a child who refuses to wear a bicycle helmet after repeated warnings is being willfully defiant.
* Allow for the child's temperament and individuality (goodness of fit). A strong-willed child needs to be raised differently from the so-called 'compliant child'.

In applying consequences, these suggestions may be helpful:

* Apply consequences as soon as possible.
* Do not enter into arguments with the child during the correction process.
* Make the consequences brief. For example, time-out (see Forms of discipline) should last one minute per year of the child's age, to a maximum of five minutes.
* Parents should mean what they say and say it without shouting at the child. Verbal abuse is no less damaging than physical punishment.
* Follow consequences with love and trust, and ensure that the child knows the correction is directed against the behaviour and not the person. Guard against humiliating the child. Model forgiveness and avoid bringing up past mistakes.

## Recommendations from the research

Routines to avoid unwanted behaviour can start from an early age. Other forms of discipline – such as praise, pre-emptive parenting and distraction – are effective from around 9 months of age. Young children generally lack the ability to understand verbal reasoning and to control their behaviour, but it is a good idea to begin to introduce a mild form of 'time-outs' from about 18 months onwards, perhaps with a parent alongside, so that by the time your child is 2 and better able to understand verbal reasoning, he will already be accustomed to this form of discipline. Being lax or harsh, yelling frequently or smacking have negative effects on children.

inconsistent parenting, changing from lax to over-reactive and all stages in between, leads to aggression in children as well as rule-breaking behaviour.

**The benefits of praise**

A key aspect of discipline is that a child develops a healthy conscience – the ability to feel empathy and an appropriate amount of remorse after he's done something wrong – so that he will not repeat the same poor behaviour again in the future. A young child has a limited capacity to understand the difference between right and wrong. Parents can influence this learning by clearly explaining wrongdoing and the consequences (time-outs), and also, by praising the child when he behaves well parents can reaffirm the notions of right and wrong and enable the development of the conscience (Volling, BL et al, 2009). When a child shows an act of kindness or shares a toy, he needs to be praised so that he learns what good behaviour is. This makes it easier for a child to understand what is meant when he is told that not sharing the toy is bad behaviour. D.J. Owen et al's research (2012) found that praise is a key part of discipline and the development of a conscience but used on its own, is likely to have limited effects. It is probably most effective when used in conjunction with other forms.

# Television viewing

Today, television is ubiquitous in most people's homes and from the day a baby will happily sit and watch television, most parents will begin to take advantage of this new found time to carry out chores or simply have a little break. The 'powers that be' in the UK have been surprisingly relaxed about challenging both the length of time children spend watching TV and the content of programmes compared to the US, Canada and Australia. The majority of research on this subject originates in these countries and their recommendations may come as a surprise to some British parents.

## CURRENT RECOMMENDATIONS

There are currently no recommendations in the UK for children viewing television, be it length of time or what is watched.

The American Academy of Paediatrics, however, recommends that children below the age of two should not watch any television and after two years of age, TV watching be limited to one to two hours a day. TV sets should not be in a child's room and all viewing should be carefully monitored by parents. Most television programmes should be educational and nonviolent and parents should use issues raised by programmes as topics for discussion.

## WHAT THE RESEARCH SAYS

### Delay in language and cognitive development

A Thai study found no effect on language development amongst toddlers allowed to watch more than two hours of television per day (Ruangdaraganon, N, 2009). However, what children watch may be more relevant. One study looked at the number of words a toddler had at age 2½ and the type of programmes he or she watched. They found that watching 'Dora the Explorer' (a cartoon in which a little girl sets out on quests) resulted in a greater vocabulary than watching 'Teletubbies' (Linebarger, DL, and Walker, D, 2005). Parents who have seen both these programmes will no doubt be able to understand this finding. Dora the Explorer has more interaction (during her travels, Dora must solve puzzles and riddles and count or use the Spanish language aided by the viewing audience) and elicits more responses from toddlers than Teletubbies, in which no real language is spoken and few responses are encouraged. It is the encouragement of responses and being actively involved that makes watching Dora the Explorer result in an increased vocabulary.

However, for children younger than 2½, language development in is not so easily influenced by educational programmes. In contrast to the above study, another investigation reported that toddlers (aged 12 to 15 months) exposed to an educational

DVD that was supposed to promote common word development, were found to have gained no new vocabulary when compared to toddlers who hadn't watched it. Instead, the researchers found that developing word usage in young children depended on the amount of time toddlers were read to (Robb, MB, et al, 2009).

Another study found that with every increased hour per day of television watching, the comprehension and memory of under-threes decreased (Zimmerman, FJ & Christakis, DA, 2005).

**Poor scholastic achievement**

The more TV a child watches, the less likely she is to achieve a university degree. Children who are permitted to watch excessive amounts of television (more than two hours every day) do not do as well in school as children with more limited exposure. They are also more likely to leave school early, particularly if over-watching continues into adolescence (Hancox, RJ, et al, 2005).

L.S. Pagani et al (2010) found that with every extra hour of television watching at age 2½, there was a 7% decrease in classroom engagement and a 6% decrease in mathematical ability later in life. Having a television in a child's own room has also been shown to decrease her academic performance (Borzekowski, DL & Robinson, TN, 2005)

D.R. Andersonet al (2001) showed that teenagers who are limited to watching 'educational' programmes before they start school (ages 2 to 4), get better grades in school, read more books, place a higher value on doing well and tend to be less aggressive – particularly if they are girls. Similarly, girls tend to be more negatively affected by exposure to violent TV programmes, getting lower grades in school. Watching fewer educational programmes results in less reading, vocabulary and maths skills, thus setting up a vicious cycle: The less skilled a young child is as a result of watching educationally poor programmes, the more likely she is to opt for educationally poor programmes later on so further exacerbating her skill base (Wright, JC, et al, 2001).

However, despite the proof that some educational programmes may be beneficial for children over the age of two, plenty of research suggests that they are not in any way beneficial for children younger than two. In fact, they may even be detrimental, no matter how educational the content (Kirkorian, H, et al, 2008). One study looking at toddlers aged between 12 and 18months instructed parents to show a DVD several times a week for one month. When compared to toddlers who hadn't been shown the DVD (the control group), the DVD-watching group had not learned any new words associated with the DVD. Children of this age learn far better from hands-on personal interaction and the researchers found that the best learning came from parents teaching their toddlers target words without watching DVDs (DeLoache, JS, et al, 2010).

**Poor social relations**

It is unclear whether watching television can be of benefit in the development of emotions and empathy. Through television watching, children may learn about emotions and may even develop feelings of empathy for certain characters and specially selected educational programmes and comedies aimed at children could increase a child's altruism and their tolerance of others (Wilson, BJ, 2008). However, a study looking at the effects of TV watching in young children found that with every extra hour of TV viewed, the children were 10% more likely to be bullied by their school-mates (Pagani, LS, et al, 2010). This might indicate that normal social responses, usually developed at a young age, may be stunted by watching too much television and this could impact on childhood relations in a school setting.

**Shorter attention spans**

It has long been believed that children who watch lots of television have shorter attention spans than children who watch less. However, the age that children first engage in regular TV watching may also be a significant factor. Rapidly changing images could be over stimulating to a young brain and this could be a factor in the development or severity of ADHD (Attention Deficit Hyperactivity Disorder), which has been associated with short attention spans. A piece of research conducted in this area found that children who were exposed to two to three hours of TV per day between the ages of one and three were more likely to have decreased attention spans at age seven (Christakis, DA, et al, 2004).

**Aggression, violence and fear**

A lot of research makes a clear link between exposure to violence from television and film watching and increased aggression and violence in children and teenagers. This effect is so noticeable that it could be considered a "threat to public health" (Huesmann, LR & Taylor, LD, 2006). However, the research is not conclusive — some researchers find a connection others not – and there is also only limited proof of the impact of viewing violence on television and crime.

Children (aged 6 to12) who watch violent television programmes are less likely to spend time playing with friends (Bickham, DS & Rich, M, 2006). Why this is so is unclear, however. Does watching violent TV cause children to be less inclined to spend time with friends or does a lack of inclination to spend time with friends result in children watching more violent TV? Quite likely, one impacts on the other. The researchers did not find any relationship between watching nonviolent TV and spending time with friends, which indicates that the influencing factor is the watching of violent TV programmes not just watching TV that impacts on time spent with friends. Also, children who spend time watching TV with friends are more likely to do other activities with them.

Most parents would seek to prevent their children watching 'violent' TV but children can routinely be exposed to violence in ways that parents may not consider (particularly if they leave the room during commercial breaks). A study looking at common TV advertisements found that a small percentage showed violent or unsafe behaviour. It is well recognised that children who watch such activities on the television are much more likely to attempt high-risk behaviours that could potentially result in injury (Potts, R, et al, 1994). Whilst such advertisements may not accompany programmes made especially for children, they are frequently screened during sporting events, and as coming attractions for films or other TV programmes (Tamburro, RF, et al, 2004).

News programmes are also a source of fear and distress for young children. One study reported that children found violence between strangers, wars, famines and natural disasters most frightening (Cantor, J & Nathanson, AI, 2006). As they get older (primary school age), children are more likely to become distressed by violence between strangers depicted on the news, although less likely to become scared by fictional stories. This shows a developing understanding of what is real and what isn't.

**Background television**

It is fairly common for parents to leave the television on during playtime or for parents or older siblings to engage in TV watching whilst a young child is playing, and this can negatively affect both the quality and quantity of normal parent-child relations. In one experiment, a game show was played in the background during a toddler's playtime. The researchers reported that even though the children barely watched the television, they spent about 25% less time focussing on play during the 30 minutes that the TV show was on because of the cumulative effect of three-second glances at the programme every few minutes (Schmidt, ME, et al, 2008). However, the effects of this type of distraction on a child's cognitive development are unknown and there may not be any long-lasting effects of enjoying a suitable programme with your child present.

A similar study found that the attention parents gave to their toddlers, in the form of playing and talking, was significantly less if the TV was on in the background even when the programme was designed for the baby (Courage, ML, et al, 2010).

**Playtime**

One study evaluated whether playtime is affected by how much television is watched. The researchers found that active play and reading (or being read to) are not affected but that a child's ability for creative play is. Effects are seen if a child (aged less than two) is permitted only one hour of television a day; he or she will spend less time doing arts and crafts, pretend playing, dressing-up and playing with toys but more active play (such as ball games and playing in a playground) will be relatively unaffected (Vandewater, EA, et al, 2006). These results beg the question as to whether children who don't watch television are more creative than children who do. One study showed that children who initially saw no TV (they were youngsters before TV appeared in an Canadian town), scored higher in creativity than those who did have a TV, but that when the non-TV-viewing children were introduced to TV, they quickly lost this capability and scored similarly to children who had access to television all along (Williams, TM, 1986).

Interestingly enough, while television decreases the amount of time in which children listen to the radio or access other media, it does not affect the amount they read (Neuman, SB, 1991). Although overall amounts of reading (either reading on one's own or being read to) are low – roughly 15 minutes per day, and this decreases with age – children who watched lots of television were neither more nor less likely to read more (Vandewater, EA, et al, 2006).

### Health

A sedentary lifestyle is as dangerous for children's health as it is for adults'. Watching too much television (two or more hours per day) leads to obesity (Tremblay, MS, et al, 2011). If children are allowed to watch more than two hours of television every day, they are more likely to become overweight adults and have higher levels of cholesterol (Hancox, RJ, et al, 2004) with a greater risk of heart disease later in life (Wijndaele, K, et al, 2011).

However, why TV watching results in obesity is not clearly understood. According to the research, it's not because TV watching entails less activity. And restricting the amount of television that children watch, does not result in them taking up more activities to fill their time (Pearce, MS, et al, 2012). Instead, current thinking is that children overeat if they do so whilst watching TV – they are less able to recognise when they are full up, or it's due to the effect of advertising, or not remembering when they last ate (Hu, FB, 2003). Either way, prolonged TV watching increases the chances of obesity and related health issues far more than diet or amount of exercise.

## Recommendations from the research

Children aged less than 2 should not watch any television; it could make them more likely to develop ADHD or, if ADHD is diagnosed later on, it could make symptoms more severe. It also limits the amount of more creative play.

Children over the age of 2 may benefit from watching educational programmes for limited amounts of time – a maximum of 2 hours per day (preferably less). A child permitted to watch more than this every day is far more likely to be overweight, do less well in school and lack social skills.

The effects of background television on children are unclear but it can mean that you spend less time with your child and may not give him your full attention even when spending time with him. While letting your child watch TV may free you to carry out chores, he should not be left unattended, particularly if watching non-children's programmes such as sport and news reports, as he may view violent or unsafe behaviour and fall prey to advertisements.

# Computer use

Public opinion tends to favour the use of computers by young children. Parents are generally proud of their young child's competency in all areas of technology and many believe that their child will start school at a disadvantage if she is not a proficient user of an iPad or other devices. There has been so much hype about the necessity of children being able to keep up in an increasingly digital world that few have really taken the time to question whether that is true or not. Nobody can deny that, at an appropriate age, there is a need for youngsters to become computer literate so that they can function well in society and school, but at what age does it become vital to begin these lessons? Or more importantly, are there issues with children using computers too early? Just what are the advantages and disadvantages of early computer use?

## CURRENT RECOMMENDATIONS

There currently exist no official guidelines on 'screen time' in the UK. However, The American Academy of Paediatrics recommends that young children's access to

media be limited to a maximum of two hours per day and those younger than two should have no access at all (see also page 00). Computers and computer games (along with televisions) should not be used in a young child's bedroom.

Developmentally Appropriate Technology in Early Childhood or DATEC (a project set up to help develop a common ICT education curriculum for children in Europe) recommends that all children should have the opportunity to explore and play with computers and other technologies from an early age but that children should be limited to programmes and applications that are predominantly educational. For three year olds, it recommends that the use of computers be limited to 10 to 20 minutes. User time can be gradually extended so that by the time a child is eight years old, she should be allowed roughly 40 minutes per day (DATEC Curriculum Guidance, 2002).

### Computer literacy

It is widely believed that young children should be encouraged to become proficient computer users so that they will succeed in the workplace as adults. For this reason, ICT (information communication technology) is now part of the National Curriculum for children aged five to seven years. Children younger than this in nursery education are also encouraged to work towards becoming competent and confident in using computers and understanding how technology is used in the world. It is relatively common for preschool children to have a significant level of computer literacy, which means they can often switch items on and off and record, store and retrieve data from computers and mobile phones.

There is some evidence that young children who have access to computers or similar technology tend to be more digitally literate when they start school compared to those who had limited access (McPake, J, et al, 2008). However, because digitally-speaking, schools contain far less accessible technology than the average home, this apparent difference in ability will remain fairly unknown to both fellow students and the teacher.

## WHAT THE RESEARCH SAYS

### Risk of obesity

Spending too much time on the computer (or watching television) has been linked with obesity, which can lead to heart disease and other health complications later in life. To avoid overweight, it is important that a child has sufficient physical activity and limited computer use (and TV watching [Tremblay, MS, et al, 2011]). Computer use, as with TV viewing, may also expose children to food advertising, which encourages them to eat more unhealthy items.

### ADHD

Playing computer games, particularly for long periods of time has been linked with the development of ADHD in children aged 6-16 years and to exacerbate the symptoms of existing ADHD (Weiss, MD, et al, 2011).

**Better academic performance**

Educational computer programmes can boost learning. V.A. Van der Kooy-Hofland et al (2012) found that children with below adequate literacy skills improved after using an educational computer programme that encouraged them to match sounds with letters. Numerous other studies show that young children who use computers at home are more school-ready and have greater cognitive abilities (Li, X, and Atkins, MS, 2004). There is also evidence that children who use computers do better academically even if the programmes they use are not directly related to skills beneficial to schooling (Blanton, WE, et al, 1997).

But not all computer use is educational. A couple of studies that looked at the use of computers in primary schools concluded that any advantages depended heavily on whether the teacher used the technology to help meet the learning needs of the child. In other words, if computers are simply used to replicate something that is better taught by a teacher, or for practice and drilling activities (doing sums or learning colours), their benefits are limited.

What may be important is not so much what your child does on the computer as how much you interact with her at the time. K. McCarrick et al (2007) found that children whose parents were very much involved with their computer use and interacted with them while the children were using one, showed better cognitive abilities (language and number skills, and being able to problem solve, learn and remember) than those left on their own. The greater the parental interaction, the more improved the child's skills.

**Social considerations**

Computer game playing, whether it be between children or with the whole family can be a social activity if done in short-to-moderate bursts. However, according to K. Subrahmanyam et al (2000), if she spends excessive amounts of time playing computer games, her social skills may be hampered. L.R. Huesmann and L.D. Taylor found that children watching or playing violent computer games, became aggressive and violent, and that the majority of computer games, including those 'suitable' for children aged 10 or over, include violence (2006). Playing violent video games has also been shown to desensitise people to the pain and suffering of others (Bushman, BJ, and Anderson, CA, 2009).

Some children may struggle to understand the difference between reality and simulation. There are numerous examples of computer games where a child takes on a character or has a virtual pet that needs to be 'cared' for. Few studies have been conducted in this area but one conducted on university students by S.L. Calvert and S.L. Tan (1994), found that those who played aggressive computer games were more likely to have aggressive thoughts than those who watched the game, which demonstrates the effects simulation can have on real life. If older students' notions of reality become blurred, the games' impact on very young children could be much greater.

Many parents find it surprising that computer use appears to elicit communication – showing another child how to do something, turn-taking and even fighting! – between peers rather than creating solitary users. A.A. Muller and M.

Perlmutter, who compared computer use with puzzle solving and the peer interaction that followed (1985), found that children were more solitary when doing a jigsaw than when using a computer.

**Other benefits of computer use**

S.W. Haugland found that when young children are encouraged to explore and make decisions, as they frequently must in computer games, they gain initiative, which increases their self-esteem (1992).

Children generally enjoy computers and C. Stephen and L. Plowman (2003) discovered that using one for a boring activity such as spelling or adding, can make the activity more exciting and palatable to a young child. It might even act as a motivator for such tasks.

L. Borecki et al (2013) showed that playing computer games can improve fine motor skills (and that computer game players also have noticeably better hand-to eye-coordination than non-players. Children who play interactive computer games develop better skills with handling objects – throwing, pushing and lifting them (known as object control skills) – but, perhaps not surprisingly, they don't develop better locomotive skills (in which the feet move the body from one place to another [Barnett, LM, et al, 2012]). However, it's not known if playing these games makes children have better object control skills or whether children who are already skilled choose these sorts of games.

**The best applications for child development**

According to DATEC, programmes used to teach maths or colours have limited benefits and those that offer some kind of reward – a funny sound or a digital 'sticker' – for right answers could negatively affect a child's motivation to learn.

## Recommendations from the research

There is still relatively little research available on the impact of computers on young children, particularly the effects of accessible multiple technologies in the long term. We do know that sedentary lifestyles can contribute to obesity, that excessive computer usage is linked to ADHD and that access to violent computer games can cause aggressiveness later on in life. Computer usage, particularly by young children, may contribute to a blurred understanding reality and simulation, the effects of which are unknown. These negatives can all be avoided by limiting the time spent using a computer and disallowing access to violent computer games.

Children who have and use computers at home have greater computer literacy, better cognitive skills and tend to do better academically. Computers can improve self-esteem, increase social interactions, motivate learners and might even help develop fine motor skills. If used in the correct way, they can be a valuable resource for parents wanting to give their children the best start in a technology-filled world. Unfortunately, there is too little research on too broad a subject and few recommendations on the appropriate age to introduce computers.

However, programmes that encourage a child to work with either a parent or another child appear to be highly beneficial, particularly those that involve working together and turn taking.

A child who sees computers being used in everyday life will not only be prepared for ICT in school but will also understand that computers play a part in many aspects of life. You can assist with this learning by showing your child that a mobile phone makes calls, that some apps play music or 'tell' stories, or that washing machines can be programmed according to the type or amount of clothes being cleaned. Or, you can show your child how to use a drawing programme or app to create a birthday card and then after its printed, to complete it using other materials, such as glitter. This will enable her to see computers and technology as tools that can be used to complete tasks.

When searching for programmes or apps for your young child to use, avoid those designed to elicit only one correct answer, as they can limit learning. Those that permit multiple correct answers – even if one is better than the others – will extend her thinking and problem-solving skills.

# Bibliography


**1. Pregnancy supplements**

Lane IR. (2011. Preventing neural tube defects with folic acid: nearly 20 years on, the majority of women remain unprotected. *Journal of Obstetrics and Gynaecology*. Oct; 31(7):581-5.

Molloy AM, Kirke PN, Brody LC, Scott JM, Mills JL. (2008). Effects of folate and vitamin B12 deficiencies during pregnancy on fetal, infant, and child development. *Food and Nutrition Bulletin*. Jun; 29(2 Supplement):S101-11; discussion S112-5.

Milne E, Greenop KR, Bower C, Miller M, van Bockxmeer FM, Scott RJ, de Klerk NH, Ashton LJ, Gottardo NG, Armstrong BK; Aus-CBT Consortium. (2012). Maternal use of folic acid and other supplements and risk of childhood brain tumors. *Cancer Epidemiology, Biomarkers & Prevention*. Nov; 21(11):1933-41.

Roth C, Magnus P, Schjølberg S, Stoltenberg C, Surén P, McKeague IW, Davey Smith G, Reichborn-Kjennerud T, Susser E. (2011)

Folic acid supplements in pregnancy and severe language delay in children. -

JAMA: The Journal of the American Medical Association. Oct 12; 306(14):1566-73.

Sharland E, Montgomery B, Granell R. (2011)

Folic acid in pregnancy - is there a link with childhood asthma or wheeze? *Australian Family Physician*. Jun; 40(6):421-4.

Vanderpump MP, Lazarus JH, Smyth PP, Laurberg P, Holder RL, Boelaert K, Franklyn JA; British Thyroid Association UK Iodine Survey Group. (2011. Iodine status of UK schoolgirls: a cross-sectional survey. *Lancet*. Jun 11; 377(9782):2007-12.

Bath SC, Button S, Rayman MP. (2012). Iodine concentration of organic and conventional milk: implications for iodine intake. *British Journal of Nutrition*. Apr; 107(7):935-40.

Helin A, Kinnunen TI, Raitanen J, Ahonen S, Virtanen SM, Luoto R. (2012). Iron intake, haemoglobin and risk of gestational diabetes: a prospective cohort study. *BMJJ: British Medical Journal* Open. Sep 25; 2(5).

Radhika MS, Bhaskaram P, Balakrishna N, Ramalakshmi BA, Devi S, Kumar BS. (2002). Effects of vitamin A deficiency during pregnancy on maternal and child health. *BJOG: An International Journal of Obstetrics and Gynaecology*. Jun; 109(6):689-93.

Steyn PS, Odendaal HJ, Schoeman J, Stander C, Fanie N, Grové D. (2003). A randomised, double-blind placebo-controlled trial of ascorbic acid supplementation for the prevention of preterm labour. *Journal of Obstetrics and Gynaecology*. Mar; 23(2):150-5.

Wagner CL, Taylor SN, Dawodu A, Johnson DD, Hollis BW. (2012). Vitamin D and its role during pregnancy in attaining optimal health of mother and fetus. *Nutrients*. Mar; 4(3):208-30.

Chappell LC, Seed PT, Briley AL, Kelly FJ, Lee R, Hunt BJ, Parmar K, Bewley SJ, Shennan AH, Steer PJ, Poston L. (1999). Effect of antioxidants on the occurrence of pre-eclampsia in women at increased risk: a randomised trial. *Lancet*. Sep 4; 354(9181):810-6.

Pelucchi C, Chatenoud L, Turati F, Galeone C, Moja L, Bach JF, La Vecchia C. (2012). Probiotics supplementation during pregnancy or infancy for the prevention of atopic dermatitis: a meta-analysis. *Epidemiology*. May; 23(3):402-14.

Ly NP, Litonjua A, Gold DR, Celedón JC. (2011). Gut microbiota, probiotics, and vitamin D: interrelated exposures influencing allergy, asthma, and obesity? *Journal of Allergy and Clinical Immunology*. May; 127(5):1087-94.

Agrawal R, Burt E, Gallagher AM, Butler L, Venkatakrishnan R, Peitsidis P. (2012). Prospective randomized trial of multiple micronutrients in subfertile women undergoing ovulation induction: a pilot study. *Reproductive Biomedicine*. 2012 Jan; 24(1):54-60.

**2. Alcohol**

Faas AE, Spontón ED, Moya PR, Molina JC. (2000). Differential responsiveness to alcohol odor in human neonates: effects of maternal consumption duringgestation. *Alcohol*. Aug; 22(1):7-17.

Robinson M, Oddy WH, McLean NJ, Jacoby P, Pennell CE, de Klerk NH, Zubrick SR, Stanley FJ, Newnham JP. (2010.) Low-moderate prenatal alcohol exposure and risk to child behavioural development: a prospective cohort study. *BJOG: an International Journal of Obstetrics and Gynaecology*. Aug; 117(9):1139-50.

Todorow M, Moore TE, Koren G. (2010). Investigating the effects of low to moderate levels of prenatal alcohol exposure on child behaviour: a critical review. *Journal of Population Therapeutics and Clinical Pharmacology*. Summer; 17(2):e323-30.

Jensen TK, Hjollund NH, Henriksen TB, Scheike T, Kolstad H, Giwercman A, Ernst E, Bonde JP, Skakkebaek NE, Olsen J. (1998). Does moderate alcohol consumption affect fertility? Follow up study among couples planning first pregnancy. *British Medical Journal* 1998 Aug 22; 317(7157):505-10.

Juhl M, Olsen J, Andersen AM, Grønbaek M. (2003. Intake of wine, beer and spirits and waiting time to pregnancy. *Human Reproduction* (Oxford, England). Sep; 18(9):1967-71.

Skogerbø Å, Kesmodel US, Wimberley T, Støvring H, Bertrand J, Landrø NI, Mortensen EL. (2012). The effects of low to moderate alcohol consumption and binge drinking in early pregnancy on executive functioning 5-year-old children. *BJOG: an International Journal of Obstetrics and Gynaecology*. Sep; 119(10):1201-10.

Kesmodel US, Bertrand J, Støvring H, Skarpness B, Denny CH, Mortensen EL; Lifestyle During Pregnancy Study Group. (2012. The effect of different alcohol drinking patterns in early to mid pregnancy on the child's intelligence, attention, and executive function. *BJOG: an International Journal of Obstetrics and Gynaecology*. Sep; 119(10):1180-90

Jaddoe VW, Bakker R, Hofman A, Mackenbach JP, Moll HA, Steegers EA, Witteman JC. (2007). Moderate alcohol consumption during pregnancy and the risk of low birth weight and preterm birth. The generation R study. *Annals of Epidemiology*. Oct; 17(10):834-40

Simpson ME, Duggal S, Keiver K. (2005). Prenatal ethanol exposure has differential effects on fetal growth and skeletal ossification. *Bone*. Mar; 36(3):521-32.

Hepper PG, Dornan JC, Lynch C, Maguire JF. (2012). Alcohol delays the emergence of the fetal elicited startle response, but only transiently. *Physiology & Behavior*. Aug 20; 107(1):76-81.

Lewis SJ, Zuccolo L, Davey Smith G, Macleod J, Rodriguez S, Draper ES, Barrow M, Alati R, Sayal K, Ring S, Golding J, Gray R. (2012). Fetal alcohol exposure and IQ at age 8: evidence from a population-based birth-cohort study. *PLoS One*. 7(11):e49407,

Andersen AM, Andersen PK, Olsen J, Grønbæk M, Strandberg-Larsen K. (2012). Moderate alcohol intake during pregnancy and risk of fetal death. *International Journal of Epidemiology*. Apr; 41(2):405-13.

Sood B, Delaney-Black V, Covington C, Nordstrom-Klee B, Ager J, Templin T, Janisse J, Martier S, Sokol RJ. (2001). Prenatal alcohol exposure and childhood behavior at age 6 to 7 years: I. dose-response effect. *Pediatrics*. Aug; 108(2):E34.

Falgreen Eriksen HL, Mortensen EL, Kilburn T, Underbjerg M, Bertrand J, Støvring H, Wimberley T, Grove J, Kesmodel US. (2012). The effects of low to moderate prenatal alcohol exposure in early pregnancy on IQ in 5-year-old children. *BJOG: an International Journal of Obstetrics and Gynaecology*. Sep; 119(10):1191-200.

Kim P, Park JH, Choi CS, Choi I, Joo SH, Kim MK, Kim SY, Kim KC, Park SH, Kwon KJ, Lee J, Han SH, Ryu JH, Cheong JH, Han JY, Ko KN, Shin CY. (2013). Effects of Ethanol Exposure During Early Pregnancy in Hyperactive, Inattentive and Impulsive Behaviors and MeCP2 Expression in Rodent Offspring. *Neurochemical Research*. Mar; 38(3):620-31

Burger PH, Goecke TW, Fasching PA, Moll G, Heinrich H, Beckmann MW, Kornhuber J. (2011). How does maternal alcohol consumption during pregnancy affect the development of attention deficit/hyperactivity syndrome in the child. *Fortschritte der Neurologie- Psychiatrie*. Sep; 79(9):500-6.

Kesmodel US, Eriksen HL, Underbjerg M, Kilburn TR, Støvring H, Wimberley T, Mortensen EL. (2012). The effect of alcohol binge drinking in early pregnancy on general intelligence in children. *BJOG: an International Journal of Obstetrics and Gynaecology.* Sep; 119(10):1222-31.

Underbjerg M, Kesmodel US, Landrø NI, Bakketeig L, Grove J, Wimberley T, Kilburn TR, Sværke C, Thorsen P, Mortensen EL. (2012. The effects of low to moderate alcohol consumption and binge drinking in early pregnancy on selective and sustained attention in 5-year-old children. *BJOG: an International Journal of Obstetrics and Gynaecology.* Sep; 119(10):1211-21.

ckstrand KL, Ding Z, Dodge NC, Cowan RL, Jacobson JL, Jacobson SW, Avison MJ. (2012. Persistent dose-dependent changes in brain structure in young adults with low-to-moderate alcohol exposure in utero. *Alcoholism, Clinical and Experimental Research.* Nov; 36(11):1892-902.

**3. Caffeine**

Crozier TW, Stalmach A, Lean ME, Crozier A. (2012). Espresso coffees, caffeine and chlorogenic acid intake: potential health implications. *Food & Function.* Jan; 3(1):30-3.

Adeney KL, Williams MA, Schiff MA, Qiu C, Sorensen TK. (2007). Coffee consumption and the risk of gestational diabetes mellitus. *Acta Obstetricia et Gynecologica Scandinavica.* 86(2):161-6.

Grant I, Cartwright JE, Lumicisi B, Wallace AE, Whitley GS. (2012. Caffeine Inhibits EGF-Stimulated Trophoblast Cell Motility through the Inhibition of mTORC2 and Akt. *Endocrinology.* Sep; 153(9):4502-10.

Bech BH, Nohr EA, Vaeth M, Henriksen TB, Olsen J. (2005). Coffee and fetal death: a cohort study with prospective data. *American Journal of Epidemiology.* Nov 15; 162(10):983-90.

Santos IS, Matijasevich A, Domingues MR. (2012). Maternal caffeine consumption and infant nighttime waking: prospective cohort study. *Pediatrics.* May; 129(5):860-8.

CARE Study Group. (2008). Maternal caffeine intake during pregnancy and risk of fetal growth restriction: a large prospective observational study. *British Medical Journal*; 3 November.

**4. Peanuts**

Strid J, Thomson M, Hourihane J, Kimber I, Strobel S. (2004). A novel model of sensitization and oral tolerance to peanut protein. *Immunology.* Nov; 113(3):293-303.

Maslova E, Granström C, Hansen S, Petersen SB, Strøm M, Willett WC, Olsen SF. (2012). Peanut and tree nut consumption during pregnancy and allergic disease in children-should mothers decreasetheir intake? Longitudinal evidence from the Danish National Birth Cohort.

*Journal of Allergy and Clinical Immunology.* Sep; 130(3):724-32.

Sicherer SH, Wood RA, Stablein D, Lindblad R, Burks AW, Liu AH, Jones SM, Fleischer DM, Leung DY, Sampson HA. (2010).Maternal consumption of peanut during pregnancy is associated with peanut sensitization in atopic infants.*Journal of Allergy and Clinical Immunology.* Dec; 126(6):1191-7.

**5. Stress**

Peters JL, Cohen S, Staudenmayer J, Hosen J, Platts-Mills TA, Wright RJ. (2012). Prenatal negative life events increases cord blood IgE: interactions with dust mite allergen and maternal atopy. *Allergy.* Apr; 67(4):545-51.

Choi H, Rauh V, Garfinkel R, Tu Y, Perera FP. (2008). Prenatal exposure to airborne polycyclic aromatic hydrocarbons and risk of intrauterine growth restriction. *Environmental Health Perspectives.* May; 116(5):658-65.

Capra L, Tezza G, Mazzei F, Boner AL. (2013. The origins of health and disease: the influence of maternal diseases and lifestyle during gestation. *Italian Journal of Pediatrics.* Jan 23; 39(1):7.

Künzle R, Mueller MD, Hänggi W, Birkhäuser MH, Drescher H, Bersinger NA. (2003). Semen quality of male smokers and nonsmokers in infertile couples. Fertility and Sterility. Feb; 79(2):287-91.

D.C. Gesink Law1,2,4, R.F. Maclehose3, M.P. Longnecker1 (2007). Obesity and time to pregnancy. *Human Reproduction.* 22 (2): 414-420.

Augood C, Duckitt K, Templeton AA. (1998). Smoking and female infertility: a systematic review and meta-analysis. *Human Reproduction.* Jun; 13(6):1532-9.

Bolúmar F, Olsen J, Rebagliato M, Sáez Lloret I, Bisanti L. (2000). Body mass index and delayed conception: a European Multicenter Study on Infertility and Subfecundity. *American Journal of Epidemiology.* Jun 1; 151(11):1072-9.

Wise LA, Rothman KJ, Mikkelsen EM, Sørensen HT, Riis AH, Hatch EE. (2012). A prospective cohort study of physical activity and time to pregnancy. *Fertility and Sterility.* May; 97(5):1136-42.e1-4.

De Coster S, van Larebeke N. (2012). Endocrine-disrupting chemicals: associated disorders and mechanisms of action. *Journal of Environmental and Public Health.* 713696.

Domar AD, Clapp D, Slawsby E, Kessel B, Orav J, Freizinger M. (2000). The impact of group psychological interventions on distress in infertile women. *Health Psychology.* Nov; 19(6):568-75.

Boivin J, Griffiths E, Venetis CA. (2011). Emotional distress in infertile women and failure of assisted reproductive technologies: meta-analysis ofprospective psychosocial studies. *BMJ: British Medical Journal (Clinical Research Edition).* Feb 23; 342:d223.

Byun JS, Lyu SW, Seok HH, Kim WJ, Shim SH, Bak CW. (2012). Sexual dysfunctions induced by stress of timed intercourse and medical treatment. *BJU International.* Oct 26.

Hassan MA, Killick SR. (2004). Negative lifestyle is associated with a significant reduction in fecundity. *Fertility and Sterility.* Feb; 81(2):384-92.

Capra L, Tezza G, Mazzei F, Boner AL. (2013). The origins of health and disease: the influence of maternal diseases and lifestyle during gestation. *Italian Journal of Pediatrics.* Jan 23; 39(1):7.

Matthiesen SM, Frederiksen Y, Ingerslev HJ, Zachariae R. (2011). Stress, distress and outcome of assisted reproductive technology (ART): a meta-analysis. *Human Reproduction.* Oct; 26(10):2763-76.

Choi H, Rauh V, Garfinkel R, Tu Y, Perera FP. (2008). Prenatal exposure to airborne polycyclic aromatic hydrocarbons and risk of intrauterine growth restriction. *Environmental Health Perspectives.* May; 116(5):658-65.

Bellinger DL, Lubahn C, Lorton D. (2008). Maternal and early life stress effects on immune function: relevance to immunotoxicology. *Journal of Immunotoxicology.* Oct; 5(4):419-44.

Ruiz RJ, Fullerton J, Dudley DJ. (2003). The interrelationship of maternal stress, endocrine factors and inflammation on gestational length. *Obstetrical & Gynecological Survey.* Jun; 58(6):415-28.

Huizink AC, Mulder EJ, Robles de Medina PG, Visser GH, Buitelaar JK. (2004). Is pregnancy anxiety a distinctive syndrome? *Early Human Development.* Sep; 79(2):81-91.

**6. Planned Caesareans**

Bragg F, et al .(2010). Variation in rates of caesarean section among English NHS trusts after accounting for maternal and clinical risk: cross sectional study. BMJ: British Medical Journal Oct 6; 341:c5065.

Koc O, Duran B. (2012. Role of elective cesarean section in prevention of pelvic floor disorders. *Current Opinion in Obstetrics & Gynecology.* Oct; 24(5):318-23.

Hay-Smith J, Mørkved S, Fairbrother KA, Herbison GP. (2008. Pelvic floor muscle training for prevention and treatment of urinary and faecal incontinence in antenatal andpostnatal women. *Cochrane Database of Systematic Reviews.* Oct 8 ;(4).

Barrett G, Peacock J, Victor CR, Manyonda I. (2005. Cesarean section and postnatal sexual health. *Birth.* Dec; 32(4):306-11.

Smith GC, Pell JP, Dobbie R. (2003). Caesarean section and risk of unexplained stillbirth in subsequent pregnancy. *Lancet.* Nov 29; 362(9398):1779-84.

Nissen E, Uvnäs-Moberg K, Svensson K, Stock S, Widström AM, Winberg J. (1996).Different patterns of oxytocin, prolactin but not cortisol release during breastfeeding in women delivered by caesarean section or by the vaginal route. *Early Human Development.* Jul 5; 45(1-2):103-18.

Madar J, Richmond S, Hey E. (1999. Surfactant-deficient respiratory distress after elective delivery at 'term'. *Acta Paediatricia* (Oslo, Norway). Nov; 88(11):1244-8.

MacKay DF, Smith GC, Dobbie R, Pell JP. (2010). Gestational age at delivery and special educational need: retrospective cohort study of 407,503 schoolchildren. *PLoS Medicine.* Jun 8; 7(6)

Christensson K, Siles C, Cabrera T, Belaustequi A, de la Fuente P, Lagercrantz H, Puyol P, Winberg J.(1993). Lower body temperatures in infants delivered by caesarean section than in vaginally delivered infants. *Acta Paediatricia* (Oslo, Norway). Feb; 82(2):128-31.

Thavagnanam S, Fleming J, Bromley A, Shields MD, Cardwell CR. (2008). A meta-analysis of the association between Caesarean section and childhood asthma. *Clinical and Experimental Allergy.* Apr; 38(4):629-33.

Yang SN, Shen LJ, Ping T, Wang YC, Chien CW. (2011). The delivery mode and seasonal variation are associated with the development of postpartum depression. *Journal of Affective Disorders.* Jul; 132(1-2):158-64.

Hall MH, Bewley S. (1999). Maternal mortality and mode of delivery. *Lancet.* Aug 28; 354(9180):776.

Al-Zirqi I, Stray-Pedersen B, Forsén L, Vangen S. (2010). Uterine rupture after previous caesarean section. *BJOG: an international journal of obstetrics and gynaecology.* Jun; 117(7):809-20.

**7. Pain relief**

Hogg MI, Wiener PC, Rosen M, Mapleson WW.(1977. Urinary excretion and metabolism of pethidine and norpethidine in the newborn. *British Journal of Anaesthesia* 49(9): 891–9.

Leap N, Sandall J, Buckland S, Huber U. (2010). Journey to confidence: women's experiences of pain in labour and relational continuity of care. *Journal of Midwifery & Women's Health.* May-Jun; 55(3):234-42.

Hodnett ED. (2002). Pain and women's satisfaction with the experience of childbirth: a systematic review. *American Journal of Obstetrics and Gynecology.* May; 186(5 Supplement Nature):S160-72.

Hodnett ED, Gates S, Hofmeyr GJ, Sakala C. (2012). Continuous support for women during childbirth.Cochrane Database of Systematic Reviews. Oct 17; 10.

Henry A, Nand SL. (2004). Women's antenatal knowledge and plans regarding intrapartum pain management at the Royal Hospital for Women. *Australian & New Zealand Journal of Obstetrics & Gynaecology.* Aug; 44(4):314-7.

Mehdizadeh A, Roosta F, Chaichian S, Alaghehbandan R. (2005). Evaluation of the impact of birth preparation courses on the health of the mother and the newborn. *American Journal of Perinatology.* Jan; 22(1):7-9.

Crowe K, von Baeyer C. (1989). Predictors of a positive childbirth experience. *Birth.* Jun; 16(2):59-63.

Fabian HM, Rådestad IJ, Waldenström U.(2005). Childbirth and parenthood education classes in Sweden. Women's opinion and possible outcomes. Acta Obstetricia et Gynecologica Scandinavica. May; 84(5):436-43.

Lawrence A, Lewis L, Hofmeyr GJ, Dowswell T, Styles C. (2009). Maternal positions and mobility during first stage labour. Cochrane Database of Systematic Reviews . Apr 15 ;( 2).

Gupta JK, Hofmeyr GJ, Shehmar M (2012). Position in the second stage of labour for women without epidural anaesthesia. Cochrane Database of Systematic Reviews May 16; 5.

Baghdadi ZD. (2000). Evaluation of audio analgesia for restorative care in children treated using electronic dental anaesthesia. *Journal of Clinical Pediatric Dentistry.* Fall; 25(1):9-12.

Ahlborg G Jr, Axelsson G, Bodin L. (1996). Shift work, nitrous oxide exposure and subfertility among Swedish midwives. *International Journal of Epidemiology.* Aug; 25(4):783-90.

Afolabi BB, Lesi FE. (2012). Regional versus general anaesthesia for caesarean section. Cochrane Database of Systematic Reviews Oct 17; 10.

**8. Spacing pregnancies**

Stamilio DM, DeFranco E, Paré E, Odibo AO, Peipert JF, Allsworth JE, Stevens E, Macones GA. (2007. Short inter-pregnancy interval: risk of uterine rupture and complications of vaginal birth after caesarean delivery. *Obstetetrics and Gynecology.* Nov; 110(5):1075-82.

King JC. (2003). The risk of maternal nutritional depletion and poor outcomes increases in early or closely spaced pregnancies. *Journal of Nutrition.* May; 133(5 Supplement 2):1732S-1736S.

Gunawardana L, Smith GD, Zammit S, Whitley E, Gunnell D, Lewis S, Rasmussen F. (2011)

Pre-conception inter-pregnancy interval and risk of schizophrenia.

British Journal of Psychiatry. Oct; 199(4):338-9.

Fuentes-Afflick E, Hessol NA. (2000)

Interpregnancy interval and the risk of premature infants.

Obstetrics and Gynecology. Mar; 95(3):383-90.

Nabukera SK, Wingate MS, Kirby RS, Owen J, Swaminathan S, Alexander GR, Salihu HM. (2008) Interpregnancy interval and subsequent perinatal outcomes among women delaying initiation of childbearing. *Journal of Obstetrics and Gynaecological Research.* Dec; 34(6):941-7.

Conde-Agudelo A, Rosas-Bermúdez A, Kafury-Goeta AC. (2006. Birth spacing and risk of adverse perinatal outcomes: a meta-analysis. *JAMA: the Journal of the American Medical Association.* Apr 19; 295(15):1809-23.

Zhu BP. (2005). Effect of inter-pregnancy interval on birth outcomes: findings from three recent US studies. *International Journal of Gynaecology and Obstetrics.* Apr; 89 Supplement 1:S25-33.

Caan B, Horgen DM, Margen S, King JC, Jewell NP. (1987). Benefits associated with WIC supplemental feeding during the inter-pregnancy interval. *American Journal of Clinical Nutrition.* Jan; 45(1):29-41.

Koch HL (1960). The Relation of Certain Formal Attributes of Siblings to Attitudes Held toward Each Other and toward Their Parents. *Monographs of the Society for Research in Child Development* Vol. 25, No. 4, pp. 1+3+5-7+9-33+35-69+71-97+99-117+119+121-124. Published by: Wiley.

Wagner ME, Schubert HJ, Schubert DS. Effects of sibling spacing on intelligence, interfamilial relations, psychosocial characteristics, and mental and physical health. *Advances in Child Development and Behaviour.*19:149-206.

Black, SE. Devereux, P J. & Salvanes, KG. (2007). Older and Wiser? Birth Order and IQ of Young Men. *IZA Discussion Papers* 3007, Institute for the Study of Labor (IZA).

Kanazawa S. (2012). Intelligence, birth order, and family size. *Personality& Social Psychology Bulletin.* Sep; 38(9):1157-64.

Rodgers JL, Cleveland HH, van den Oord E, Rowe DC. (2000. Resolving the debate over birth order, family size, and intelligence. *American Psychologist.* Jun; 55(6):599-612.

Harold T. Christensen (1968). Children in the Family: Relationship of Number and Spacing to Marital Success. *Journal of Marriage and Family* Vol. 30, No. 2, Family Planning and Fertility Control (May, 1968), pp. 283-289 Published by: National Council on Family Relations.

Kasey S. Buckles and Elizabeth L. Munnich (2012). Birth Spacing and Sibling Outcomes. *Journal of Human Resources.* 47 no. 3613-64.2

**9. Breastfeeding**

Lönnerdal B. (2003). Nutritional and physiologic significance of human milk proteins. *American Journal of Clinical Nutrition.* Jun; 77(6):1537S-1543S.

Leunissen RW, Kerkhof GF, Stijnen T, Hokken-Koelega A. (2009).Timing and tempo of first- year rapid growth in relation to cardiovascular and metabolic risk profile in early adulthood. *JAMA: The Journal of the American Medical Association.* Jun 3; 301(21):2234-42.

Kramer MS, et at (2001). PROBIT Study Group (Promotion of Breastfeeding Intervention Trial). Promotion of Breastfeeding Intervention Trial (PROBIT): a randomized trial in the Republic of Belarus. *JAMA: The Journal of the American Medical Association.* Jan 24-31; 285(4):413-20.

Savino F, Lupica MM. (2006). Breast milk: biological constituents for health and well-being in infancy. *Recenti Progressi in Medicina.* Oct; 97(10):519-27.

Wegienka G, Ownby DR, Havstad S, Williams LK, Johnson CC. (2006). Breastfeeding history and childhood allergic status in a prospective birth cohort. *Annals of Allergy, Asthma& Immunology.* Jul; 97(1):78-83.

Purvis DJ, Thompson JM, Clark PM, Robinson E, Black PN, Wild CJ, Mitchell EA.(2005). Risk factors for atopic dermatitis in New Zealand children at 3.5 years of age. *British Journal of Dermatology.* Apr; 152(4):742-9.

Kramer MS et al PROBIT Study Group. (2007). Effects of prolonged and exclusive breastfeeding on child height, weight, adiposity, and blood pressure at age 6.5y: evidence from a large randomized trial. *American Journal of Clinical Nutrition.* Dec; 86(6):1717-21.

Kelly DA, Phillips AD, Elliott EJ, Dias JA, Walker-Smith JA. (1989. Rise and fall of coeliac disease 1960-85. *Archives of Disease in Childhood.* Aug; 64(8):1157-60.

Wilson AC, Forsyth JS, Greene SA, Irvine L, Hau C, Howie PW. (1998). Relation of infant diet to childhood health: seven year follow up of cohort of children in Dundee infant feeding study. *BMJ: British Medical Journal* (clinical research edition). Jan 3; 316(7124):21-5.

Mårild S, Hansson S, Jodal U, Odén A, Svedberg K. (2004). Protective effect of breastfeeding against urinary tract infection. *Acta Paediatrica* (Oslo, Norway). Feb; 93(2):164-8.

Fewtrell MS, Lucas A, Morgan JB. (2003). Factors associated with weaning in full term and preterm infants. *Archives of Disease in Childhood. Fetal and Neonatal Edition.* Jul; 88(4):F296-301.

Owen CG, Martin RM, Whincup PH, Davey-Smith G, Gillman MW, Cook DG. (2005). The effect of breastfeeding on mean body mass index throughout life: a quantitative review of published and unpublished observational evidence. *American Journal of Clinical Nutrition.* Dec; 82(6):1298-307.

Kramer MS, Matush L, Vanilovich I, Platt RW, Bogdanovich N, Sevkovskaya Z, Dzikovich I, Shishko G, Collet JP, Martin RM, Smith GD, Gillman MW, Chalmers B,Hodnett E, Shapiro S. (2009). A randomized breast-feeding promotion intervention did not reduce child obesity in Belarus. *Journal of Nutrition.* Feb; 139(2):417S-21S.

Gouveri E, Papanas N, Hatzitolios AI, Maltezos E. (2011). Breastfeeding and diabetes. *Current Diabetes Review.* Mar; 7(2):135-42.

Kühn T, Kroke A, Remer T, Schönau E, Buyken AE. (2012). Is breastfeeding related to bone properties? A longitudinal analysis of associations between breast feeding duration and pQCT parameters in children and adolescents. *Maternal & Child Nutrition.* Aug 22.

Martin RM, Gunnell D, Owen CG, Smith GD. (2005). Breast-feeding and childhood cancer: A systematic review with metaanalysis. *International Journal of Cancer.* Dec 20; 117(6):1020-31.

Hansen TS, Jess T, Vind I, Elkjaer M, Nielsen MF, Gamborg M, Munkholm P. (2011). Environmental factors in inflammatory bowel disease: a case-control study based on a Danish inception cohort. *Journal of Crohn's & Colitis.* Dec; 5(6):577-84.

Moon RY, Fu L. (2012). Sudden infant death syndrome: an update. *Pediatrics in Review* / American Academy of Pediatrics. Jul; 33(7):314-20.

Kramer MS, et al (2008). Breastfeeding and child cognitive development: new evidence from a large randomized trial. *Archives of General Psychiatry.* May; 65(5):578-84.

Jedrychowski W, et al (2012). Effect of exclusive breastfeeding on the development of children's cognitive function in the Krakow prospective birth cohort study. *European Journal of Pediatrics.* Jan; 171(1):151-8.

Weaver IC, et al (2004). Epigenetic programming by maternal behavior. *Nature Neuroscience.* Aug; 7(8):847-54.

Daniels MC, Adair LS. (2005). Breast-feeding influences cognitive development in Filipino children. *Journal of Nutrition.* Nov; 135(11):2589-95.

Rea MF. (2004). [Benefits of breastfeeding and women's health].*Jornal de Pediatria* (Rio de Janero). Nov; 80(5 Supplement):S142-6.

Wiklund PK, et al (2012). Lactation is associated with greater maternal bone size and bone strength later in life. *Osteoporosis International.* Jul; 23(7):1939-45.

Lahiri M, Morgan C, Symmons DP, Bruce IN. (2012). Modifiable risk factors for RA: prevention, better than cure? *Rheumatology* (Oxford, England). Mar; 51(3):499-512.

Motil KJ, Sheng HP, Kertz BL, Montandon CM, Ellis KJ. (1998). Lean body mass of well-nourished women is preserved during lactation. *American Journal of Clinical Nutrition.* Feb; 67(2):292-300.

Barbosa C, Vasquez S, Parada MA, Gonzalez JC, Jackson C, Yanez ND, Gelaye B, Fitzpatrick AL (2009). The relationship of bottle feeding and other sucking behaviors with speech disorder in Patagonian pre-schoolers. *BMC Pediatrics.* Oct 21; 9:66.

Ferguson M, Molfese PJ. (2007). Breast-fed infants process speech differently from bottle-fed infants: evidence from neuroelectrophysiology. *Developmental Neuropsychology.* 31(3):337-47.

Broad FE. (1975). Further studies on the effects of infant feeding on speech quality. *New Zealand Medical Journal.* Dec 10; 82(553):373-6.

Labbok MH, Hendershot GE. (1987). Does breast-feeding protect against malocclusion? An analysis of the 1981 Child Health Supplement to the National Health Interview Survey. *American Journal of Preventative Medicine.* Jul-Aug; 3(4):227-32.

Romero CC, Scavone-Junior H, Garib DG, Cotrim-Ferreira FA, Ferreira RI. (2011). Breastfeeding and non-nutritive sucking patterns related to the prevalence of anterior open bite in primary dentition. *Journal of Applied Oral Science.* Apr; 19(2):161-8.

Ngom PI, Diagne F, Samba Diouf J, Ndiaye A, Hennequin M. (2008). [Prevalence and factors associated with non-nutritive sucking behavior. Cross sectional study among 5- to 6-year-old Senegalese children]. *L'Orthodontie Francaise.* Jun; 79(2):99-106.

Lehtonen J, Könönen M, Purhonen M, Partanen J, Saarikoski S, Launiala K. (1998). The effect of nursing on the brain activity of the newborn. *Journal of Pediatrics.* Apr; 132(4):646-51.

Kim P, Feldman R, Mayes LC, Eicher V, Thompson N, Leckman JF, Swain JE (2011). Breastfeeding, brain activation to own infant cry, and maternal sensitivity. *Journal of Child Psychology and Psychiatry, and Allied Disciplines.* Aug; 52(8):907-15.

Fall CH, Barker DJ, Osmond C, Winter PD, Clark PM, Hales CN. (1992). Relation of infant feeding to adult serum cholesterol concentration and death from ischaemic heart disease. *BMJ: British Medical Journal* (Clinical Research Edition). Mar 28; 304(6830):801-5.

Leeson CP, Kattenhorn M, Deanfield JE, Lucas A. (2001). Duration of breast feeding and arterial distensibility in early adult life: population based study. *BMJ: British Medical Journal* (Clinical Research Edition). Mar 17; 322(7287):643-7.

Leon DA, Ronalds G (2009.Breast-feeding influences on later life--cardiovascular disease.*Advances in Experimental Medicine and Biology.* 639:153-66.

Armstrong KL, Van Haeringen AR, Dadds MR, Cash R. (1998). Sleep deprivation or postnatal depression in later infancy: separating the chicken from the egg. *Journal of Paediatric and Child Health.* Jun; 34(3):260-2.

Montgomery-Downs HE, Clawges HM, Santy EE. (2010). Infant feeding methods and maternal sleep and daytime functioning. Pediatrics. Dec; 126(6):e1562-8.

Thomas KA. (2000). Differential effects of breast- and formula-feeding on preterm infants' sleep-wake patterns. *Journal of Obstetric, Gynecologic and Neonatal Nursing.* Mar-Apr; 29(2):145-52.

Lauzon-Guillain Bd, Wijndaele K, Clark M, Acerini CL, Hughes IA, Dunger DB, Wells JC, Ong KK. (2012). Breastfeeding and infant temperament at age three months. *PLoS One.* ; 7(1):e29326

MDykes F. (2005) Government funded breastfeeding peer support projects: implications for practice. Maternal & Child Nutrition. Jan; 1(1):21-31.

Amir LH. (2006). Breastfeeding--managing 'supply' difficulties. *Australian Family Physician.* Sep; 35(9):686-9.

McFadden A, Toole G. (2006). Exploring women's views of breastfeeding: a focus group study within an area with high levels of socio-economic deprivation. *Maternal & Child Nutrition.* Jul; 2(3):156-68.

Brown JD, Peuchaud SR (2008. Media and breastfeeding: friend or foe? *International Breastfeeding Journal.* Aug 4; 3:15.

Cox CR, Goldenberg JL, Arndt J, Pyszczynski T. (2007. Mother's milk: an existential perspective on negative reactions to breast-feeding. *Personality & Social Psychology Bulletin.* Jan; 33(1):110-22.

Kim P, Feldman R, Mayes LC, Eicher V, Thompson N, Leckman JF, Swain JE. (2011). Breastfeeding, brain activation to own infant cry, and maternal sensitivity. *Journal of Child Psychology and Psychiatry, and Allied Disciplines.* Aug; 52(8):907-15.

Virginia Schmied1, Deborah Lupton (2001. Blurring the boundaries: breastfeeding and maternal subjectivity. Sociology of Health & Illness, Mar; 23(2):234–250.

McInnes RJ, Chambers JA. (2008).Supporting breastfeeding mothers: qualitative synthesis. *Journal of Advanced Nursing.* May; 62(4):407-27.

**10. Circumcision**

Simforoosh N, Tabibi A, Khalili SA, Soltani MH, Afjehi A, Aalami F, Bodoohi H. (2012). Neonatal circumcision reduces the incidence of asymptomatic urinary tract infection: a large prospective study with long-term follow up using Plastibell. *Journal of Pediatric Urology.* Jun; 8(3):320-3.

The American Academy of Pediatrics (2012) Circumcision Policy Statement Task force on circumcision. *Pediatrics* 130(3) September 1, pp. 585 -586

The American Academy of Pediatrics (2012) Technical Report: Male Circumcision Task force on circumcision. *Pediatrics* 130(3):e758-e785.

Jagannath VA, Fedorowicz Z, Sud V, Verma AK, Hajebrahimi S. (2012). Routine neonatal circumcision for the prevention of urinary tract infections in infancy. Cochrane Database of Systematic Reviews.Nov 14; 11

Siegfried N, Muller M, Deeks JJ, Volmink J. (2009. Male circumcision for prevention of heterosexual acquisition of HIV in men. Cochrane Database of Systematic Reviews. Apr 15 ;( 2).

Tobian AA, Gray RH, Quinn TC (2010). Male circumcision for the prevention of acquisition and transmission of sexually transmitted infections: the case for neonatal circumcision. *Archives of Pediatric and Adolescent Medicine.* Jan; 164(1):78-84

Schoen EJ, Oehrli M, Colby Cd, Machin G. (2000). The highly protective effect of newborn circumcision against invasive penile cancer. *Pediatrics.* Mar; 105(3):E36.

Brady-Fryer B, Wiebe N, Lander JA. (2004). Pain relief for neonatal circumcision.Cochrane Database of Systematic Reviews (online). Oct 18 ;( 4)

Weiss HA, Larke N, Halperin D, Schenker I. (2010). Complications of circumcision in male neonates, infants and children: a systematic review. BMC Urology. Feb 16; 10:2.

Pieretti RV, Goldstein AM, Pieretti-Vanmarcke R. (2010). Late complications of newborn circumcision: a common and avoidable problem. *Pediatric Surgery International.* May; 26(5):515-8.

Morris BJ, Bailey RC, Klausner JD, Leibowitz A, Wamai RG, Waskett JH, Banerjee J, Halperin DT, Zoloth L, Weiss HA, Hankins CA. (2012). Review: a critical evaluation of arguments opposing male circumcision for HIV prevention in developed countries. *AIDS Care.* 24(12):1565-75.

Kigozi G, Watya S, Polis CB, Buwembo D, Kiggundu V, Wawer MJ, Serwadda D, Nalugoda F, Kiwanuka N, Bacon MC, Ssempijja V, Makumbi F, Gray RH. (2008). The effect of male circumcision on sexual satisfaction and function, results from a randomized trial of male circumcision for human immunodeficiency virus prevention, Rakai, Uganda. *BJU International.* 2008 Jan; 101(1):65-70.

Fink KS, Carson CC, DeVellis RF. (2002. Adult circumcision outcomes study: effect on erectile function, penile sensitivity, sexual activity and satisfaction. *Journal of Urology.* May; 167(5):2113-6.

**11. Nappies**

Carlsen E, Giwercman A, Keiding N, Skakkebaek NE. (1992). Evidence for decreasing quality of semen during past 50 years. *British Medical Journal* Sep 12; 305(6854):609-13.

Partsch CJ, Aukamp M, Sippell WG. (2000). Scrotal temperature is increased in disposable plastic lined nappies. *Archives of Disease in Childhood.* Oct; 83(4):364-8.

Campbell RL, Seymour JL, Stone LC, Milligan MC. (1987). Clinical studies with disposable diapers containing absorbent gelling materials: evaluation of effects on infantskin condition. *Journal of the American Academy of Dermatology.* Dec; 17(6):978-87.

Erasala GN, Romain C, Merlay I. (2011). Diaper area and disposable diapers. *Current Problems in Dermatology.* 40:83-9.

Akin F, Spraker M, Aly R, Leyden J, Raynor W, Landin W. (2001). Effects of breathable disposable diapers: reduced prevalence of Candida and common diaper dermatitis. *Pediatric Dermatology.* Jul-Aug; 18(4):282-90.

Kamat M, Malkani R. (2003. Disposable diapers: a hygienic alternative. *Indian Journal of Pediatrics.* Nov; 70(11):879-81.

Wong DL, Brantly D, Clutter LB, De Simone D, Lammert D, Nix K, Perry KA, Smith DP, White KH. (1992). Diapering choices: a critical review of the issues. *Pediatric Nursing.* Jan-Feb; 18(1):41-54.

Lintner K, Genet V. (1998). A physical method for preservation of cosmetic products. *International Journal of Cosmetic Science.* =Apr;20(2):103-15.

DeVito MJ, Schecter A. (2002. Exposure assessment to dioxins from the use of tampons and diapers. *Environmental Health Perspectives.* Jan; 110(1):23-8.

Anderson RC, Anderson JH. (1999. Acute respiratory effects of diaper emissions. *Archives of Environmental Health.* Sep-Oct; 54(5):353-8.

Alberta L, Sweeney SM, Wiss K. (2005).Diaper dye dermatitis. *Pediatrics.* Sep; 116(3):e450-2.

**12. Infant supplements**

Gregory JR, Collins DL, Davies PSW, Hughes JM, Clarke PC. (2010)

Health Service Executive (Eire) (2010) Vitamin D Supplementation for Infants - Information for Health Professionals. *National Diet and Nutrition Survey: children aged 1 ½ to 4 ½ years.* Volume 1: Report of the diet and nutrition survey London: HMSO, 1995.

Lauer B, Spector N. (2012). Vitamins. *Pediatrics in Review/American Academy of Pediatrics.* Aug; 33(8):339-51.

Marshall I, Mehta R, Petrova A. (2012). Vitamin D in the maternal-fetal-neonatal interface: clinical implications and requirements for supplementation. *Journal of Maternal-Fetal & Neonatal Medicine.* Dec 14.

Litonjua AA. (2012). Vitamin D deficiency as a risk factor for childhood allergic disease and asthma. *Current Opinion in Allergy and Clinical Immunology.* Apr; 12(2):179-85.

Kneller K. (2012). [Vitamins in pediatrics]. [Article in French]. *Revue Medicale de Bruxelles.* Sep; 33(4):339-45.

More J. (2007). Who needs vitamin supplements? *Journal of Family Health Care.* 17(2):57-60.

Leaf AA; RCPCH Standing Committee on Nutrition. (2007). Vitamins for babies and young children. *Archives of Disease in Childhood.* Feb; 92(2):160-4.

Osborn DA, Sinn JK. (2007). Prebiotics in infants for prevention of allergic disease and food hypersensitivity. Cochrane Database of Systematic Reviews. Oct 17;( 4).

Rautava S, Kainonen E, Salminen S, Isolauri E. (2012). Maternal probiotic supplementation during pregnancy and breast-feeding reduces the risk of eczema in the infant. *Journal of Allergy and Clinical Immunology.* Dec; 130(6):1355-60.

Guandalini S. (2011). Probiotics for prevention and treatment of diarrhoea. *Journal of Clinical Gastroenterology.* Nov; 45 Supplement: S149-53.

Johnston BC, Goldenberg JZ, Vandvik PO, Sun X, Guyatt GH. (2011). Probiotics for the prevention of pediatric antibiotic-associated diarrhea. Cochrane Database of Systematic Reviews. Nov 9 ;( 11).

Niittynen L, Pitkäranta A, Korpela R. (2012). Probiotics and otitis media in children. *International Journal of Pediatric Otorhinolaryngology.* Apr; 76(4):465-70.

Williams G, Craig JC (2009). Prevention of recurrent urinary tract infection in children. *Current Opinion in Infectious Diseases.* Feb; 22(1):72-6.

Vouloumanou EK, Makris GC, Karageorgopoulos DE, Falagas ME. (2009). Probiotics for the prevention of respiratory tract infections: a systematic review. *International Journal of Antimicrobial Agents.* Sep; 34(3):197.e1-10.

Hao Q, Lu Z, Dong BR, Huang CQ, Wu T. (2011). Probiotics for preventing acute upper respiratory tract infections. Cochrane Database of Systematic Reviews). Sep 7 ;( 9).

Romieu I, Torrent M, Garcia-Esteban R, Ferrer C, Ribas-Fitó N, Antó JM, Sunyer J. (2007. Maternal fish intake during pregnancy and atopy and asthma in infancy. *Clinical and Experimental Allergy.* Apr; 37(4):518-25.

D'Vaz N, Meldrum SJ, Dunstan JA, Martino D, McCarthy S, Metcalfe J, Tulic MK, Mori TA,

Prescott SL.(2012). Postnatal fish oil supplementation in high-risk infants to prevent allergy: randomized controlled trial. *Pediatrics.* Oct; 130(4):674-82.

Meldrum SJ, D'Vaz N, Simmer K, Dunstan JA, Hird K, Prescott SL. (2012). Effects of high-dose fish oil supplementation during early infancy on neurodevelopment and language: a randomised controlled trial. *British Journal of Nutrition.* Oct 28; 108(8):1443-54.

Willatts P, Forsyth JS, DiModugno MK, Varma S, Colvin M. (1998). Effect of long-chain polyunsaturated fatty acids in infant formula on problem solving at 10 months of age. *Lancet.* Aug 29; 352(9129):688-91.

Simmer K, Patole SK, Rao SC. (2011). Long-chain polyunsaturated fatty acid supplementation in infants born at term. Cochrane Database of Systematic Reviews (online). Dec 7; (12).

Waddell L. (2012). The power of vitamins. *Journal of Family Health Care.* Jan-Feb; 22(1):14, 16-20, 22-5.

Briefel R, Hanson C, Fox MK, Novak T, Ziegler P. (2006). Feeding Infants and Toddlers Study: do vitamin and mineral supplements contribute to nutrient adequacy or excess among US infants and toddlers? *Journal of the American Dietetic Association.* Jan; 106(1 Supplement 1):S52-65.

**13. Hygiene and cleanliness**

Bodner C, Godden D, Seaton A. (1998). Family size, childhood infections and atopic diseases. The Aberdeen WHEASE Group. *Thorax.* Jan; 53(1):28-32.

Strachan DP. (1989. Hay fever, hygiene, and household size. *BMJ: British Medical Journal.* Nov 18; 299(6710):1259-60.

Karmaus W, Arshad SH, Sadeghnejad A, Twiselton R. (2004). Does maternal immunoglobulin E decrease with increasing order of live offspring? Investigation into maternal immune tolerance. *Clinical and Experimental Allergy.* Jun; 34(6):853-9.

Ball TM, Castro-Rodriguez JA, Griffith KA, Holberg CJ, Martinez FD, Wright AL. (2000).

Siblings, day-care attendance, and the risk of asthma and wheezing during childhood. *The New England Journal of Medicine.* Aug 24; 343(8):538-43.

Svanes C, Jarvis D, Chinn S, Burney P. (1999). Childhood environment and adult atopy: results from the European Community Respiratory Health Survey. *Journal of Allergy and Clinical Immunology.* Mar; 103:415-20.

Ben-Shoshan M, Harrington DW, Soller L, Fragapane J, Joseph L, Pierre YS, Godefroy SB, Elliott SJ, Clarke AE. (2012. Demographic predictors of peanut, tree nut, fish, shellfish, and sesame allergy in Canada. *Journal of Allergy*(Cairo). ; 2012:858306.

Sherriff A, Golding J; Alspac Study Team. (2002). Factors associated with different hygiene practices in the homes of 15 month old infants. *Archives of Disease in Childhood.* Jul; 87(1):30-5.

Nickmilder M, Carbonnelle S, Bernard A. (2007). House cleaning with chlorine bleach and the risks of allergic and respiratory diseases in children. *Pediatric Allergy and Immunology.* Feb; 18(1):27-35.

Scott E, Bloomfield SF, Barlow CG. (1984). Evaluation of disinfectants in the domestic environment under 'in use' conditions. *Journal of Hygiene* (London). Apr; 92(2):193-203.

Farooqi IS, Hopkin JM. (1998). Early childhood infection and atopic disorder. *Thorax.* Nov; 53(11):927-32.

Celedón JC, Fuhlbrigge A, Rifas-Shiman S, Weiss ST, Finkelstein JA. (2004. Antibiotic use in the first year of life and asthma in early childhood. *Clinical and Experimental Allergy.* Jul; 34(7):1011-6.

Hesselmar B, Aberg N, Aberg B, Eriksson B, Björkstén B. (1999. Does early exposure to cat or dog protect against later allergy development?. *Clinical and Experimental Allergy.* May; 29(5):611-7.

Bloomfield SF, Stanwell-Smith R, Crevel RW, Pickup J. (2006). Too clean, or not too clean: the hygiene hypothesis and home hygiene. *Clinical and Experimental Allergy.* Apr; 36(4):402-25.

Pelosi U, Porcedda G, Tiddia F, Tripodi S, Tozzi AE, Panetta V, Pintor C, Matricardi PM. (2005). The inverse association of salmonellosis in infancy with allergic rhinoconjunctivitis and asthma at school-age: a longitudinal study. *Allergy.* May; 60(5):626-30.

Yazdanbakhsh M, Matricardi PM. (2004). Parasites and the hygiene hypothesis: regulating the immune system?. Clinical Reviews in Allergy & Immunology. Feb; 26(1):15-24.

Pelucchi C, Chatenoud L, Turati F, Galeone C, Moja L, Bach JF, La Vecchia C. (2012. Probiotics supplementation during pregnancy or infancy for the prevention of atopic dermatitis: a meta-analysis. *Epidemiology.* May; 23(3):402-14.

Rook GA, Adams V, Hunt J, Palmer R, Martinelli R, Brunet LR. (2004). Mycobacteria and other environmental organisms as immunomodulators for immunoregulatory disorders. *Springer Seminars in Immunopathology.* Feb; 25(3-4):237-55.

Camporota L, Corkhill A, Long H, Lordan J, Stanciu L, Tuckwell N, Cross A, Stanford JL, Rook GA, Holgate ST, Djukanovic R. (2003). The effects of Mycobacterium vaccae on allergen-induced airway responses in atopic asthma. *European Respiratory Journal.* Feb; 21(2):287-93.

Griffith C, Worsfold D, Mitchell R. (1998). Food preparation, risk communication and the consumer. *Food Control*; 9:225–32.

**14. Sleep interventions**

Bell, S. M., & Ainsworth, M. D. S. (1972). Infant crying and maternal responsiveness. *Child Development,* 43, 117l-1190.

Cook F, Bayer J, Le HN, Mensah F, Cann W, Hiscock H. (2012. Baby Business: a randomised controlled trial of a universal parenting program that aims to prevent early infant sleep and cry problems and associated parental depression. *BMC Pediatrics.* Feb 6; 12:13.

Mindell JA, Kuhn B, Lewin DS, Meltzer LJ, Sadeh A; American Academy of Sleep Medicine. (2006). Behavioral treatment of bedtime problems and night wakings in infants and young children. *Sleep.* Oct; 29(10):1263-76.

Rickert VI, Johnson CM. (1988).Reducing nocturnal awakening and crying episodes in infants and young children: a comparison between scheduled awakenings and systematic ignoring. *Pediatrics.* Feb; 81(2):203-12.

Morgenthaler TI, Owens J, Alessi C, Boehlecke B, Brown TM, Coleman J Jr, Friedman L, Kapur VK, Lee-Chiong T, Pancer J, Swick TJ; American Academy of Sleep Medicine. (2006). Practice parameters for behavioral treatment of bedtime problems and night wakings in infants and young children. *Sleep.* Oct; 29(10):1277-81.

Zuckerman B, Stevenson J, Bailey V. (1987). Sleep problems in early childhood: continuities, predictive factors, and behavioral correlates. *Pediatrics.* Nov; 80(5):664-71.

Hiscock H, Bayer J, Gold L, Hampton A, Ukoumunne OC, Wake M. (2007). Improving infant sleep and maternal mental health: a cluster randomised trial. *Archives of Disease in Childhood.* Nov; 92(11):952-8

Papousek M, von Hofacker N.(1998). Persistent crying in early infancy: a non-trivial condition of risk for the developing mother-infant relationship. *Child: Care, Health and Development.* Sep; 24(5):395-424.

Hiscock H, Wake M. (2001). Infant sleep problems and postnatal depression: a community-based study. *Pediatrics.* Jun; 107(6):1317-22.

Morrell JM. (1999). The role of maternal cognitions in infant sleep problems as assessed by a new instrument, the maternal cognitions about infant sleep questionnaire. *Journal of Child Psychology and Psychiatry, and Allied Disciplines.* Feb; 40(2):247-58.

Waters E, Cummings EM. (2000. A secure base from which to explore close relationships. *Child Development.* Jan-Feb; 71(1):164-72.

Murray L, Ramchandani P. (2007). Might prevention be better than cure? *Archives of Disease in Childhood.* Nov; 92(11):943-4.

Price AM, Wake M, Ukoumunne OC, Hiscock H. (2012). Five-year follow-up of harms and benefits of behavioral infant sleep intervention: randomized trial. *Pediatrics.* Oct; 130(4):643-51

Hiscock H, Bayer JK, Hampton A, Ukoumunne OC, Wake M. (2008). Long-term mother and child mental health effects of a population-based infant sleep intervention: cluster-randomized, controlled trial. *Pediatrics.* Sep; 122(3):e621-7.

Middlemiss W, Granger DA, Goldberg WA, Nathans L. (2012). Asynchrony of mother-infant hypothalamic-pituitary-adrenal axis activity following extinction of infant crying responses induced during the transition to sleep. *Early Human Development.* Apr; 88(4):227-32.

Larson MC, White BP, Cochran A, Donzella B, Gunnar M. (1998). Dampening of the cortisol response to handling at 3 months in human infants and its relation to sleep, circadian cortisol activity, and behavioral distress. *Developmental Psychobiology.* Dec; 33(4):327-37.

Gunnar MR, Donzella B. (2002). Social regulation of the cortisol levels in early human development. *Psychoneuroendocrinology.* Jan-Feb; 27(1-2):199-220.

Blunden SL, Thompson KR, Dawson D. (2011). Behavioural sleep treatments and night time crying in infants: challenging the status quo. *Sleep Medicine Reviews.* Oct; 15(5):327-34.

Allan n. Schore (2001). Effects of a secure attachment relationship on right brain development, affect regulation, and infant mental health. *Infant Mental Health Journal*, 22(1–2): 7–66.

Beijers R, Riksen-Walraven JM, de Weerth C. (2012). Cortisol regulation in 12-month-old human infants: Associations with the infants' early history of breastfeeding and co-sleeping. *Stress.* Nov 29.

St James-Roberts I, Sleep J, Morris S, Owen C, Gillham P. (2001). Use of a behavioural programme in the first 3 months to prevent infant crying and sleeping problems. *Journal of Paediatrics and Child Health.* Jun; 37(3):289-97.

Goodlin-Jones BL, Burnham MM, Gaylor EE, Anders TF. (2001). Night waking, sleep-wake organization, and self-soothing in the first year of life. *Journal of Developmental and Behavioral Pediatrics.* Aug; 22(4):226-33.

**15. MMR**

Grabenstein JD. (1999). Moral considerations with certain viral vaccines. *Christian Pharmacology;* 2(2): 3–6.

Wakefield AJ, Murch SH, Anthony A, Linnell J, Casson DM, Malik M, Berelowitz M, Dhillon AP, Thomson MA, Harvey P, Valentine A, Davies SE, Walker-Smith JA. (1998). Ileal-lymphoid-nodular hyperplasia, non-specific colitis, and pervasive developmental disorder in children.*Lancet.* Feb 28; 351(9103):637-41.

O'Leary JJ, Uhlmann V, Wakefield AJ. (2000). Measles virus and autism. *Lancet.* Aug 26; 356(9231):772.

Uhlmann V, Martin CM, Sheils O, Pilkington L, Silva I, Killalea A, Murch SB, Walker-Smith J, Thomson M, Wakefield AJ, O'Leary JJ. (2002). Potential viral pathogenic mechanism for new variant inflammatory bowel disease. *Molecular Pathology.* Apr; 55(2):84-90.

Hornig M, Briese T, Buie T, Bauman ML, Lauwers G, Siemetzki U, Hummel K, Rota PA, Bellini WJ, O'Leary JJ, Sheils O, Alden E, Pickering L, Lipkin WI. (2008). Lack of association between measles virus vaccine and autism with enteropathy: a case-control study. *PLoS One.* Sep 4; 3(9):e3140.

Hensley E, Briars L. (2003). Closer look at autism and the measles-mumps-rubella vaccine. *Journal of the American Pharmacists Association* (2003). Nov-Dec; 50(6):736-41.

Deer B. (2011). How the case against the MMR vaccine was fixed. *BMJ: British Medical Journal* Jan 5; 342:c5347.

Pérez-Vilar S et al. (2012). Suspected adverse events to measles, mumps and rubella vaccine reported to the Community of Valencia Pharmacovigilance Centre. *Anales Pediatria* (Barcelona, Spain). Sep 13. pii: S1695-4033(12)00340-2.

Pearce A, Law C, Elliman D, Cole TJ, Bedford H; Millennium Cohort Study Child Health Group. (2008). Factors associated with uptake of measles, mumps, and rubella vaccine (MMR) and use of single antigen vaccines in a contemporary UK cohort: prospective cohort study. *BMJ: British Medical Journal* Apr 5; 336(7647):754-7

Halsey NA. (2001). Combination vaccines: defining and addressing current safety concerns. *Clinical Infectious Diseases.* Dec 15; 33 Supplement 4:S312-8.

Brown KF, Long SJ, Ramsay M, Hudson MJ, Green J, Vincent CA, Kroll JS, Fraser G, Sevdalis N. (2012). U.K. parents' decision-making about measles-mumps-rubella (MMR) vaccine 10 years after the MMR-autism controversy: a qualitative analysis. *Vaccine.* Feb 27; 30(10):1855-64.

Stowe J, Andrews N, Taylor B, Miller E. (2009.No evidence of an increase of bacterial and viral infections following Measles, Mumps and Rubella vaccine. *Vaccine.* Feb 25; 27(9):1422-5.

Kirkland A. (2012). Credibility battles in the autism litigation. *Social Studies of Science.* Apr; 42(2):237-61.

Schneeweiss B, Pfleiderer M, Keller-Stanislawski B. (2008). Vaccination safety update. *Deutsche Arzteblatt International.* Aug;105(34-35):590-5.

**16. Dummies**

Jaafar SH, Jahanfar S, Angolkar M, Ho JJ. (2012). Effect of restricted pacifier use in breastfeeding term infants for increasing duration of breastfeeding. Cochrane Database Systematic Reviews Jul 11; 7:CD007202.

Jackson JM, Mourino AP. (1999). Pacifier use and otitis media in infants twelve months of age or younger. *Pediatric Dentistry.* Jul-Aug; 21(4):255-60.

Niemelä M, Pihakari O, Pokka T, Uhari M. (2000). Pacifier as a risk factor for acute otitis media: A randomized, controlled trial of parental counselling. *Pediatrics.* Sep; 106(3):483-8.

Comina E, Marion K, Renaud FN, Dore J, Bergeron E, Freney J. (2006). Pacifiers: a microbial reservoir. *Nursing & Health Sciences.* Dec; 8(4):216-23.

da Silveira LC, Charone S, Maia LC, Soares RM, Portela MB. (2009. Biofilm formation by Candida species on silicone surfaces and latex pacifier nipples: an in vitro study. *Journal of Clinical Pediatric Dentistry.* Spring; 33(3):235-40.

Franco Varas V, Gorritxo Gil B. (2012). Pacifier sucking habit and associated dental changes. Importance of early diagnosis. *Anales Pediatria (Barcelona, Spain).* Dec; 77(6):374-80.

Adair SM, Milano M, Dushku JC. (1992). Evaluation of the effects of orthodontic pacifiers on the primary dentitions of 24- to 59-month-old children: preliminary study. *Pediatric Dentistry.* Jan-Feb; 14(1):13-8.

Peressini S. (2003). Pacifier use and early childhood caries: an evidence-based study of the literature. *Journal (Canadian Dental Association).* Jan; 69(1):16-9.

Ollila P, Niemelä M, Uhari M, Larmas M. (1997). Risk factors for colonization of salivary lactobacilli and Candida in children. *Acta Odontologica Scandinavica.* Jan; 55(1):9-13.

Barbosa C, Vasquez S, Parada MA, Gonzalez JC, Jackson C, Yanez ND, Gelaye B, Fitzpatrick AL.(2009). The relationship of bottle feeding and other sucking behaviors with speech disorder in Patagonian preschoolers. *BMC Pediatrics.* Oct 21; 9:66

Neville HL, Huaco J, Vigoda M, Sola JE. (2008). Pacifier-induced bowel obstruction-not so soothing. *Journal of Pediatric Surgery.* Feb; 43(2):e13-5.

Rubi SC, Garcia M, Leiva A, Plaza F. (1990). Intestinal obstruction caused by a pacifier. *Annales de Pediatrie* (Paris). Oct; 37(8):543-5.

Moon RY, Fu L. (2012). Sudden infant death syndrome: an update. *Pediatrics in Review/American Academy of Pediatrics.* Jul; 33(7):314-20.

Johnson CM. (1991). Infant and toddler sleep: a telephone survey of parents in one community. *Journal of Developmental and Behavioral Pediatrics.* Apr; 12(2):108-14.

Pinelli J, Symington A. (2000). How rewarding can a pacifier be? A systematic review of non-nutritive sucking in preterm infants. *Neonatal Network: NN.* Dec; 19(8):41-8.

Carbajal R, Chauvet X, Couderc S, Olivier-Martin M. (1999). Randomised trial of analgesic effects of sucrose, glucose, and pacifiers in term neonates. *British Medical Journal* Nov 27; 319(7222):1393-7.

Ludington-Hoe SM, Cong X, Hashemi F. (2002). Infant crying: nature, physiologic consequences, and select interventions. Neonatal Network. Mar;21(2):29-36.

**17. Colic**

Räihä H, Lehtonen L, Korhonen T, Korvenranta H. (1996). Family life 1 year after infantile colic.

*Archives of Pediatrics & Adolescent Medicine.* Oct; 150(10):1032-6.

Savino F, Palumeri E, Castagno E, Cresi F, Dalmasso P, Cavallo F, Oggero R. (2006). Reduction of crying episodes owing to infantile colic: A randomized controlled study on the efficacy of a new infant formula. *European Journal of Clinical Nutrition.* Nov; 60(11):1304-10.

Lothe L, Lindberg T, Jakobsson I. (1990). Macromolecular absorption in infants with infantile colic. *Acta Paediatrica Scandinavica.*1990 Apr; 79(4):417-21.

Miller-Loncar C, Bigsby R, High P, Wallach M, Lester B.(2004. Infant colic and feeding difficulties. *Archives of Disease in Childhood.* 2004 Oct; 89(10):908-12.

Savino F, Cordisco L, Tarasco V, Locatelli E, Di Gioia D, Oggero R, Matteuzzi D. (2011). Antagonistic effect of Lactobacillus strains against gas-producing coliforms isolated from colicky infants.

*BMC Microbiology.* Jun 30; 11:157.

Matheson I. (1995). [Infantile colic--what will help?].[Article in Norwegian] *Tidsskrift for den Norske Laegeforening.* Aug 20; 115(19):2386-9.

Milidou I, Henriksen TB, Jensen MS, Olsen J, Søndergaard C. (2012). Nicotine replacement therapy during pregnancy and infantile colic in the offspring. *Pediatrics.* Mar; 129(3):e652-8.

White BP, Gunnar MR, Larson MC, Donzella B, Barr RG. (2000. Behavioral and physiological responsivity, sleep, and patterns of daily cortisol production in infants with and without colic.*Child Development.* Jul-Aug; 71(4):862-77.

Evans K, Evans R, Simmer K (1995). Effect of the method of breast feeding on breast engorgement, mastitis and infantile colic. *Acta Paediatrica* (Oslo, Norway). Aug; 84(8):849-52.

Metcalf TJ, Irons TG, Sher LD, Young PC. (1994). Simethicone in the treatment of infant colic: a randomized, placebo-controlled, multicenter trial. *Pediatrics.* Jul; 94(1):29-34.

Sas D, Enrione MA, Schwartz RH. (2004). Pseudomonas aeruginosa septic shock secondary to "gripe water" ingestion. *Pediatric Infectious Diseases Journal.* Feb; 23(2):176-7.

Lucassen P. (2010). Colic in infants. *Clinical Evidence.* Feb 5.

Savino F, Brondello C, Cresi F, Oggero R, Silvestro L. (2002). Cimetropium bromide in the treatment of crisis in infantile colic. *Journal of Pediatric Gastroenterology and Nutrition.* Apr; 34(4):417-9.

Hill DJ, Roy N, Heine RG, Hosking CS, Francis DE, Brown J, Speirs B, Sadowsky J, Carlin JB. (2005). Effect of a low-allergen maternal diet on colic among breastfed infants: a randomized, controlled trial. *Pediatrics.* Nov; 116(5):e709-15.

Kanabar D, Randhawa M, Clayton P. (2001). Improvement of symptoms in infant colic following reduction of lactose load with lactase. *Journal of Human Nutrition and Dietetics.* Oct; 14(5):359-63.

Bocquet A, Bresson JL, Briend A, Chouraqui JP, Darmaun D, Dupont C, Frelut ML, Ghisolfi J, Goulet O, Putet G, Rieu D, Turck D, Vidailhet M; Comité de Nutrition de la Société Française de Pédiatrie.(2001). Infant formulas and soy protein-based formulas: current data. *Archives de Pediatrie.* Nov; 8(11):1226-33.

Savino F, Pelle E, Palumeri E, Oggero R, Miniero R. (2007). Lactobacillus reuteri (American Type Culture Collection Strain 55730) versus simethicone in the treatment of infantile colic: a prospective randomized study. *Pediatrics.* Jan; 119(1):e124-30.

Akçam M, Yilmaz A. (2006). Oral hypertonic glucose solution in the treatment of infantile colic. *Pediatrics International.* Apr; 48(2):125-7.

Barr RG, Pantel MS, Young SN, Wright JH, Hendricks LA, Gravel R. (1999). The response of crying newborns to sucrose: is it a "sweetness" effect? *Physiol ogy & Behavior.* May; 66(3):409-17.

Savino F, Cresi F, Castagno E, Silvestro L, Oggero R. (2005). A randomized double-blind placebo-controlled trial of a standardized extract of Matricariae recutita, Foeniculum vulgare and Melissa officinalis (ColiMil) in the treatment of breastfed colicky infants. *Phytotherapy Research.* Apr; 19(4):335-40.

St James-Roberts I, Goodwin J, Peter B, Adams D, Hunt S. (2003. Individual differences in responsivity to a neurobehavioural examination predict crying patterns of 1-week-old infants at home. *Devopmental Medicine and Child Neurology.* Jun; 45(6):400-7.

St James-Roberts I, Alvarez M, Csipke E, Abramsky T, Goodwin J, Sorgenfrei E. (2006). Infant crying and sleeping in London, Copenhagen and when parents adopt a "proximal" form of care. *Pediatrics.* Jun; 117(6):e1146-55.

Kheir AE. (2012). Infantile colic, facts and fiction. *Italian Journal of Pediatrics.* Jul 23; 38:34.

**18. Co-sleeping**

ResMcKenna JJ, McDade T. (2005). Why babies should never sleep alone: a review of the co-sleeping controversy in relation to SIDS, bedsharing and breast feeding. *The Journal of Paediatrics.* Jun; 6(2):134-52.

Mace S. (2006). Where should babies sleep? *Community Practitioner: the journal of the Community Practitioners' & Health Visitors' Association.* Jun; 79(6):180-3.

Nakamura S, Wind M, Danello MA. (1999). Review of hazards associated with children placed in adult beds. *Archives of Pediatric & Adolescent Medicine.* Oct; 153(10):1019-23.

Shaefer SJ. (2012). Review finds that bed sharing increases risk of sudden infant death syndrome. *Evidence-based Nursing.* Oct; 15(4):115-6.

Baddock SA, Galland BC, Bolton DP, Williams SM, Taylor BJ. (2012). Hypoxic and hypercapnic events in young infants during bed-sharing. *Pediatrics.* Aug; 130(2):237-44.

Kelmanson IA. (2010). Sleep disturbances in two-month-old infants sharing the bed with parent(s). *Minerva Pediatrica.* Apr; 62(2):161-9.

Mosko S, Richard C, McKenna J, Drummond S, Mukai D. (1997). Maternal proximity and infant CO2 environment during bedsharing and possible implications for SIDS research. *American Journal of Physical Anthropology.* Jul; 103(3):315-28.

Blair PS, Heron J, Fleming PJ. (2010). Relationship between bed sharing and breastfeeding: longitudinal, population-based analysis. *Pediatrics.* Nov; 126(5):e1119-26.

Santos IS, Mota DM, Matijasevich A, Barros AJ, Barros FC. (2009). Bed-sharing at 3 months and breast-feeding at 1 year in southern Brazil. *The Journal of Pediatrics.* Oct; 155(4):505-9

Ball HL. (2003). Breastfeeding, bed-sharing, and infant sleep. *Birth.* Sep; 30(3):181-8.

Barajas RG, Martin A, Brooks-Gunn J, Hale L. (2011. Mother-child bed-sharing in toddlerhood and cognitive and behavioral outcomes. *Pediatrics.* Aug; 128(2):e339-47.

Keller MA, Goldberg WA. (2004). Co-sleeping: Help or hindrance for young children's independence. *Infant and Child Development* Dec; 13(5):369–388.

**19. Baby-led versus parent-led**

St James-Roberts I, Alvarez M, Csipke E, Abramsky T, Goodwin J, Sorgenfrei E. (2006). Infant crying and sleeping in London, Copenhagen and when parents adopt a "proximal" form of care. *Pediatrics.* Jun; 117(6):e1146-55.

St James-Roberts I, Hurry J, Bowyer J, Barr RG. (1995). Supplementary carrying compared with advice to increase responsive parenting as interventions to prevent persistent infant crying. *Pediatrics.* Mar; 95(3):381-8.

Alvarez M. (2004). Caregiving and early infant crying in a Danish community. *Journal of Developmental and Behavioral Pediatrics.* 2004 Apr; 25(2):91-8.

Nikolopoulou M, St James-Roberts I. (2003). Preventing sleeping problems in infants who are at risk of developing them. *Archives of Disease in Childhood.* Feb; 88(2):108-11.

Touchette E, Petit D, Paquet J, Boivin M, Japel C, Tremblay RE, Montplaisir JY. (2005. Factors associated with fragmented sleep at night across early childhood. *Archives of Pediatrics & Adolescent Medicine.* Mar; 159(3):242-9.

Anders TF, Halpern LF, Hua J. (1992). Sleeping through the night: a developmental perspective. *Pediatrics.* Oct; 90(4):554-60.

Adams LA, Rickert VI. (1989). Reducing bedtime tantrums: comparison between positive routines and graduated extinction. *Pediatrics.* Nov; 84(5):756-61.

Rickert VI, Johnson CM. (1988). Reducing nocturnal awakening and crying episodes in infants and young children: a comparison between scheduled awakenings and systematic ignoring. *Pediatrics.* Feb; 81(2):203-12.

Renfrew MJ, Lang S, Martin L, Woolridge M. (2000). Interventions for influencing sleep patterns in exclusively breastfed infants. Cochrane Databaseof Systematic Reviews. 2000;(2).

Hayama J, Adachi Y, Nishino N, Ohryoji F. (2008). [Impact of parenting behavior relevant to infant's sleep on maternal sleep and health].[Article in Japanese]Nihon Koshu Eisei Zasshi: *Japanese Journal of Public Health.* Oct; 55(10):693-700.

**20. Carseats**

Kapoor T, Altenhof W, Snowdon A, Howard A, Rasico J, Zhu F, Baggio D. (2011). A numerical investigation into the effect of CRS misuse on the injury potential of children in frontal and side impact crashes. *Accident Analysis and Prevention.* Jul; 43(4):1438-50.

Lowne R, Roy P, Paton I. Child Occupant Protection 2nd Symposium Proceedings. Warrendale, PA: SAE; (1997). A Comparison of the Performance of Dedicated Child Restraint Attachment Systems (ISOFIX). *SAE* 973302; pp. 71–86.

Bilston LE, Brown J, Kelly P. (2005). Improved protection for children in forward-facing restraints during side impacts. *Traffic Injury Prevention.* Jun; 6(2):135-46.

**21. Introducing solids**

European Food Safety Authority (EFSA) Panel on Dietetic Products, Nutrition and Allergies (NDA), (2009). Scientific Opinion on the appropriate age for introduction of complementary feeding of infants. *EFSA Journal* 7(12): 1423.

Kramer MS, Kakuma R. (2012). Optimal duration of exclusive breastfeeding. Cochrane Database of Systematic Reviews. Aug 15; 8.

Quigley MA, Kelly YJ, Sacker A. (2009). Infant feeding, solid foods and hospitalisation in the first 8 months after birth. *Archives of Disease in Childhood.* Feb; 94(2):148-50.

Chantry CJ, Howard CR, Auinger P. (2006). Full breastfeeding duration and associated decrease in respiratory tract infection in US children. *Pediatrics.* Feb; 117(2):425-32.

Forsyth JS, Ogston SA, Clark A, Florey CD, Howie PW. (1993. Relation between early introduction of solid food to infants and their weight and illnesses during the first two years of life. BMJ: British Medical Journal Jun 12; 306(6892):1572-6.

Kramer MS, Vanilovich I, Matush L, Bogdanovich N, Zhang X, Shishko G, Muller-Bolla M, Platt RW.(2007). The effect of prolonged and exclusive breast-feeding on dental caries in early school-age children. New evidence from a large randomized trial. *Caries Research.* 41(6):484-8.

Reilly JJ, Wells JC. (2005). Duration of exclusive breast-feeding: introduction of complementary feeding may be necessary before 6 months of age. *British Journal of Nutrition.* Dec; 94(6):869-72.

Schwartz C, Chabanet C, Laval C, Issanchou S, Nicklaus S. (2012). Breast-feeding duration: influence on taste acceptance over the first year of life. British Journal of Nutrition. Jul 4:1-8.

Mennella JA, Griffin CE, Beauchamp GK. (2004). Flavor programming during infancy. *Pediatrics.* Apr; 113(4):840-5.

Nicklaus S. (2011). Children's acceptance of new foods at weaning. Role of practices of weaning and of food sensory properties. *Appetite.* Dec; 57(3):812-5.

Cohen RJ, Rivera LL, Canahuati J, Brown KH, Dewey KG. (1995). Delaying the introduction of complementary food until 6 months does not affect appetite or mother's report of food acceptance of breast-fed infants from 6 to 12 months in a low income, Honduran population. *Journal of Nutrition.* Nov; 125(11):2787-92.

Yilmaz G, Gürakan B, Cakir B, Tezcan S. (2002). Factors influencing sleeping pattern of infants. *Turkish Journal of Pediatrics.* Apr-Jun; 44(2):128-33.

Robertson RM. (1974). Letter: Solids and "sleeping through". *British Medical Journal.* Feb 2; 1(5900):200.

Norris JM, Barriga K, Hoffenberg EJ, Taki I, Miao D, Haas JE, Emery LM, Sokol RJ, Erlich HA, Eisenbarth GS, Rewers M. (2005). Risk of celiac disease autoimmunity and timing of gluten introduction in the diet of infants at increased risk of disease. *Journal of the American Medical Association.* May 18; 293(19):2343-51.

Guandalini S. (2007). The influence of gluten: weaning recommendations for healthy children and children at risk for celiac disease. *Nestle Nutrition Workshop Series, Pediatric Programmme.* 2007; 60:139-51; discussion 151-5.

Ludvigsson JF, Fasano A. (2012). Timing of introduction of gluten and celiac disease risk. *Annals of Nutrition & Metabolism.* 60 Suppl 2:22-9.

Szajewska H, Chmielewska A, Pie cik-Lech M, Ivarsson A, Kolacek S, Koletzko S, Mearin ML, Shamir R, Auricchio R, Troncone R; PREVENTCD Study Group. (2012). Systematic review: early infant feeding and the prevention of coeliac disease. *Alimentary Pharmacology & Therapeutics.* Oct;36(7):607-18.

Filipiak B, Zutavern A, Koletzko S, von Berg A, Brockow I, Grübl A, Berdel D, Reinhardt D, Bauer CP, Wichmann HE, Heinrich J; GINI-Group (2007). Solid food introduction in relation to eczema: results from a four-year prospective birth cohort study. *Journal of Pediatrics.* Oct; 151(4):352-8

Grimshaw KE, Allen K, Edwards CA, Beyer K, Boulay A, van der Aa LB, Sprikkelman A, Belohlavkova S, Clausen M, Dubakiene R, Duggan E, Reche M, Marino LV, Nørhede P, Ogorodova L, Schoemaker A, Stanczyk-Przyluska A, Szepfalusi Z, Vassilopoulou E, Veehof SH, Vlieg-Boerstra BJ, Wjst M, Dubois AE. (2009). Infant feeding and allergy prevention: a review of current knowledge and recommendations. A EuroPrevall state of the art paper. *Allergy.* Oct; 64(10):1407-16.

Wilson AC, Forsyth JS, Greene SA, Irvine L, Hau C, Howie PW. (1998). Relation of infant diet to childhood health: seven year follow up of cohort of children in Dundee infant feeding study. *British Medical Journal.* Jan 3; 316(7124):21-5.

Wahlberg J, Vaarala O, Ludvigsson J; ABIS-study group. (2006). Dietary risk factors for the emergence of type 1 diabetes-related autoantibodies in 21/2 year-old Swedish children. *British Journal of Nutrition.* Mar; 95(3):603-8.

Mehta KC, Specker BL, Bartholmey S, Giddens J, Ho ML (1998). Trial on timing of introduction to solids and food type on infant growth. *Pediatrics.* Sep; 102(3 Pt 1):569-73.

Chomtho S, Wells JC, Williams JE, Davies PS, Lucas A, Fewtrell MS. (2008). Infant growth and later body composition: evidence from the 4-component model. *American Journal of Clinical Nutrition.* Jun; 87(6):1776-84.

Schack-Nielsen L, Sørensen TIa, Mortensen EL, Michaelsen KF. (2010). Late introduction of complementary feeding, rather than duration of breastfeeding, may protect against adult overweight. *American Journal of Clinical Nutrition.* Mar; 91(3):619-27.

Robinson S, Fall C. (2012). Infant nutrition and later health: a review of current evidence. *Nutrients.* Aug; 4(8):859-74.

Owen CG, Martin RM, Whincup PH, Smith GD, Cook DG. (2006). Does breastfeeding influence risk of type 2 diabetes in later life? A quantitative analysis of published evidence. *American Journal of Clinical Nutrition.* Nov; 84(5):1043-54.

Butte NF, Lopez-Alarcon MG, Garza C (2002): Nutrient adequacy of exclusive breastfeeding for the term infant during the first six months of life. World Health Organisation, Geneva.

Leunissen RW, Kerkhof GF, Stijnen T, Hokken-Koelega A. (2009). Timing and tempo of first-year rapid growth in relation to cardiovascular and metabolic risk profile in early adulthood. *Journal of the American Medical Association.* Jun 3; 301(21):2234-42.

Salmenperä L, Perheentupa J, Siimes MA. (1985). Exclusively breast-fed healthy infants grow slower than reference infants. *Pediatric Research.* Mar; 19(3):307-12.

Fewtrell M, Wilson DC, Booth I, Lucas A. (2010). Six months of exclusive breast feeding: how good is the evidence? *BMJ: British Medical Journal.* Jan 13; 342:c5955.

Krebs NF, Hambidge KM. (2007). Complementary feeding: clinically relevant factors affecting timing and composition. *American Journal of Clinical Nutrition.* Feb; 85(2):639S-645S.

**22. Baby-led weaning**

Brown A, Lee M. (2011). An exploration of experiences of mothers following a baby-led weaning style: developmental readiness for complementary foods. *Maternal & Child Nutrition.* Nov 28.

Wright CM, Cameron K, Tsiaka M, Parkinson KN. (2011). Is baby-led weaning feasible? When do babies first reach out for and eat finger foods? *Maternal & Child Nutrition.* Jan; 7(1):27-33.

Brown A, Lee M. (2011). An exploration of experiences of mothers following a baby-led weaning style: developmental readiness for complementary foods. *Maternal & Child Nutrition.* Nov 28.

Northstone K, Emmett P, Nethersole F; ALSPAC Study Team. Avon Longitudinal Study of Pregnancy and Childhood (2001). The effect of age of introduction to lumpy solids on foods eaten and reported feeding difficulties at 6 and 15 months. *Journal of Human Nutrition and Dietetics.* Feb; 14(1):43-54.

Rowan H, Harris C. (2012). Baby-led weaning and the family diet. A pilot study. *Appetite.* Jun; 58(3):1046-9.

Townsend E, Pitchford NJ. (2012). Baby knows best? The impact of weaning style on food preferences and body mass index in early childhood in a case-controlled sample. *BMJ: British Medical Journal.* Feb 6; 2(1):e000298.

Cameron SL, Heath AL, Taylor RW (2012). How feasible is baby-led weaning as an approach to infant feeding? A review of the evidence. *Nutrients.* Nov 2; 4(11):1575-609.

**23. Jarred baby food**

Yeung DL, Pennell MD, Leung M, Hall J. (1982). Commercial or homemade baby food. *Canadian Medical Association Journal.* Jan 15; 126(2):113.

Ferguson EL, Darmon N. (2007). Traditional foods vs. manufactured baby foods. Nestle Nutrition Workshop Series, Pediatric Programme. 2007; 60:43-61.

Zand N, Chowdhry BZ, Pollard LV, Pullen FS, Snowden MJ, Zotor FB. (2012). Commercial 'ready-to-feed' infant foods in the UK: macro-nutrient content and composition. *Maternal & Child Nutrition.* Oct 1.

van den Boom S, Kimber AC, Morgan JB. (1997). Nutritional composition of home-prepared baby meals in Madrid. Comparison with commercial products in Spain and home-made meals in England. *Acta Paediatrica* (Oslo, Norway). Jan; 86(1):57-62.

Zand N, Chowdhry BZ, Wray DS, Pullen FS, Snowden MJ. (2012). Elemental content of commercial 'ready to-feed' poultry and fish based infant foods in the UK. *Food Chemistry.* Dec 15; 135(4):2796-801.

Zand N, Chowdhry BZ, Pullen FS, Snowden MJ, Tetteh J. (2012). Simultaneous determination of riboflavin and pyridoxine by UHPLC/LC-MS in UK commercial infant meal food products. *Food Chemistry.* Dec 15; 135(4):2743-9. .

Hurley KM, Black MM. (2010). Commercial baby food consumption and dietary variety in a state-wide sample of infants receiving benefits from the special supplemental nutrition program for women, infants, and children. *Journal of the American Dietetic Association.* Oct; 110(10):1537-41.

Bennett AE, O'Connor AL, Canning N, Kenny A, Keaveney E, Younger K, Flynn MA. (2012). Weaning onto solid foods: some of the challenges. *Irish Medical Journal.* Sep; 105(8):266-8.

Mondello L, Zoccali M, Purcaro G, Franchina FA, Sciarrone D, Moret S, Conte L, Tranchida PQ. (2012). Determination of saturated-hydrocarbon contamination in baby foods by using on-line liquid-gas chromatography and off-line liquid chromatography-comprehensive gas chromatography combined with mass spectrometry. *Journal of Chromatography.* Oct 12;1259:221-6.

**24. Going organic**

Palupi E, Jayanegara A, Ploeger A, Kahl J (2012). Comparison of nutritional quality between conventional and organic dairy products: a meta-analysis. *Journal of the Science of Food and Agriculture,* Nov 92(14):2774-81.

Vanderpump MP, Lazarus JH, Smyth PP, Laurberg P, Holder RL, Boelaert K, Franklyn JA; British Thyroid Association UK Iodine Survey Group. (2011). Iodine status of UK schoolgirls: a cross-sectional survey. *The Lancet.* Jun 11; 377(9782):2007-12.

Alvarez-Fernández E, Domínguez-Rodríguez J, Capita R, Alonso-Calleja CJ (2012). Influence of housing systems on microbial load and antimicrobial resistance patterns of Escherichia coli isolates from eggs produced for human consumption. *Journal of Food Protection.* May :5(5):847-53.

Raigón MD, Rodríguez-Burruezo A, Prohens J (2012). Effects of organic and conventional cultivation methods on composition of eggplant fruits. *Journal of Agricultural and Food Chemistry.* June 9;58(11):6833-40.

Smith-Spangler C, Brandeau ML, Hunter GE, Bavinger JC, Pearson M, Eschbach PJ, Sundaram V, Liu H, Schirmer P, Stave C, Olkin I, Bravata DM. (2012). Are organicfoods safer or healthier than conventional alternatives?: a systematic review. *Annals of Internal Medicine.* Sept 4;157(5):348-66.

Lester GE, Manthey JA, Buslig BS. (2007). Organic verses conventionally grown Rio Red whole grapefruit and juice: comparison of production, inputs, market quality, consumer acceptance, and human health-bioactive compounds. *Journal of Agricultural and Food Chemistry.* May 30; 55(11):4474-80.

Carbonaro M, Mattera M, Nicoli S, Bergamo P, Cappelloni M. (2002). Modulation of antioxidant compounds in organic vs conventional fruit (peach, Prunus persica L., and pear, Pyruscommunis L.) *Journal of Agricultural and Food Chemistry.* Sept 11; 50(19):5458-62.

Stracke BA, Rüfer CE, Weibel FP, Bub A, Watzl B. (2009). Threeyear comparison ofthe polyphenol contents and antioxidant capacities in organically and conventionally produced apples (Malus domestica Bork. Cultivar 'Golden Delicious'). *Journal of Agricultural and Food Chemistry.* Jun 10;57(11):4598-605 and Feb 25; 57(4):1188-94.

Juroszek P, Lumpkin HM, Yang RY, Ledesma DR, Ma CH. (2011). Fruit quality and bioactive compounds with antioxidant activity of tomatoes grown on-farm: comparison oforganic and conventional management systems. *Clinical and Experimental Dermatology.* Aug; 36(6):573-7; quiz 577-8.

Shams K, Grindlay DJ, Williams HC. (2005). What's new in atopic eczema? An analysis of systematic reviews published in 2009-2010. *Food Additives and Contaminants.* May; 22(5):437-42.

Piemontese L, Solfrizzo M, Visconti A. Occurrence of patulin in conventional and organic fruit products in Italy and subsequent exposure assessment. *Food additives and Contaminants.* May;22(5):437-42.

Eskenazi B, Bradman A, Castorina R. (1999). Exposures of children to organophosphate pesticides and their potential adverse health effects. *Environmental Health Perspectives.* June; 107 Suppl 3:409-19.

Ellis KA, Innocent G, Grove-White D, Cripps P, McLean WG, Howard CV, Mihm M. (2006). Comparing the fatty acid composition of organic and conventional milk. *Journal of Dairy Science.* June; 89(6):1938-50.

O'Donnell AM, Spatny KP, Vicini JL, Bauman DE (2010). Survey ofthe fatty acid composition of retail milk differing in label claims based on production management practices. *Journal of Dairy Science.* May;93(5):1918-25.

Dr. Alan Dangour (lead), Ms. Andrea Aikenhead, Ms. Arabella Hayter, Dr. Elizabeth Allen, Dr. Karen Lock, Professor Ricardo Uauy (2009). Nutrition and Public Health Intervention Research Unit London School of Hygiene & Tropical Medicine. Comparison of putative health effects of organically and conventionally produced foodstuffs: a systematic review. *Report for the Food Standards Agency* July.

**25. Fussy eaters**

Pliner P, Loewen ER. (1997). Temperament and food neophobia in children and their mothers. *Appetite.* Jun; 28(3):239-54.

Cooke LJ, Haworth CM, Wardle J. (2007). Genetic and environmental influences on children's food neophobia. *American Journal of Clinical Nutrition.* Aug; 86(2):428-33.

Tsuji M, Nakamura K, Tamai Y, Wada K, Sahashi Y, Watanabe K, Ohtsuchi S, Ando K, Nagata C. (2012). Relationship of intake of plant-based foods with 6-n-propylthiouracil sensitivity and food neophobia in Japanese preschool children. *European Journal of Clinical Nutrition.* Jan; 66(1):47-52.

Pidamale R, Sowmya B, Thomas A, Jose T. (2012). Genetic sensitivity to bitter taste of 6-n Propylthiouracil: A useful diagnostic aid to detect early childhood caries in pre-school children. *Indian Journal of Human Genetics.* Jan; 18(1):101-5.

Galloway AT, Lee Y, Birch LL. (2003). Predictors and consequences of food neophobia and pickiness in young girls. *Journal of American Dietetic Association.* Jun; 103(6):692-8.

Fisher JO, Birch LL, Smiciklas-Wright H, Picciano MF. (2000). Breast-feeding through the first year predicts maternal control in feeding and subsequent toddler energy intakes. *Journal of American Dietetic Association.* Jun; 100(6):641-6.

Mennella JA, Jagnow CP, Beauchamp GK (2001). Prenatal and postnatal flavor learning by human infants. *Pediatrics.* Jun; 107(6):E88..

Wardle J, Herrera ML, Cooke L, Gibson EL. (2003). Modifying children's food preferences: the effects of exposure and reward on acceptance of an unfamiliar vegetable. *European Journal of Clinical Nutrition.* Feb; 57(2):341-8.

Carruth BR, Skinner JD (2000). Revisiting the picky eater phenomenon: neophobic behaviors of young children. *Journal of American College of Nutrition.* Nov-Dec; 19(6):771-80.

Addessi E, Galloway AT, Visalberghi E, Birch LL. (2005). Specific social influences on the acceptance of novel foods in 2-5-year-old children. *Appetite.* Dec; 45(3):264-71.

Hendy HM, Raudenbush B. (2000). Effectiveness of teacher modelling to encourage food acceptance in preschool children. *Appetite.* Feb; 34(1):61-76.

Brown R, Ogden J. (2004). Children's eating attitudes and behaviour: a study of the modelling and control theories of parental influence. *Health Education Research.* Jun; 19(3):261-71.

Orrell-Valente JK, Hill LG, Brechwald WA, Dodge KA, Pettit GS, Bates JE. (2007) "Just three more bites": an observational analysis of parents' socialization of children's eating at mealtime. *Appetite.* Jan; 48(1):37-45.

Birch LL, McPhee L, Steinberg L, Sullivan S. Conditioned flavor preferences in young children. *Physiology & Behaviour.* Mar; 47(3):501-5.

Birch LL, Fisher JO. (1998). Development of eating behaviors among children and adolescents. *Pediatrics.* Mar; 101(3 Pt 2):539-49.

Birch LL. (1999). Development of food preferences. *Annual Review of Nutrition*; 19:41-62.

Carruth BR, Skinner JD. (2000). Revisiting the picky eater phenomenon: neophobic behaviors of young children. *Journal of the American College of Nutrition.* Nov-Dec; 19(6):771-80.

Falciglia GA, Couch SC, Gribble LS, Pabst SM, Frank R. (2000). Food neophobia in childhood affects dietary variety. *Journal of American Dietetics Association.* Dec; 100(12):1474-81.

Taylor JP, Evers S, McKenna M. (2005). Determinants of healthy eating in children and youth. [Article in English, French] *Canadian Journal of Public Health.* 2005 Jul-Aug; 96 Supplement 3:S20-6, S22-9.

**26. Restricting the bottle**

Early childhood caries–facts and prevention. *Therapeutische Umschau. Revue Therapeutique.* Feb; 65(2):75-82.

Tiberia MJ, Milnes AR, Feigal RJ, Morley KR, Richardson DS, Croft WG, Cheung WS. (2007). Risk factors for early childhood caries in Canadian preschool children seeking care. *Pediatric Dentistry.* May-Jun; 29(3):201-8.

Mohebbi SZ, Virtanen JI, Vahid-Golpayegani M, Vehkalahti MM. (2008). Feeding habits as determinants of early childhood caries in a population where prolonged breastfeeding is the norm. *Community Dentistry and Oral Epidemiology*. Aug; 36(4):363-9.

Schwartz SS, Rosivack RG, Michelotti P. (1993). A child's sleeping habit as a cause of nursing caries. *ASDC Journal of Dentistry for Children*. Jan-Feb; 60(1):22-5.

Febres C, Echeverri EA, Keene HJ. (1997). Parental awareness, habits, and social factors and their relationship to baby bottle tooth decay. *Pediatric Dentistry*. Jan-Feb; 19(1):22-7.

Smith PJ, Moffatt ME. (1998). Baby-bottle tooth decay: are we on the right track? *International Journal of Circumpolar Health*. 57 Supplement 1:155-62.

Pavlov MI, Naulin-Ifi C. (1999). [Plea for prevention and early management of baby bottle tooth decay syndrome].[Article in French] *Archives Pediatrie*. Feb; 6(2):218-22.

Bonuck KA, Kahn R. (2002). Prolonged bottle use and its association with iron deficiency anemia and overweight: a preliminary study. *Clinical Pediatrics* (Phila). Oct; 41(8):603-7.

Brotanek JM, Halterman JS, Auinger P, Flores G, Weitzman M. (2005). Iron deficiency, prolonged bottle-feeding, and racial/ethnic disparities in young children. *Archives of Pediatric & Adolescent Medicine*. Nov; 159(11):1038-42.

Gooze RA, Anderson SE, Whitaker RC. (2011). Prolonged bottle use and obesity at 5.5 years of age in US children. *Journal of Pediatrics*. Sep; 159(3):431-6.

Bonuck K, Kahn R, Schechter C. (2004). Is late bottle-weaning associated with overweight in young children? Analysis of NHANES III data. *Clinical Pediatrics (Philadelphia)*. Jul-Aug; 43(6):535-40.

Keim SA, Fletcher EN, TePoel MR, McKenzie LB.(2012). Injuries associated with bottles, pacifiers, and sippy cups in the United States, 1991-2010. *Pediatrics*. Jun; 129(6):1104-10.

Tiberia MJ, Milnes AR, Feigal RJ, Morley KR, Richardson DS, Croft WG, Cheung WS. (2007). Risk factors for early childhood caries in Canadian preschool children seeking care. *Pediatric Dentistry*. May-Jun; 29(3):201-8.

Khadra-Eid J, Baudet D, Fourny M, Sellier E, Brun C, François P. (2012). [Development of a screening scale for children at risk of baby bottle tooth decay]. [Article in French] *Archives de Pediatrie*. Mar; 19(3):235-41.

**27. Toddler milk**

Horrocks LA, Yeo YK (1999). Health benefits of docosahexaenoic acid (DHA). *Pharmacological Research* (Italy). Sep; 40(3):211-25.

Minns LM, Kerling EH, Neely MR, Sullivan DK, Wampler JL, Harris CL, Berseth CL, Carlson SE. (2010). Toddler formula supplemented with docosahexaenoic acid (DHA) improves DHA status and respiratory health in a randomized, double-blind, controlled trial of US children less than 3 years of age. *Prostaglandins, Leukotrienes and Essential Fatty Acids*. Apr-Jun; 82(4-6):287-93.

Thorisdottir AV, Ramel A, Palsson GI, Tomassson H, Thorsdottir I. (2012). Iron status of one-year-olds and association with breast milk, cow's milk or formula in late infancy. *European Journal of Nutrition*. Dec 2

Soh P, Ferguson EL, McKenzie JE, Skeaff S, Parnell W, Gibson RS. (2002). Dietary intakes of 6-24-month-old urban South Island New Zealand children in relation to biochemical iron status. *Public Health Nutrition*. Apr; 5(2):339-46.

Sachdev H, Gera T, Nestel P. (2006). Effect of iron supplementation on physical growth in children: systematic review of randomised controlled trials. *Public Health Nutrition*. Oct; 9(7):904-20.

Maldonado Lozano J, Baró L, Ramírez-Tortosa MC, Gil F, Linde J, López-Huertas E, Boza JJ,

Gil A.(2007). [Intake of an iron-supplemented milk formula as a preventive measure to avoid low iron status in 1-3 year-olds].[Article in Spanish]. *Anales de Pediatria* (Barcelona, Spain). Jun; 66(6):591-6.

Lozoff B, Castillo M, Clark KM, Smith JB. (2012). Iron-fortified verses low-iron infant formula: developmental outcome at 10 years. *Archives of Pediatrics & Adolescent Medicine*. Mar; 166(3):208-15.

Berry NJ, Jones SC, Iverson D. (2012). Circumventing the WHO Code? An observational study. *Archives of Disease in Childhood*. Apr; 97(4):320-5.

**28. Discipline**

Canadian Paediatric Society. Psychological Paediatrics Committee (2004). Effective discipline for children. *Paediatric Child Health*. 9:37–41

Knox, M. (2010). On hitting children: A review of corporal punishment in the United States. *Journal of Pediatric Health Care*. 24, 103-107.

H L MacMillan, M H Boyle, M Y Wong, E K Duku, J E Fleming, and C A Walsh (1999). Slapping and spanking in childhood and its association with lifetime prevalence of psychiatric disorders in a general population sample. *CMAJ*. October 5; 161(7): 805–809.

Byford, M., Kuh, D., and Richards, M.( 2012). Parenting practices and intergenerational associations in cognitive ability. *International Journal of Epidemiology;* 41(1): 263-272.

Blum NJ, Williams GE, Friman PC, Christophersen ER. (1995). Disciplining young children: the role of verbal instructions and reasoning. *Pediatrics. Aug; 96(2 Pt 1):336-41.*

Howard BJ. (1991). Discipline in early childhood. *Pediatric Clinics of North America*. Dec; 38(6):1351-69.

Simeon D, Guralnik O, Schmeidler J, Sirof B, Knutelska M. (2001). The role of childhood interpersonal trauma in depersonalization disorder. *American Journal of Psychiatry*. Jul; 158(7):1027-33.

Dowling CB, Smith Slep AM, O'Leary SG. (2009). Understanding pre-emptive parenting: relations with toddlers' misbehavior, over-reactive and lax discipline, and praise. *Journal of Clinical Child and Adolescent Psychology*. Nov; 38(6):850-7.

Bayer JK, Ukoumunne OC, Mathers M, Wake M, Abdi N, Hiscock H. (2012). Development of children's internalising and externalising problems from infancy to five years of age. *Australian and New Zealand Journal of Psychiatry*. Jul; 46(7):659-68.

Tung I, Li JJ, Lee SS. (2012). Child sex moderates the association between negative parenting and childhood conduct problems. *Aggressive Behavior*. May-Jun; 38(3):239-51.

Volling BL, Mahoney A, Rauer AJ. (2009). Sanctification of Parenting, Moral Socialization, and Young Children's Conscience Development. *Psychology of Religion and Spirituality*. Feb; 1(1):53-68.

Owen DJ, Slep AM, Heyman RE. (2012). The effect of praise, positive nonverbal response, reprimand, and negative nonverbal response on child compliance: a systematic review. *Clinical Child and Family Psychology Review*. Dec; 15(4):364-85.

**29. TV**

Wilson BJ. (2008). Media and children's aggression, fear, and altruism. *The Future of Children. Spring*; 18(1):87-118.

Pagani, L. S., Fitzpatrick, C. F., Barnett, T. A., & Dubow, E. (2010). Prospective associations between early childhood televiewing and later mental and physical health. *Archives of Pediatric and Adolescent Medicine*, 164, 425–431.

Hu FB. (2003). Sedentary lifestyle and risk of obesity and type 2 diabetes. *Lipids*. 2003 Feb; 38(2):103-8.

Cantor J, Nathanson AI. (1996). Children's Fright Reactions to Television News. *Journal of Communication;* Dec;46(4): 139–152.

Bickham DS, Rich M. (2006); Is television viewing associated with social isolation? Roles of exposure time, viewing context, and violent content. *Archives of Pediatrics & Adolescent Medicine*. 2006 Apr; 160(4):387-92.

Huesmann, LR & Taylor, LD (2006). The role of the mass media in violent behavior. In R. C. Brownson et al. (Eds.). *Annual Review of Public Health*, 26. Palo Alto, CA: Annual Reviews Publishers.

Ruangdaraganon N, Chuthapisith J, Mo-suwan L, Kriweradechachai S, Udomsubpayakul U, Choprapawon C. (2009). Television viewing in Thai infants and toddlers: impacts to language development and parental perceptions. *BMC Pediatrics*. May 22; 9:34.

Deborah L. Linebarger, Dale Walker (2005). Infants' and Toddlers' Television Viewing and Language Outcomes. *American Behavioral Scientist*; Jan; 48(5):624-645

Robb MB, Richert RA, Wartella EA. (2009). Just a talking book? Word learning from watching baby videos. *British Journal of Developmental Psychology*. Mar; 27(Pt 1):27-45.

Zimmerman FJ, Christakis DA. (2005). Children's television viewing and cognitive outcomes: a longitudinal analysis of national data. *Archives of Pediatric & Adolescent Medicine*. Jul; 159(7):614-8.

Hancox RJ, Milne BJ, Poulton R. (2005). Association of television viewing during childhood with poor educational achievement. *Archives of Pediatric & Adolescent Medicine*. Jul; 159(7):607-13.

Borzekowski DL, Robinson TN. (2001). The remote, the mouse, and the no. 2 pencil: the household media environment and academic achievementamong third grade students. *Monographs of the Society for Research in Child Development.* 66(1):I-VIII, 1-147.

Anderson DR, Huston AC, Schmitt KL, Linebarger DL, Wright JC. (2001). Early childhood television viewing and adolescent behavior: the recontact study. *Child Development.* Sep-Oct; 72(5):1347-66.

Wright JC, Huston AC, Murphy KC, St Peters M, Piñon M, Scantlin R, Kotler J. (2008). The relations of early television viewing to school readiness and vocabulary of children from low-income families: the early window project. *The Future of Children.* Spring; 18(1):39-61.

Kirkorian HL, Wartella EA, Anderson DR. (2008). Media and young children's learning. *The Future of Children.* 18, No. 1:39-61.

Christakis DA, Zimmerman FJ, DiGiuseppe DL, McCarty CA. (2004). Early television exposure and subsequent attentional problems in children. *Pediatrics.* Apr; 113(4):708 -713.

DeLoache JS, Chiong C, Sherman K, Islam N, Vanderborght M, Troseth GL , Strouse GA,

O'Doherty K. ( 2010). Do Babies Learn From Baby Media? *Psychological Science;* 21(11):1570 –1574.

Huesmann LR, Taylor LD. (2006). The role of media violence in violent behavior. *Annual Review of Public Health.* 27:393-415.

Neuman, S. B. (1991). Literacy in the television age: The myth of the TV effect. Norwood, New Jersey: Ablex Publishing Corporation

Potts R, Doppler M, Hernandez M. (1994). Effects of television content on physical risk-taking in children. *Journal of Experimental Child Psychology.* Dec; 58(3):321-31.

Tamburro RF, Gordon PL, D'Apolito JP, Howard SC. (2004) Unsafe and violent behavior in commercials aired during televised major sporting events. Pediatrics. Dec; 114(6):e694-8.

Joanne Cantor1, Amy I. Nathanson2 (1996). Children's Fright Reactions to Television News. *Journal of Communication;* Dec; 46( 4): 139–152.

Schmidt ME, Pempek TA, Kirkorian HL, Lund AF, Anderson DR. (2008). The effects of background television on the toy play behavior of very young children. *Child Development.* Jul-Aug;79(4):1137-51.

Courage ML, Murphy AN, Goulding S, Setliff AE. (2006). When the television is on: the impact of infant-directed video on 6- and 18-month-olds' attention during toy play and on parent-infant interaction. *Infant Behavior & Development.* Apr; 33(2):176-88.

Vandewater EA, Bickham DS, Lee JH. (2006). Time well spent? Relating television use to children's free-time activities. *Pediatrics.* Feb; 117(2):e181-91.

Williams, TM, (1979), "The Impact of Television: A Natural Experiment Involving Three Communities," Paper presented at the Annual Meeting of the International Communication Association (Philadelphia, Pennsylvania, May 1-5, 1979).

Vandewater EA, Bickham DS, Lee JH. (2006). Time well spent? Relating television use to children's free-time activities. *Pediatrics.* Feb; 117(2):e181-91.

Tremblay MS, LeBlanc AG, Kho ME, Saunders TJ, Larouche R, Colley RC, Goldfield G,

Connor Gorber S. (2011). Systematic review of sedentary behaviour and health indicators in school-aged children and youth. *International Journal Behavioral Nutritional and Physical Activity.* Sep 21; 8:98.

Hancox RJ, Milne BJ, Poulton R. (2004). Association between child and adolescent television viewing and adult health: a longitudinal birth cohort study. *Lancet.* Jul 17-23; 364(9430):257-62.

Wijndaele K, Brage S, Besson H, Khaw KT, Sharp SJ, Luben R, Bhaniani A, Wareham NJ,

Ekelund U.(2011). Television viewing and incident cardiovascular disease: prospective associations and mediation analysis in theEPIC Norfolk Study. *PLoS One*; 6(5):e20058.

Hu FB. (2003).Sedentary lifestyle and risk of obesity and type 2 diabetes. *Lipids.* Feb; 38(2):103-8.

Pearce MS, Basterfield L, Mann KD, Parkinson KN, Adamson AJ, Reilly JJ; Gateshead Millennium Study Core Team. (2012). Early predictors of objectively measured physical activity and sedentary behaviour in 8-10 year old children: the Gateshead Millennium Study. *PLoS One*; 7(6):e37975.

**30. Computer**

Weiss MD, Baer S, Allan BA, Saran K, Schibuk H. (2011). The screens culture: impact on ADHD. *Attention Deficit and Hyperactivity Disorders.* Dec;3(4):327-34.

McPake, J., Stephen, C., Plowman, L. & Berch-Heyman, S. (2008). Developing Digital Literacy at Home : The Impact of Parents' Attitudes and Preschool Children's Preferences. Paper given at the American Educational Research Association Annual Meeting, New York (March 24-28).

Tremblay MS, LeBlanc AG, Kho ME, Saunders TJ, Larouche R, Colley RC, Goldfield G, Connor Gorber S (2011). Systematic review of sedentary behaviour and health indicators in school-aged children and youth. International Journal of Behavioral Nutrition and Physical Activity. Sep 21;8:98.

Van der Kooy-Hofland VA, Bus AG, Roskos K.(2012). Effects of a brief but intensive remedial computer intervention in a sub-sample of kindergartners with early literacy delays. *Reading and Writing.* Aug;25(7):1479-1497.

Li X, Atkins MS. (2004). Early childhood computer experience and cognitive and motor development. *Pediatrics.* Jun;113(6):1715-22.

Blanton, W, E, Moorman, G.B., Hayes, B.A., & Warner, M. L. (1997). Effects of Participation in the Fifth Dimension on Far Transfer. *Journal of Education Computing Research,*16: 371-396.

McCarrick, K., & Xiaoming, (2007). Buried treasure: The impact of computer use on young children's social, cognitive, language development and motivation. AACE Journal,15(1), 73-95.

Subrahmanyam K, Kraut RE, Greenfield PM, Gross EF. (2000). The impact of home computer use on children's activities and development. *The Future of Children.* Fall-Winter; 10(2):123-44.

L. R, Huesmann and L. D. Taylor (2006). The role of media violence in violent behaviour. *Annual Review of Public Health.* 27:1.1–1.23

Bushman, B. J., & Anderson, C. A. (2009). Comfortably numb: Desensitizing effects of violent media on helping others. *Psychological Science*, 20, 273-277.

Calvert, S.L. & Tan, S.L.(1994). Impact of virtual reality on young adults' physiological arousal and aggressive thoughts: Interaction versus observation. *Journal of Applied Developmental Psychology,* 15(1):125-139.

Muller, A.A., and Perlmutter, M. (1985). Preschool children's problem-solving interactions at computers and jigsaw puzzles. *Journal of Applied Developmental Psychology,*6: 173–186.

Stephen, C. & Plowman, L. (2003). Information and Communication Technologies in Pre-School Settings: A Review of the Literature. *International Journal of Early Years Education,* 11(3): 224-234.

Haugland, S.W. (1992). The effects of computer software on preschool children's developmental gains. *Journal of Computing in Childhood Education.* 3(1), 15-30.

Lukasz Borecki, Katarzyna Tolstych, Mieczyslaw Pokorski. (2013). Computer Games and Fine Motor Skills. *Advances in Experimental Medicine and Biology;* 755:343-348.

Barnett LM, Hinkley T, Okely AD, Hesketh K, Salmon J. (2012). Use of electronic games by young children and fundamental movement skills? *Perceptual and Motor Skills.* Jun; 114(3):1023-34.

# Index

## Q,R



## S

# Acknowledgments

**From the author:**
I would firstly like to thank my husband, Gavin, for always believing that I could write this book. I can honestly say that I would have listened to the rather loud voices of doubt and without his constant encouragement, this book would never have been written. I also want to thank my little girl, to whom this book is dedicated. Without a desire to do my best for her, I would never had had the inclination or motivation to find out what the research says and put it into action.

Thanks also to my parents for allowing me the time to write, and especially to my Mum for proof reading and formatting the bibliographies – not the most riveting of tasks! Thanks to my sister, Jo, for her expert medical advice.
Thanks to Zelda, Hannah, Amy and Beth for thinking it was all possible.

A big thank you to Amy Carroll for taking a chance on an unsolicited email and for all her help in making the complexities of research evidence available to all.

And finally: Soli Deo Gloria.

**Carroll & Brown would like to thank the following for supplying images:**
Car seat p112 Courtesy of Dorel UK
Farmers' market p129 by Gareth Jones. Courtesy of FARMA, The National Farmers' Retail & Markets Association, 12 Southgate Street, Winchester, SO23 9EF
Image p139 Courtesy of BabyBjörn, www.babybjorn.com